Advances in Asian Human-Environmental Research

Series Editor
Prof. Marcus Nüsser, South Asia Institute, University of Heidelberg, Germany

Editorial Board
Prof. Eckart Ehlers, University of Bonn, Germany
Prof. Harjit Singh, Jawaharlal Nehru University, New Delhi, India
Prof. Hermann Kreutzmann, Freie Universität Berlin, Germany
Prof. Kenneth Hewitt, Waterloo University, Canada
Prof. Urs Wiesmann, University of Bern, Switzerland
Prof. Sarah J. Halvorson, University of Montana, USA
Dr. Daanish Mustafa, King's College London, UK

Aims and Scope

The series aims at fostering the discussion on the complex relationships between physical landscapes, natural resources, and their modification by human land use in various environments of Asia. It is widely acknowledged that human-environment interactions become increasingly important in area studies and development research, taking into account regional differences as well as bio-physical, socioeconomic and cultural particularities.

The book series seeks to explore theoretic and conceptual reflection on dynamic human-environment systems applying advanced methodology and innovative research perspectives. The main themes of the series cover urban and rural landscapes in Asia. Examples include topics such as land and forest degradation, glaciers in Asia, mountain environments, dams in Asia, medical geography, vulnerability and mitigation strategies, natural hazards and risk management concepts, environmental change, impacts studies and consequences for local communities. The relevant themes of the series are mainly focused on geographical research perspectives of area studies, however there is scope for interdisciplinary contributions.

More information about this series at http://www.springer.com/series/8560

Saloni Gupta

Contesting Conservation

Shahtoosh Trade and Forest Management in Jammu and Kashmir, India

Saloni Gupta
Udaipur, Rajasthan, India

ISSN 1879-7180   ISSN 1879-7199   (electronic)
Advances in Asian Human-Environmental Research
ISBN 978-3-319-72256-6   ISBN 978-3-319-72257-3   (eBook)
https://doi.org/10.1007/978-3-319-72257-3

Library of Congress Control Number: 2017961763

© Springer International Publishing AG 2018
This work is subject to copyright. All rights are reserved by the Publisher, whether the whole or part of
the material is concerned, specifically the rights of translation, reprinting, reuse of illustrations, recitation,
broadcasting, reproduction on microfilms or in any other physical way, and transmission or information
storage and retrieval, electronic adaptation, computer software, or by similar or dissimilar methodology
now known or hereafter developed.
The use of general descriptive names, registered names, trademarks, service marks, etc. in this publication
does not imply, even in the absence of a specific statement, that such names are exempt from the relevant
protective laws and regulations and therefore free for general use.
The publisher, the authors and the editors are safe to assume that the advice and information in this book
are believed to be true and accurate at the date of publication. Neither the publisher nor the authors or the
editors give a warranty, express or implied, with respect to the material contained herein or for any errors
or omissions that may have been made. The publisher remains neutral with regard to jurisdictional claims
in published maps and institutional affiliations.

Cover image: Nomads near Nanga Parbat, 1995. Copyright © Marcus Nüsser (used with permission)

Printed on acid-free paper

This Springer imprint is published by Springer Nature
The registered company is Springer International Publishing AG
The registered company address is: Gewerbestrasse 11, 6330 Cham, Switzerland

For
Pabi and Barepapa

Acknowledgements

I have received support from several people in initiating and completing this research. First and foremost, I would like to thank the Commonwealth Scholarship Commission for their financial support in pursuing this project. I am also indebted to all the *shahtoosh* workers and residents of the villages who confided in me and shared their time and thoughts. The truth and genuineness of their demands have provided me with the motivation to voice their concerns loudly in this book. I extend my thanks to the members of the Wildlife Trust of India and officials of various government departments who found time within their busy schedules to speak to me.

I would like to thank my PhD supervisor, Dr. Subir Sinha, who had always motivated me with the novelty of the topic of my research and provided insightful comments on my thesis drafts. I am also grateful to Jens Lerche, Mahesh Rangarajan, Gabriela Lichtenstein, Deniz Kandiyoti, Neil Thin, Richard Axelby, George Kunnath, Bhaskar Vira and Ashish Aggarwal for their suggestions on my earlier drafts.

I would also like to acknowledge the support that I received from my family. I thank my parents, Sh. Sudesh Kumar Gupta and Smt. Sarishta Gupta, who have been a great source of inspiration to me. In initiating this research project and writing this book, I have constantly been reminded of my school days when, on returning from a hard day's work, my father would spend three hours every day working with me, not letting me finish until satisfied. Writing this book has caused me to experience again the joy of striving for satisfaction. I thank my mother for her silent prayers that have continuously turned things in my favour.

For Saurabh, I have no words. I saw him in different roles and capacities, from a friend who has constantly encouraged me to a guide who has contributed immensely in shaping and presenting my ideas. I have also felt enthusiastic about my work on seeing him overjoyed whenever I have succeeded in putting complex arguments across. I yearn to reciprocate the support and comfort that I have received from him while writing this book. For Arghya, my son, I cannot thank him enough for his happy 'bye-byes' which have motivated me to work peacefully in his sleeping hours. What a pleasure to see him sleeping at this moment as I write this note of appreciation for him!

Finally, I would like to express my deepest feelings of gratitude to my great grandmother, Smt. Rajan Devi, and my grandfather, Sh. Hans Raj Gupta, who took great pride in me doing this research and who showed keen interest in my topic by enquiring the various questions I ask in the field! Both of them are no longer in this world and so will not be able to see the completion of this work, but their memories will always be the strongest source of strength to me.

Contents

1 Introduction 1
   1.1 Wildlife and Forest Conservation in J&K: An Introduction 3
      1.1.1 The Ban on *Shahtoosh* Trade 3
      1.1.2 National Afforestation Programme 5
   1.2 Political Ecology: Approaches and Analytical Framework 8
   1.3 Theoretical Positioning 9
   1.4 Contesting Conservation: What This Study Contributes 14
   1.5 Chapter Layout 15
   References 17

**2 Jammu and Kashmir: Contextualising Conservation
in Specific Sites** 23
   2.1 Research Setting 24
      2.1.1 Jammu and Kashmir: An Introduction 24
      2.1.2 Field Locations 28
   2.2 Research Methods 29
      2.2.1 Ethnography of Conservation Interventions 29
      2.2.2 Research in the Context of Violent Conflict 32
   References 34

3 Tibetan Antelope and *Shahtoosh* Shawl: A Brief History 37
   3.1 Tibetan Antelope 38
      3.1.1 Chiru and Its Habitat 38
      3.1.2 The *Shahtoosh* Wool: Myths and Realities 39
   3.2 *Shahtoosh* Shawl 43
      3.2.1 From Raw Wool to Finished Shawl:
          The Production Process 43
      3.2.2 *Shahtoosh* Workers: Population and Distribution 51

3.3	Kashmir Shawl Industry		51
	3.3.1	Origin and Development of the Shawl Industry	51
	3.3.2	Marginalisation and Exploitation of the Shawl Workers: Pre-independence	56
	3.3.3	*Shahtoosh* Workers and the New State: Post-independence	58
3.4	Conclusion		60
References			61

4 The Ban on *Shahtoosh*: Sustainability for Whom? 63

4.1	Ban on *Shahtoosh*		65
	4.1.1	A Chronology of Events	65
	4.1.2	The Prospects of *chiru* Farming: Observations of the 'Expert Group'	73
4.2	State and Shawl Workers		78
	4.2.1	Weak Enforcement and Split Role of the State	78
	4.2.2	Shawl Workers Response to the Ban: Protest and Politics	81
4.3	Illegality and Conflict		85
	4.3.1	The Trade Continues: Illegality and Shadow Networks of *shahtoosh*	85
	4.3.2	Militancy and *shahtoosh*: Exploring the Connections	87
4.4	Conclusion		89
References			91

5 The Micropolitics of the Ban on *Shahtoosh*: Costs and Reparations 95

5.1	Impacts of Banning		96
	5.1.1	The Origin of Wool and the Unpopularity of the Ban	97
	5.1.2	Different Categories, Differential Impact	98
	5.1.3	Machines and Adulteration	103
	5.1.4	Decreasing Wages, Increasing Prices: Strategies of Labour Exploitation and Control	106
	5.1.5	Declining Social Prestige and Cultural Heritage	111
5.2	Rehabilitation and Alternative Livelihoods: Accountability of Whom?		113
5.3	Conclusion		117
References			119

6 Forests, State and People: A Historical Account of Forest Management and Control in J&K 121

6.1	Forest Management in Early Colonial Period	122
6.2	Local Access Versus Commercial Interests: The Politics of Scientific Forestry in the Late Colonial Period	124

Contents   xi

    6.3   National Interests Versus Local Needs: The Politics
          of Forest Management in the Post-Colonial Period   133
    6.4   Conclusion   139
    References   140

**7   Joint Management of Forests and Split Role of the State:
      The Politics of Forest Conservation in J&K**   143
    7.1   Joint Management of Forests: New Arenas of 'Partnership'
          and 'Participation'   145
          7.1.1   National Afforestation Programme: Facilitating
                  JFM Through 'Decentralisation'   148
    7.2   Setting the Scene: Interplay Between Centre,
          State and Non-state Actors   153
          7.2.1   FD and JFMCs   153
          7.2.2   SFC and FD   158
          7.2.3   MoEF and FD   159
          7.2.4   NGOs and FD   160
    7.3   The Politics of Forest Resource Control   162
          7.3.1   Navni and Chinnora: A Brief Introduction   162
          7.3.2   Our Forests, Their Timber   163
          7.3.3   Split Role of the Field-Staff: Forest Regulations
                  vis-a-vis Local Needs   166
    7.4   Conclusion   169
    References   170

**8   The Micropolitics of Forest Use and Control: New Spaces
      for Cooperation and Conflict**   173
    8.1   From Centralisation to Decentralisation:
          Do Blockages Disappear?   175
    8.2   Panchayat and JFMC: Conflicting Powers and Functions   180
    8.3   Increased Biomass, Reduced Access   185
    8.4   Illegal Timber Felling: What If Fence Eats the Grass?   194
    8.5   Conclusion   199
    References   200

**9   On Conservation Politics: Cooperation,
      Conflicts and Contestations**   205
    9.1   Power as Dispersed and Fluid   207
    9.2   Between Cooperation and Conflict: Spaces for Contestation   208
          9.2.1   Science, Law and Politics   208
          9.2.2   Uneasy Partnerships and Collaborations   210
          9.2.3   Split Role of the State   211
          9.2.4   Differential Impact, Differential Abilities   212
          9.2.5   Rationality of Rule   213

	9.2.6	On Illegality	214
	9.2.7	Limited Space for Protest	215
9.3	Who Is Accountable?		216
9.4	Practical Implications		217
9.5	Conclusion		218
References			218

Bibliography 223

Index 237

Acronyms

AKKA	Almi Khudai Khidmatgar Association
BSF	Border Security Force
CCF	Chief Conservator of Forests
CF	Conservator of Forests
CITES	Convention on International Trade in Endangered Species of Wild Fauna and Flora
COP	Conference of Parties
DFO	Divisional Forest Officer
DIG	Deputy Inspector General
FD	Forest Department
FDA	Forest Development Agency
GI	Geographical Indication
GoJ&K	Government of Jammu and Kashmir
IFAW	International Fund for Animal Welfare
IFS	Indian Forest Service
IUCN	International Union for the Conservation of Nature
J&K	Jammu and Kashmir
JBIC	Japan Bank for International Cooperation
JFM	Joint Forest Management
JFMC	Joint Forest Management Committee
MoEF	Ministry of Environment and Forests
MoU	Memorandum of Understanding
NAEB	National Afforestation and Eco-Development Board
NAP	National Afforestation Programme
NC	National Conference
NGO	Non-Governmental Organisation
NTFP	Non-Timber Forest Produce
PCCF	Principal Chief Conservator of Forests
PDP	People's Democratic Party
PIL	Public Interest Litigation
RO	Range Officer

SD	Sustainable Development
SFC	State Forest Corporation
SIDA	Swedish International Development Agency
SL	Sustainable Livelihoods
VFC	Village Forest Committee
WPSI	Wildlife Protection Society of India
WTI	Wildlife Trust of India
WWF	World Wide Fund for Nature

Units of Conversion

1 pound sterling (GBP) = 83 Indian rupees (Rs.) [average value in 2006–2007]
1 lakh = 100,000
1 crore = 10,000,000
1 kanal = 0.125 acres
1 kilograms (kg) = 1000 grams
1 kilometres (km) = 1000 metres
1 cubic meter = 35.3 cubic feet [approximately]

List of Figures

Fig. 3.1   Production process of the *shahtoosh* shawl   44
Fig. 3.2   Spinning yarn with a traditional hand-driven wheel   46
Fig. 3.3   A weaver weaving a pashmina shawl in a karkhana in Srinagar ...   48

Fig. 7.1   Actors in Forest Administration and Management, J&K   149

Fig. 8.1   A ring peeled pine in the closure of Navni   196
Fig. 8.2   A burnt deodar trunk in the closure of Chinnora   196

List of Table and Map

Table 2.1 Occupational distribution of *shahtoosh* workers 31

Map 2.1 Field locations in Jammu and Kashmir 24

Chapter 1
Introduction

Abstract This book examines and analyses the politics of recent wildlife and forest conservation interventions in the Indian state of Jammu and Kashmir (hereafter J&K). With a long history of militant separatist movement and violence, J&K has drawn the attention of scholars interested in the issues of strategic affairs, autonomy, secession, terrorism, ethnicity, identity based conflicts, displacement and forced migration. Over the last few decades, these problems relating to political instability in the region have eclipsed other issues of significant research interest, despite the fact that amidst conflict, interactions between social, political, economic and environmental forces have simultaneously been taking place in J&K (as in other states of India). The problems pertaining to natural resources and livelihoods have also been unfolding in J&K as in the rest of India, but have not received due consideration by the scholars of the region. Owing to the fact that the vast majority of the population of the state is dependent on natural resources for livelihoods and sustenance, and that power relations largely determine access to and use of resources, the field of environmental politics provides a significant area of research on J&K. In this book I discuss the agendas, interests and power of different actors involved in two nature conservation interventions i.e. ban on *shahtoosh* trade and National Afforestation Programme, and analyse how environmental policies permeate different layers of politics from macro- to microlevel in the process of implementation.

Keywords Nature conservation · Political ecology · Jammu and Kashmir · *Shahtoosh* · Forest management

This book examines and analyses the politics of recent wildlife and forest conservation interventions in the Indian state of Jammu and Kashmir (hereafter J&K). With a long history of militant separatist movement and violence, J&K has drawn the attention of scholars interested in the issues of strategic affairs, autonomy, secession, terrorism, ethnicity, identity based conflicts, displacement and forced migration. Over the last few decades, these problems relating to political instability in the region have eclipsed other issues of significant research interest, despite the fact that amidst conflict, interactions between social, political, economic and environmental

© Springer International Publishing AG 2018
S. Gupta, *Contesting Conservation*, Advances in Asian Human-Environmental
Research, https://doi.org/10.1007/978-3-319-72257-3_1

forces have simultaneously been taking place in J&K (as in other states of India). The problems pertaining to natural resources and livelihoods have also been unfolding in J&K as in the rest of India, but have not received due consideration by the scholars of the region. Owing to the fact that the vast majority of the population of the state is dependent on natural resources for livelihoods and sustenance, and that power relations largely determine access to and use of resources, the field of environmental politics provides a significant area of research on J&K. It is with this interest and motivation that I have undertaken this study on the politics of nature conservation, aiming to fill the gap in the available literature on the region.

In 2002, two significant nature conservation interventions took place in the state of J&K. First, an international ban on the production of *shahtoosh* (wool derived from the Tibetan antelope, *Pantholops hodgsonii*) was enforced on the state following pressure from the international conservation community as well as the Indian government. The banning of *shahtoosh* resulted in the loss of traditional occupations for a large number of shawl workers in Kashmir. Second, the central Ministry of the Environment and Forests sponsored National Afforestation Programme (hereafter NAP) was initiated in J&K for promoting forest regeneration and protection with the help of local communities. In line with the dominant global forest management discourse of the time, the NAP aimed to facilitate the functioning of 'joint forest management' through devolution of project funds to the village forest committees, resulting in new avenues for cooperation and conflict between the forest users and managers. Apart from making significant impact on the livelihoods and sustenance of populations dependent on natural resources (*shahtoosh* wool and forest produce) in J&K, these two processes have also generated various forms of politics within local communities on the one hand, and between conservationist groups, central government, provincial government, local state and resource users on the other, in relation to control over and access to resources. In this book I discuss the agendas, interests and power of different actors involved in the two processes, and analyse how environmental policies permeate different layers of politics from macro- to microlevel in the process of implementation.[1]

In order to understand the recent politics of nature conservation in J&K, I address the following four analytical questions. First, how does power determine access to and control over natural resources? Second, how are global conservation interventions understood, accepted and reshaped by various actors? Third, what has been the impact of these recent interventions on different categories of resource users? Fourth, in what ways do these conservation processes converge with the attempts of historically powerful actors (state and local elites) to dominate the poor and marginalised populations, specifically shawl workers and forest dependent communities? While answering these questions, my purpose is not to challenge the arguments put forward by conservationists in relation to the urgent need for conserving natural

---

[1] In this book, the terms macro and micro are used to refer to the sites of study as well as relative scales. While the term micro is used to define the local or the specific site where the policy is enforced, macro refers to sites outside the local, including international, national, state and divisional levels, and also the linkages between them.

resources, both forests and wildlife, but to question the dominance of conservation interests at the cost of livelihoods and subsistence needs of the local communities. In this book, I demonstrate that in the absence of alternative livelihoods or rehabilitation support to the affected populations, the conservation policies are unlikely to meet desired goals.

1.1 Wildlife and Forest Conservation in J&K: An Introduction

A brief introduction to the two recent nature conservation interventions selected for this study deserves a mention before unravelling the key theoretical and analytical issues with which this book engages. The two interventions presented below provide interesting cases for our understanding of the larger politics of wildlife and forest conservation, especially in conflict regions.

1.1.1 The Ban on Shahtoosh Trade

As mentioned above, the international ban on trade involving *shahtoosh* was extended to J&K in 2002 to protect the 'near threatened' species of the Tibetan antelope, commonly known as *chiru*.[2] This antelope is killed for its wool, known as *shahtoosh*, which is smuggled into Kashmir from Tibet and woven into shawls and scarves.[3] While the source of *shahtoosh* is Tibet, the manufacturing of shawls takes place exclusively in the Kashmir valley. The shawls are very expensive and have remained in great demand as a luxury product in high-end fashion markets worldwide.[4] Over the years, the increasing demand for *shahtoosh* shawls resulted in the large scale poaching of the antelope in Tibet by local herdsmen.[5]

---

[2] In previously published Red List assessments of the International Union for the Conservation of Nature (IUCN), *chiru* appeared as 'vulnerable' in 1996 and 'endangered' from 2000 onwards. In the latest assessment of IUCN 2016, *chiru* is listed as 'near threatened' (IUCN 2016)

[3] The origin of shawl weaving in Kashmir is obscure. The local people of the Kashmir valley associate its origin with Shah Hamdani, a Sufi saint who introduced this skill from Persia to Kashmir centuries ago. I deal with the myths and legends associated with its origin at length in chapter three of the book.

[4] Before the ban, the price of a plain *shahtoosh* shawl would range from Rs. 15,000 to Rs. 40,000 depending upon the quality of the wool used. The price of an intricately embroidered *shahtoosh* shawl could even go up to eight to ten *lakhs* of rupees. After the ban, the price of *shahtoosh* shawls has increased manifold. I explain the continued production of *shahtoosh* shawls and its illegal trade in chapters four and five of this book.

[5] For centuries, people in Kashmir had clung to the age old belief that *shahtoosh* wool is shed by an animal seasonally and collected from rocks and bushes by the herdsmen. The truth about the connection between the massacre of the Tibetan antelope, and *shahtoosh* shawls was revealed to

There are 21 different categories of *shahtoosh* workers involved in the various stages of production and marketing of *shahtoosh* shawls.[6] *Bhotias*, the pastoral nomads of western Himalayan region, bring the unprocessed wool from Tibet to Kashmir to be sold to local wool dealers and shawl manufacturers (locally known as *vastas*).[7] The agents of manufacturers (*poiywans*) distribute the raw wool to separators (*charun wajens*) and spinners (*katun wajens*) for spinning, and then to weavers (*vovers*).[8] The woven shawls are then taken to clippers (*puruzgars*), darners (*raffugars*), and washers (*dhobis*) for finishing, and finally to the designers (*naqqashs*), printers (*chappawals*) and embroiderers (ragbars) for designing and embroidery. These shawls are then sold by hawkers (*pheriwalas*) and exporters in domestic and international markets. It is noteworthy that the wool dealers, manufacturers, agents and exporters have reaped the maximum benefits from the *shahtoosh* trade while the skilled workers involved in the production have suffered exploitation at the hands of these powerful actors.

The antelope has been included in Appendix I of the Convention on International Trade in Endangered Species of Wild Fauna and Flora (CITES) since its inception in July 1975, making any trade in *chiru* body parts and its derivatives illegal (CITES 1999). In India, the species was listed under Schedule I of the Wildlife Protection Act (1977), making any harm to the animal, or trade in its products, an illegal act. The state of J&K is the only state in India which has a separate Wildlife Protection Act. Here, *chiru* appeared in Schedule II of the Wildlife Protection Act (1978), which permitted trade, however, 'controlled' by the state. Owing to the declining number of *chirus*, evidenced by an American wildlife biologist, George Schaller, the Wildlife Protection Society of India (WPSI), a New Delhi based animal rights organisation, filed a 'Public Interest Litigation' (PIL) in the J&K High Court in 1998, calling for a complete ban on the manufacture and sale of *shahtoosh* shawls.[9]

---

the scientific and conservation communities only in 1988 by George Schaller, an American biologist who saw herdsmen in the village of Gerze plucking the *shahtoosh* wool from the hides of the antelope to sell it to wool traders (see Schaller 1993).

[6] A detailed description of all the different categories of workers and the production process is presented in chapter three. In this book, I have used the term '*shahtoosh* workers' to refer generally to all the different categories of workers engaged in *shahtoosh* shawl production. At some places, this term specifically refers to the poorer artisans from within the broader community of people involved in the shawl production, especially when comparing the interests of the poorer workers with those of powerful actors such as manufacturers, hawkers, agents and traders. Also, while using the term 'poor *shahtoosh* workers', I refer mainly to their economically poor conditions.

[7] It is to be noted that manufacturers, unlike traders or merchants, are themselves involved in the actual production process of shawls such as embroidering or weaving, alongside controlling the entire production process, employing subordinate workers and owning the finished products.

[8] In a gendered division of labour, spinning is done exclusively by women and weaving by men.

[9] The WPSI is a non-profit organisation of India founded by a wildlife photographer Belinda Wright in 1994 with the aim of providing support and information to combat wildlife poaching and illegal trade of wildlife parts. The WPSI is involved in all of India's major wildlife conservation issues and have been in the forefront of media campaigns to highlight the importance of wildlife protection. It is funded by a wide range of international and national donors including the British High Commission, New Delhi, Ford Foundation, HSBC Ltd. and American Museum of Natural History. The Society's members include leading conservationists and business people.

1.1 Wildlife and Forest Conservation in J&K: An Introduction

After a 4 year long legal battle between animal rights organisations and the J&K government, the J&K High Court in 2002, ordered the J&K state to comply with the regulations of CITES to which India is a signatory. Following this, and also due to the pressures of the central government and international conservationist groups, the J&K legislative assembly amended its Wildlife Protection Act and moved the Tibetan antelope from Schedule II to Schedule I, giving it complete protection but resulting in the loss of traditional occupations of a large number of shawl workers in the Kashmir valley.

This book examines the process of banning *shahtoosh* in J&K and the surrounding politics between environmentalist groups, central and state governments, and *shahtoosh* workers. It analyses how the international ban on *shahtoosh* trade was received, resisted and negotiated at different levels from global to local. The ban on *shahtoosh* trade is examined in the light of the following questions: (a) What are the agendas and interests of the diverse set of actors (international, national, state and local) associated with the ban on *shahtoosh* trade? (b) What is the status of *shahtoosh* trade in J&K state since its ban in 2002? (c) How can the state be understood in its dual role, acting as an agency to impose the ban and also contesting the ban in view of the demands of the people or even allowing illegal production and the trade of *shahtoosh* to continue? (d) How has the ban affected the livelihoods of different categories of *shahtoosh* workers? (e) How did the *shahtoosh* workers respond to the ban and to what extent were their interests served by the state and non-state agencies? (f) How do local power relations within the *shahtoosh* worker community determine the control over shawl production and trade after the imposition of the ban? These questions can help us analysing how global wildlife conservation interventions unfold from national to local levels, and highlight how such interventions can seep into existing relations of domination and subordination at the microlevel. An integrated analysis of macro- and microforms of politics and the differential impact of the ban on *shahtoosh* within the local communities can enhance our understanding of multiple ways through which the global meets with the local.

1.1.2 National Afforestation Programme

The forest use, control and access in India has been characterised by issues of cooperation as well as conflict between multiple actors including local users, timber contractors, state forest departments, the central Ministry of the Environment and Forests, and international development and conservation communities. From the colonial period until the late 1980s, forest management and conservation in J&K has largely remained under the exclusive control of forest bureaucracy, as in other parts of India.[10] Yet, some attempts to incorporate the participation of local forest

---

[10] I provide a detailed historical account of forest management and use during this period in chapter six of the book.

users in the projects sponsored by the Forest Department (hereafter FD) and international donors have been made in discrete locations within J&K from the 1970s onwards. The 'farm forestry' programme was initiated in 1972 to 'afforest' rural areas and supplement the incomes of farmers, and a World Bank funded 'social forestry' project was started in 1982 with the main objective of tree plantation on common lands so as to combat the global problem of 'deforestation'. Under these projects, village forest committees were formed in some villages of the state for the purpose of implementation but they lacked any effective powers and remained dormant after project completion. With the change in international development thinking in favour of 'decentralised natural resource management' by the late 1980s and the rising demands for local control of local forest resources by environmental activists in India, the Ministry of the Environment and Forests (MoEF), Government of India in 1990 initiated an ambitious programme, the 'Joint Forest Management' (hereafter JFM), to involve village communities in the protection and development of degraded forest areas surrounding villages. The JFM programme was implemented in J&K in 1992, and required formation of 'joint forest management committees' (hereafter JFMCs) with an understanding that the local communities would get better access to forest resources in return for their shared responsibility in forest protection and regeneration. However, the JFMCs lacked financial powers, which still remained within the FD.

Reviewing the functioning of the JFM in the first decade of its implementation, the MoEF, in 2002, initiated the National Afforestation Programme (NAP) to facilitate the functioning of the JFM, and to provide a firm and sustainable mechanism for the devolution of funds to the JFMCs for effective project implementation. The new scheme of the NAP was to be implemented through a two-tier decentralised structure including Forest Development Agencies or FDAs (at the forest division level) and JFMCs (at the village level).[11] Under the NAP, in contrast to previous JFM projects, the funds were to be transferred directly by the MoEF to the FDAs headed by the territorial Conservator of Forests (bypassing the state FD headquarters) which were then to be released by the FDAs to individual JFMCs.

This study of 'joint forest management' under the NAP in J&K provides an opportunity to analyse the politics of forest conservation and management involving various actors, including the MoEF, J&K Forest Department, forest field-staff, JFMCs and village communities. It engages with the following analytical questions: (a) How can we understand the changing systems of forest management and control in J&K and their implications for the relationship between forest resources and dependent populations? (b) What potential does the NAP have for sustainable management of forest resources through decentralisation, partnership and local participation? (c) To what extent has the programme been able to address the problems of access to timber, fuelwood and fodder for village residents? (d) How can the 'local state' be understood in its divided role of implementing the programme and also, at times, contravening the forest laws and regulations? (e) How do the differential

---

[11] A detailed discussion on the objectives of the NAP, the flow of funds, and the structure and composition of the FDAs and JFMCs is presented in chapter seven of this book.

1.1 Wildlife and Forest Conservation in J&K: An Introduction

power and abilities of various actors influence their transgression of forest regulations and indulgence in illegal timber harvesting? This study can enhance our understanding of the various layers of politics from macro- to microlevels, through which forest conservation and management policies are shaped and reshaped. It can demonstrate the centrality of power relations in defining the outcomes of forest conservation interventions. It can also unravel the various mechanisms by which the conservation interests can converge with the attempts of historically powerful actors to maintain control over forest resources while also creating new possibilities of cooperation, alliances and conflicts at different levels, from national to village.

In order to understand the wider politics of nature conservation, and to capture the political, cultural and regional diversities within the J&K state that could influence the outcome of conservation interventions, it was logical to select more than one nature conservation intervention for the purpose of this research. Hence, I chose one such intervention in the field of wildlife and another in the arena of forest conservation. The banning of *shahtoosh*, while affecting its trade worldwide, is of the utmost significance to the people of Kashmir valley, as the entire production of shawls is concentrated in this region. The impact of the ban on *shahtoosh* in the Jammu region is inconsequential apart from the general decline in trade. Moreover, the political dynamics of Kashmir valley are very different from that of the Jammu region, making it difficult to draw any generalisations on the politics of nature conservation based on a single study.[12] Therefore, I selected one intervention (ban on *shahtoosh*) from Kashmir, and another (on forest conservation) from the Jammu region. Also, since *shahtoosh* shawl production is an urban based activity, mainly concentrated in Srinagar city, the summer capital of the state, while the JFM programme is implemented in rural areas, the choice of two conservation interventions was made, keeping in view the various ways that local communities, both rural and urban, amidst the fragile economy of the state, can respond to nature conservation policies affecting their livelihoods and access to natural resources.

The selection of two recent processes of nature conservation in J&K was also influenced by the fact that while there is a direct conflict between the interests of the affected community, the state and those of the conservationist groups in the case of *shahtoosh*, there is no overt conflict between the interests of forest dependent populations and conservation goals in the case of the NAP. It was, therefore, worth examining how the nature conservation policies unfold under situations of both converging and diverging interests of the different players involved, for getting a holistic picture of nature conservation in J&K.

A common denominator in both cases was the requirement to evaluate the power, agendas and roles of different actors at multiple levels – international, national, state and local. Both the interventions also address the question of livelihoods and subsistence needs of the local communities in relation to the conservation policies. Also it is important to mention that although the focus of my research was not to investigate the illegal trading of *shahtoosh* as well as timber resources in J&K, I saw commonalities in both cases in terms of the dual role of the state and its nexus with powerful

---

[12] I discuss the political dynamics of J&K and introduce my field sites in the next chapter.

8 1 Introduction

actors (e.g. shawl manufacturers, timber contractors and local elites) which largely influence the outcome of nature conservation interventions.

1.2 Political Ecology: Approaches and Analytical Framework

The two processes of wildlife and forest conservation in J&K are analysed using a political ecology approach as power in resource use and access is the central component of this study. Scholars have emphasised the significance of 'political ecology' in understanding various forms of struggles involved in controlling natural resources (for a review, see Robbins 2004; Forsyth 2003; Jayal 2001; Watts 2000; Greenberg and Park 1994 and Blaikie 1985), and also in gaining insights into the responses of local actors to the challenges and opportunities posed by global discourses of conservation, environmentalism, sustainability and development (see for example, Dressler et al. 2010; Shackleton et al. 2010; Singh 2008; Springate and Blaikie 2007; Sundar et al. 2001 and Stott and Sullivan 2000).

Political ecology has been described in various ways: Watts (2000: 257) argues that political ecology understands the complex relationship of nature with society through a careful study of 'forms of access and control over resources and their implications for environmental health and sustainable livelihoods'. Also, Bryant and Bailey (1997: 28) observe that political ecologists accept the idea that 'costs and benefits associated with environmental change are for the most part distributed among actors unequally'. Robbins (2004: 12) suggests that political ecology is able to critically explain 'what is wrong with dominant accounts of environmental change, while at the same time, explore alternatives [...] in the face of mismanagement and exploitation'. Others have defined political ecology as a framework that links macrolevel political economic processes with microlevel aspects of human ecology (Dodds 1998); or 'a synthesis [...] to link the distribution of power with productive activity and ecological analysis' (Greenberg and Park 1994: 1). It can also be seen as an approach to identify 'political circumstances that forced people into activities which caused environmental degradation in the absence of alternative possibilities' (Stott and Sullivan 2000: 4).

In the context of developing countries, Moore (1993: 381) alleges that political ecology approaches have given little attention to the 'rich micropolitics that condition environmental conflicts' (see also Campbell et al. 2001; Moore 1996). It is argued that political ecology analyses should focus on the relationship of class, ethnic, and gender structures to conflicts over access to resources, the interrelations among local resource users and groups of society who affect resource use, and diversity in the decisions of local resource managers (see Rocheleau et al. 1996; Stonich 1993). Some scholars have also cautioned against the understanding of local communities as fairly autonomous and homogeneous (e.g. Painter and Durham 1995; Agrawal and Gibson 1999) and to ignore the internal complexities of the state while analysing natural resource management problems (see Baviskar 2007; Vasan 2006; Vira 1999; Gupta 1995; Moore 1993).

While recognising the importance of microlevel politics, Bryant and Bailey (1997) suggest that the third world political ecology research actually suffers from an 'exaggerated emphasis on local activities at the cost of regional and global analyses', and that there is an 'anthropological style preoccupation with local problems even when those issues are linked to structural forces' (ibid.: 6). A more appropriate course for political ecology analyses, Bryant and Bailey (1997: 196) suggest would be a 'political engagement on several fronts, at different scales and with various actors'. Along the same lines, Robbins (2004: 10) sees political ecology as evaluating the 'influence of variables acting on a number of scales, each nested within another, with local decisions influenced by regional and global policies'. Some other scholars have also acknowledged the role of historical analysis in understanding the complex relationship between social, political and ecological processes (see Agrawal 2005; Mosse 2003; Peluso 1993; Guha 1990; Blaikie and Brookfield 1987; Blaikie 1985). It is also suggested that the management of natural resources, inherently political in nature, requires inter- and transdisciplinary research rather than being 'disciplinarily divided' (Mollinga 2008: 8/2010).

Keeping in view the significance of interdisciplinarity as well as the need to have a wider canvas for multi-scale analysis, the political ecological framework adopted in this book draws from historical, political and anthropological approaches, and incorporates the study of role, agendas and interests of multiple actors at different levels – from global to local – to understand the two recent nature conservation interventions in J&K. I use interdisciplinary approach in order to examine how various historical, political, social, economic and ecological factors combine to shape nature conservation policies and their effects. The integrative multi-level and multi-actor framework of the study also takes into account the differentiated concerns of various social groups within the affected communities as well as the internal complexities within the state while examining the banning of *shahtoosh* and the implementation of NAP.

1.3 Theoretical Positioning

The issues of sustainability, livelihoods, participation and decentralised natural resource management are central in the policy and practice of the mainstream institutions in the arena of nature conservation. One of the earliest documents of the mainstream, *World Conservation Strategy* (IUCN 1980, para. 20.6) reads: 'Conservation is entirely compatible with the growing demand for "people-centred" development that achieves a wider distribution of benefits to whole populations'. Adams (2001: 270) argues that initially, the conservationist groups believed that conservation could sustain development and also that development could be reconfigured to achieve conservation goals. The mainstream conservation community advanced 'sustainable development' (SD) as a solution to the tensions between environment and poverty issues.[13] By the late 1980s, with the popularity of the concept of SD, conservationists began to claim that these objectives could be achieved at all levels: globally, nationally and locally (Adams 2001). Conservation is portrayed as something that, if properly organised, could be made to meet the needs of the local people, and sustainable development and conservation are always easier to achieve when they bring positive economic benefits to communities affected by these endeavours (Tisdell 1999; see also Dryzek 2005). There is also a strong belief that poverty is the root of environmental degradation. Scholars such as Chambers (1987: 1) challenge this viewpoint by arguing that 'the poor are not the problem, they are the solution for sustainable development'. The 'sustainable livelihoods' (SL) approach was then advanced as a solution to bridge the gap between conservation goals and the interests of local populations, and to better understand the relationship between poverty and the environment.[14]

Wildlife conservation and forestry, however, provide significant challenges to realising the goals of sustainable management of natural resources along with the sustainable livelihoods of dependent populations. Despite the celebration of the SD and SL approaches, the concept of 'sustainability' is ambiguous and does not clearly suggest what is to be sustained and for whom. It will also not be false to state that most discussion on sustainability is at the 'high concept level' rather than 'practical problem-solving level' (Hempel 1999). Yearley (1996) states that the goal of SD fails to provide an uncontested, universal discourse because the very concept is difficult to define (see also Carswell et al. 1997). Sachs (1993) argues that SD fails to address the difficult questions of what kinds of needs should be satisfied and how available resources should be distributed. The SL approach synthesises many issues into a single framework, but like SD also remains ambiguous and conveys a somewhat 'cleansed', 'neutral' approach to power issues (Ashley and Carney 1999; see also Barraclough 1997). The central importance of 'political factors' and the harsh realities of power relations in determining people's access to local resources are also largely neglected in both SD and SL literatures (Baumann and Sinha 2000). Lele (1991) argues that lacking faith in local participation, the SD approaches have led to

---

[13] The most common definition of SD reads: 'development that meets the needs of the present without compromising the ability of future generations to meet their own needs' (WCED 1987). In other words, SD suggests undertaking 'environmental planning and management in a way that does minimum damage to ecological processes without putting a brake on human aspirations for economic and social improvement' (Redclift 1987: 39).

[14] The 'Sustainable Livelihoods' approach assumes that people can pursue a range of livelihood outcomes i.e. more income, increased well being, reduced vulnerability and a sustainable resource base. This framework identifies five capital assets available to the poor: natural capital (the natural resource stocks- soil, water, air, genetic resources etc.), physical capital (ecological cycles from which resources flow and services useful for livelihoods are derived), economic or financial capital (the capital base- cash, credit/debt, savings, and other economic assets, including basic infrastructure and production equipment and technologies), human capital (the skills, knowledge, ability to work, good health and physical capability), social capital (the social networks of trust, social relations, affiliations and associations) (Chambers 1988: 7). For a review on the SL framework, see Ellis (2000), Bebbington (1999), Scoones (1998), Carney (1998) and Chambers and Conway (1992).

1.3 Theoretical Positioning

inadequacies and contradictions in policy-making, resulting in their failure to address the problems of poverty and environmental degradation. These approaches also ignore illegal activities and the issues of corruption in natural resource management (see for example, Collins 2008; Nordstrom 2004; Robbins 2000 and Peluso 1993). I found the issues of power relations as well as corruption and illegality as central in both wildlife and forest conservation processes in J&K, an issue inadequately addressed in SD approaches. While noting the contributions of the SD and SL approaches, in this study, I have examined to what extent the conservation goals are compatible with the subsistence needs of resource users and dependent populations.

The mainstream conservation and development communities have responded to the critiques in relation to the exclusion of local communities in conservation programmes by incorporating ideals of 'participation' and 'decentralisation' (especially since the late 1980s) for realising the goals of sustainable natural resource management. It is emphasised that active and willing participation of the local communities is necessary for any nature regeneration programme to succeed (Sunderland and Campbell 2008; Hulme and Murphree 2001; Gadgil 1998). It is also argued that the decentralised management over natural resources would ensure the long term sustainability of resources as well as the livelihoods of resource dependent populations (see D'Silva and Nagnath 2002; Saigal 2000 and Lynch and Talbot 1995). Adams (1996: 9) suggests that 'large scale creative conservation can involve both the restoration of nature and the reinvigoration of local communities'. Similar arguments for conservation through local cooperation have been made by Shepherd (1998) and Chambers et al. (1989). Some critics have cautioned that participation may not always alter the unequal power relations between resource users and the state as well as within local communities (e.g. Mabee and Hoberg 2006; Ballabh et al. 2002; Balland and Platteau 1996). Others have expressed scepticism on the potential of decentralised resource management for securing better outcomes for the poor as well as preserving environmental health (see Sarap and Sanrangi 2009; Sundar 2004; Ribot 1998). I shall provide a more detailed discussion on the claims of both optimists and sceptics in relation to participation, partnership and decentralisation while presenting the empirical findings in the later part of the book.

Contrary to the 'win-win' impulse of the SD and SL approaches based on cooperation between different stakeholders, the alternative stream of thinking sees global environmental concerns as a potential ground for conflict between the differently placed actors at various levels from global to local. One set of scholars highlight these conflicts at the global level between the North and the South. For example, Gupta (1998) views global environmental treaties as a 'new form of governmentality' while Peluso (1993) sees the global conservation agenda as inherently 'coercive' in nature. Others have defined global conservation interventions through concepts such as 'ecological imperialism' (Calvert and Calvert 1999), 'fortress conservation' (Hulme and Murphree 1999), 'green neo-liberalism' or 'eco-governmentality' (Goldman 1996), 'ecototalitarianism' (Dietz 1996) and 'conservation imperialism' (Guha and Martinez-Alier 1997). Calvert and Calvert (1999) argue that the aggressive support for forest conservation in tropical countries is linked to the interests of developed countries in the North. Likewise, Buttel (1992) observes that concerns about 'our common future' are, in practice, more parochially focused on the future of the wealthy industrial countries.

While I find the arguments presented above convincing to a certain extent, I also see a lack of attention paid by these scholars to understand the various forms of politics generated in response to conservation policies imposed from 'above'. Also, I problematise the assumption in the arguments of these scholars that the rationality disseminated by international treaties potentially alters the behaviour of people at the local level and changes their relationship with nature. In this study, I have argued that the conservation interventions do not go unchallenged but are contested, negotiated and reshaped by the interests of diverse actors at different levels from national to local. I shall deal with this point while analysing wildlife and forest conservation policies in J&K later in the book.

Another set of scholars highlight the increasing use of conservation ideology to justify coercive interventions at national and sub-national levels. For example, Peluso (1993) argues that in the guise of resource control, the state uses 'legitimate violence' to control people, especially recalcitrant regional groups or minority groups who challenge the authority of the state in exclusive control of resources. Bryant and Bailey (1997) maintain that environmental conservation is rarely seen by states as a goal in itself, but rather as a means to various political and economic ends (see also Blaikie 2006). Some other scholars have also explained the use of conservation ideology in systematic state intervention for controlling natural resource use by local communities (see for example, Lohmann 1996; Kothari et al. 1995; Ghimire 1994; Utting 1993; Neumann 1992 and Scott 1985).

Critics have also discussed the various ways and methods through which the state is able to exercise its control over local natural resources. For example, Agrawal (2005) analyses how forest conservation efforts are realised by the state and suggests that it is done through the mechanism of 'intimate government' (by forming forest councils and dispersing its rule) and creation of 'environmental subjects' (local forest users imbibing conservation ideals). Li maintains that by portraying deficiencies in communities in relation to effective resource management, state forest agencies 'retain their customary role as the party that produces policies, plans and regulations, prescribing and enforcing the proper relations between people and forests' (Li 2007: 280). She goes on to argue that often the state adopts the language of community forest management but treats it as 'a contract between villagers and the forest department in which the department dictates the terms' (ibid.: 280). Dean (1999: 209) argues that rights to resource use are made conditional on the performance of local communities, adding a coercive element to governmental strategies (see also Sundar 2001; Poffenberger and Singh 1996). Apart from these mechanisms, some scholars suggest that the state (and conservationists) also uses 'scientific rationality' as a means of establishing its privileged position of legitimate controller of local natural resources (see Forsyth and Walker 2008; Saberwal and Rangarajan 2003). Let us keep these viewpoints in mind while examining the role of state and conservationist groups in redefining the relationship between local communities and natural resources in J&K, in the following chapters.

1.3 Theoretical Positioning

This book complements the emerging literature on the politics of contingency and contestation in the context of natural resources (e.g. Gupta 2009; Li 2007; Rangan 2004 and 1995; Sundar et al. 2001).[15] My research, while acknowledging the centrality of politics in resource control and access, also recognises the dispersed and fluid nature of power and its exercise at multiple levels by conservationist groups, the heterogeneous state and local elites for retaining their hegemony and domination. Such an understanding of power motivates us to problematise the notion of the state as a unified or monolithic entity, and also to appreciate the diverse interests and abilities of differently positioned social groups within the local communities (cf. Midgal 1988). Below, I note the arguments of some other scholars who have influenced my understanding of how global conservation interventions shape local realities and get reshaped in the process of implementation.

Rose (1999: 51) argues that governance is not a process in which rule extends itself unproblematically across a territory but 'a matter of fragile relays, contested locales and fissiparous affiliations'. Along the same lines, Moore (2000: 43) suggests that international development needs to be understood 'not as a machine that secures fixed and determined outcomes but rather as a site of contestation'. Gupta notes that development, a powerful tool of domination in the postcolonial era, has also been appropriated and reshaped by subaltern groups, particularly in the context of Third World agrarian populism (Gupta 1998: 320). It is also argued that very often, global encroachments on the ecology of a nation or encroachments of a predatory state on the life and livelihoods of the local communities result in indigenous or regionalist expressions, a kind of reaction which Cederlof and Sivaramakrishnan (2006) call 'ecological nationalisms'. Such indigenous claims are not merely acts of strategic cultural identity politics but also claims to territory, its resources and the desire to maintain subsistence livelihoods (ibid.). Martinez-Alier (2003) explains how local settler communities in Guatemala have learnt to defend their interests and manoeuvre through the language of indigenous territorial rights and sustainable development. Moreover, Rossi (2004: 26) argues that relative distance from the sources of development rationality increases the room for manoeuvre available to the recipients of policies and interventions. In addition, Rangan (1997) observes that an analytical distinction needs to be made between 'property' and 'control'

---

[15] The approach in this research is comparable to some recent studies in the field of natural resource management. For example, Gupta (2014) analyses the interests, agendas and power of various actors involved in 'watershed development' interventions in India, ranging from international development and donor agencies to the state, as well as different types of grassroots organisations, and presents complex processes of cooperation, conflict and negotiation between the various 'agents' and 'recipients' of watershed development programmes. Li (2007), in the context of the community forestry programme in Indonesia, examines the role of a range of parties beyond the state that attempt to govern forest resources including NGOs and donor agencies with their teams of expert consultants, social reformers, scientists and research institutions. She argues that community forest management is the assemblage that has emerged in the space of struggle between villagers and forest bureaucracies on the forest edge. Sundar et al. (2001), in their study of the JFM in five states of India, provide a detailed ethnographic work at the village level with a survey of NGOs and forest departments at state and national levels, and argue that in the process of implementing participatory forestry, communities and their needs are being reconfigured.

over natural resources and emphasises that both proposals for state ownership or local communities' ownership of forest resources are flawed because they assume that 'ownership' status determines the ways in which resources are used and managed. In reality, at the ground level, practices reveal several contestations and negotiations between the state and local communities rather than exhibiting a monolithic imposition of proprietary rights (ibid.). A common denominator in the arguments of these various scholars is that they contest the idea that policies from 'above' have the sole power to transform the local realities. In this book, I have demonstrated this point in the context of both the banning of *shahtoosh* and forest conservation intervention in J&K.

1.4   Contesting Conservation: What This Study Contributes

The principal originality of the research on which this book is based lies in its empirical focus on understanding the politics surrounding the banning of *shahtoosh* trade, and the functioning of joint forest management in J&K. The available literature on *shahtoosh* in particular and the shawl industry in general, mainly consists of historical accounts by travellers (such as Younghusband 1909; Schonsberg 1853; Moorcroft 1820) and the works of art historians (e.g. Rehman and Jafri 2006; Ahad 1987; Ames 1986; Irwin 1973). These accounts do not suggest any connections between *shahtoosh* and the Tibetan antelope; the origin of wool remains unclear although *shahtoosh* has been described as the finest quality wool in comparison to other varieties such as *pashmina*.[16] Apart from these historical accounts of the shawl industry, there is no academic work specifically on *shahtoosh* shawl production before or after the ban. The project reports produced by wildlife conservation organisations from the late 1990s onwards are the only secondary materials available on this subject. In the case of forest conservation, there is no comprehensive account of forest management in the state during the colonial period apart from the annual assessment reports and working plans produced by the J&K Forest Department. There is also no academic work on the performance of the JFM or NAP in the state. This study is, therefore, the first systematic account of the recent politics of wildlife and forest conservation in J&K.

The literature on J&K is also preoccupied with issues relating to political conflict, militancy and forced migration in the post-independence period. The field of ecology and environment has largely remained unexplored in the literature on this region, with the available studies mainly focusing on the natural science aspects of the environment, such as wetlands conservation, flora and fauna diversity, environmental pollution, deforestation etc. (see for example, Bhatt 2004; Chadha 1991). There is a lack of attention to social and political aspects of the environment and natural resources management. This study is an attempt in the field of political

---

[16] *Pashmina*, popularly known as *cashmere* (outside India) is another fine quality wool, derived from a reared goat found in Ladakh.

ecology in J&K which can also contribute to the emerging literature on environmental politics in conflict situations (see Le Billon 2005; Korf 2004; Peluso and Watts 2001; Homer-Dixon 1999).

This research also makes a significant contribution to understanding the illegality and corruption in natural resource use and management. While illegal wildlife and the timber trade were not the primary objectives of my research, I was able to gather information on these issues while collecting data on the impact of the ban on *shahtoosh* workers and interacting with village residents and forest field-staff. Despite widespread evidence of bribery and illegal exchange in natural resource management, corruption is largely unexplored and undertheorised in the literature (Robbins 2000; Omara-Ojungu 1992). By explaining some new mechanisms adopted by various actors in the illegal production and trade of *shahtoosh* as well as illegal timber harvesting, my findings add to the existing literature on illegality and corruption in the context of resource use (see for example, Bloomer 2009; Wardell and Lund 2006; Wiggins et al. 2004; Leon 1994; Peluso 1993).

This study can also enhance our knowledge of how conservation interventions are understood, contested and reshaped at various levels. The dispersed and fluid notion of power advanced in the book can be particularly useful in understanding the politics of contingency and contestation in the arena of natural resource governance. This book also contributes to the debates on the role of local power relations in determining access to forest resources; the differential impacts of conservation policies on resource dependent populations; the dual role of implementing agencies; and the relationship between law, science and politics in nature conservation interventions. Besides contributing to academic literature and debates on nature conservation, this research could also provide inputs on policy-making. It suggests that issues of local access to forest resources and the rehabilitation of affected communities need to be adequately addressed in the policies to make conservation work. This becomes even more significant in conflict regions and fragile economies, as in the case of J&K, where issues of violence, exclusion and poverty are intertwined.

1.5 Chapter Layout

The book is divided into nine chapters. Following this introductory chapter, I present research setting and note on fieldwork in Chap. 2. The next six chapters provide in-depth analyses of the two interventions chosen for this study. The empirical findings and discussion on the ban on the *shahtoosh* trade are presented in Chaps. 3, 4 and 5 while those of forest conservation are presented in Chaps. 6, 7 and 8 of the book. Chapter 9 presents the main conclusions on the basis of the two processes studied. Below, I provide the summary of contents in each chapter.

Chapter 2 gives an overview of Jammu and Kashmir state with specific reference to the political dynamics since the year of independence. I also introduce my field locations in J&K and describe my research methods and design for data collection.

Towards the end of the chapter, I present the specific challenges faced while collecting data for my research.

Chapter 3 begins by providing background information on *chiru* and its habitat, and traces the diverse myths and legends relating to the origin of *shahtoosh*. I describe the production process of the *shahtoosh* shawl, and the tasks performed by different categories of shawl workers. I explain the growth and development of the shawl industry since the medieval era until the 1980s, and highlight the poor conditions of shawl workers in the Valley who have experienced exploitation at the hands of both rulers and shawl manufacturers under different regimes. I also examine the changing and overlapping roles of various categories of workers in the shawl industry. Chapter 4 explains the chronology of events that led to the imposition of the ban on *shahtoosh* and analyses the legal battle between the conservationist groups and J&K state. I also present the arguments of various actors with regard to prospects of *chiru* breeding. While examining the response of the *shahtoosh* workers to the ban, I highlight the limited space of protest available to them. I also problematise the role of the state both in enforcing the ban and allowing the now illegal trade to continue (a process I have termed 'split role' of the state). The chapter ends by sketching shadow networks of *shahtoosh* and presenting diverse viewpoints on the connection between militancy and the *shahtoosh* trade in Kashmir. Chapter 5 highlights various forms of micropolitics surrounding the ban on *shahtoosh*. I discuss the perpetuation of myths regarding the origin of wool and the rationale behind the banning of trade, the differential impact of the ban on various categories of *shahtoosh* workers, their changing socio-cultural relations and new mechanisms of exploitation adopted by manufacturers and other powerful actors. I illustrate the ongoing illegal production of *shahtoosh* shawls in the Valley and explore a phenomenon wherein powerful actors delegate illegal tasks to their subordinate workers (a process I have termed 'delegated illegality'). The chapter ends by examining the initiatives of state and non-state organisations to support the ban affected shawl workers.

In Chap. 6, I turn to the study of forest conservation and management. I begin by a brief description of forest management in the early colonial period and provide a detailed analysis of the politics of 'scientific forestry' in J&K from the late nineteenth century to the 1970s. Following this, I highlight the shift in forest management practices from 'custodial' (in the colonial and post-colonial period) to 'social' (by the 1980s) and also the changing relationship between forest managers and forest users over the years. Chapter 7 analyses the rationale and objectives of the NAP, and the prospects for achieving sustainable management of forests through local participation and devolution of power, funds and responsibilities. I examine the interplay between the MoEF, the FD and the State Forest Corporation to suggest repeated re-centralisation in the guise of decentralisation of forest management and control. Following the discussion on the politics of policy-making and implementation, I provide background information on the twin villages chosen for the study. The focus then turns to analysing the micropolitics of resource use, involving the forest field-staff, villagers and timber contractors. Chapter 8 extends the discussion on micropolitics further by examining the interplay between the JFMC, forest field-staff and *panchayat*. I highlight the areas of tension as well as new areas of cooperation between these actors and analyse the differential impact of forest conservation intervention on differently positioned rural social groups with regard to their access to forest resources. The chapter also demonstrates the dual role of the forest field-staff in enforcing forest regulations and allowing illegal timber harvesting by various players including the para-military forces deployed in the state, affecting the outcome of forest conservation policies.

Chapter 9 concludes by reflecting on the broader findings of the study and relocating its contribution to the existing literature on the politics of natural resource conservation as well as on the state of Jammu and Kashmir. I also outline the broad implications of my findings for policy-making in the arena of natural resource management.

References

Adams W (1996) Future nature: a vision for conservation. Earthscan, London
Adams WM (2001) Green development: environment and sustainability in the third world. Routledge, London
Agrawal A (2005) Environmentality: technologies of government and political subjects. Duke University Press, London
Agrawal A, Gibson C (1999) Enchantment and disenchantment: the role of community in natural resource conservation. World Dev 27(4):629–649
Ahad A (1987) Kashmir to Frankfurt: a study of arts and crafts. Rima Publishing House, New Delhi
Ames F (1986) The Kashmir shawl. Antique Collectors Club, Suffolk
Ashley C, Carney C (1999) Sustainable livelihoods: lessons from early experience. Department for International Development, London
Ballabh V, Balooni K, Dave S (2002) Why local resource management institutions decline: a comparative analysis of van (forest) panchayats and forest protection committees in India. World Dev 30(12):2153–2167
Balland JM, Platteau JP (1996) Halting degradation of natural resources: is there a role for rural communities? Oxford University Press, Oxford
Barraclough S (1997) Rural development and the environment: towards ecologically and socially sustainable development in rural areas. United Nations Research Institute for Social Development, Geneva
Baumann P, Sinha S (2000) Panchayati raj institutions and natural resource management. Overseas Development Institute, London
Baviskar A (ed) (2007) Waterscapes: the cultural politics of a natural resource. Orient Longman, New Delhi
Bebbington A (1999) Capitals and capabilities: a framework for analysing peasant viability, rural livelihood and poverty. World Dev 27(12):2021–2044
Bhatt S (2004) Kashmir ecology and environment: new concerns and strategies. APH Publishers, New Delhi
Blaikie P (1985) The political economy of soil erosion in developing countries. Longman, London
Blaikie P (2006) Is small really beautiful? Community based natural resource management in Malawi and Botswana. World Dev 34(11):1942–1957
Blaikie P, Brookfield H (1987) Land degradation and society. Methuen, London
Bloomer J (2009) Using a political ecology framework to examine extra-legal livelihood strategies: a Lesotho-based case study of cultivation of and trade in cannabis. J Polit Ecol 16:49–69

Bryant R, Bailey S (1997) Third world political ecology. Routledge, London

Buttel FH (1992) Environmentalisation: origins, processes and implications for rural social change. Rural Sociol 57:127–146

Calvert P, Calvert S (1999) The south, the north and the environment. Pinter, London

Campbell BM, de Jong W, Luckert M, Mandondo A, Matose F, Nemarundwe N, Sithole B (2001) Challenges to proponents of CPR systems: despairing voices from the social forests of Zimbabwe. World Dev 29(4):589–600

Carney D (1998) Sustainable rural livelihoods, Regional development dialogue 8. Department for International Development, London

Carswell G, Hussain G, McDowell K, Wolmer W (1997) Sustainable livelihoods: a conceptual approach. IDS. Mimeo, Brighton

Cederlof G, Sivaramakrishnan K (2006) Introduction: ecological nationalisms: claiming nature for making history. In: Cederlof G, Sivaramakrishnan K (eds) Ecological nationalisms. Permanent Black, New Delhi, pp 1–40

Chadha SK (1991) Kashmir: ecology and environment. Mittal Publications, New Delhi

Chambers R (1987) Sustainable livelihoods, environment and development: putting poor rural people first, Paper No. 240. Institute of Development Studies, Sussex

Chambers R (1988) Sustainable rural livelihoods: a strategy for people, environment and development. In: Conroy C, Litvinoff M (eds) The greening of aid: sustainable livelihoods in practice. Earthscan, London, pp 1–46

Chambers R, Conway G (1992) Sustainable rural livelihoods: practical concepts for the twenty-first century. In: IDS discussion paper 296. Institute of Development Studies, Sussex

Chambers R, Saxena NC, Shah T (1989) To the hands of the poor: water and trees. Immediate Technology Publications, London

CITES (1999) Conservation of and control of trade in the Tibetan antelope. www.cites.org/eng/res/11/11-08R13.shtml. Accessed 1 Feb 2006

Collins TW (2008) The political ecology of hazard vulnerability: marginalisation, facilitation and the production of differential risks to urban wildfires in Arizona's White Mountains. J Polit Ecol 15:21–43

D'Silva E, Nagnath B (2002) Behroonguda: a rare success story in joint Forest management. Econ Polit Wkly 9:551–557

Dean M (1999) Governmentality: power and rule in modern society. Sage, London

Dietz AJ (1996) Entitlements to natural resources: contours of political environmental geography. International Books, Utrecht

Dodds D (1998) Lobster in the rain forest: the political ecology of miskito wage labor and agricultural deforestation. J Polit Ecol 5:83–108

Dressler W, Buscher B, Schoon M, Brockington D, Hayes T, Kull C, McCarthy J, Shrestha K (2010) From hope to crisis and back again? A critical history of the global CBNRM narrative. Environ Conserv 37(1):5–15

Dryzek J (2005) Deliberative democracy in divided societies: alternatives to agonism and analgesia. Polit Theo 33:218–242

Ellis F (2000) Rural livelihoods and diversity in developing countries. Oxford University Press, Oxford

Forsyth T (2003) Critical political ecology: the politics of environmental science. Routledge, London

Forsyth T, Walker A (2008) Forest guardians, forest destroyers: the politics of environmental knowledge in northern Thailand. University of Washington Press, Seattle

Gadgil M (1998) Grassroots conservation practices: revitalising the traditions. In: Kothari A, Pathak N, Anuradha R, Taneja B (eds) Communities and conservation: natural resource management in south and Central Asia. Sage, New Delhi, pp 219–238

Ghimire KB (1994) Parks and people: livelihood issues in national parks management in Thailand and Madagascar. Dev Chang 25:195–229

Goldman M (1996) Eco-governmentality and other transnational practices of a "green" World Bank. In: Peet R, Watts M (eds) Liberation ecologies. Routledge, New York, pp 166–192

References

Greenberg J, Park T (1994) Political ecology. J Polit Ecol 2(1):1–12
Guha RC (1990) The unquiet woods: ecological change and peasant resistance in the Himalayas. University of California Press, Berkeley
Guha RC, Martinez-Alier J (1997) Varieties of environmentalism: essays north and south. Earthscan, London
Gupta A (1995) Blurred boundaries: the discourse of corruption, the culture of politics and the imagined state. Am Ethnol 22(2):375–402
Gupta A (1998) Postcolonial developments in the making of modern India. Duke University Press, Durham
Gupta S (2009) The politics of development in rural Rajasthan, India: evidence from water conservation and watershed development programmes since the early 1990s. PhD thesis. University of London
Gupta S (2014) Worlds apart? Challenges of multi-agency partnership in participatory watershed development in Rajasthan, India. Dev Stud Res 1(1):100–112
Hempel L (1999) Conceptual and analytical challenges in building sustainable communities. In: Mazmanian D, Kraft M (eds) Towards sustainable communities: transition and transformations in environmental policy. The MIT Press, Cambridge, MA, pp 33–62
Homer-Dixon T (1999) Environment scarcity and violence. Princeton University Press, New Jersey
Hulme D, Murphree M (1999) Policy arena: communities, wildlife and the "new conservation" in Africa. J Int Dev 11(2):277–285
Hulme D, Murphree M (2001) African wildlife and livelihoods: the promise and performance of community conservation. James Currey, London
Irwin J (1973) The Kashmir shawl. Her Majesty's Stationary Office, London
IUCN (1980) World conservation strategy: living resource conservation for sustainable development. http://data.iucn.org/dbtw-wpd/edocs/WCS-004.pdf. Accessed 20 Feb 2011
IUCN (2016) IUCN red list of threatened species. www.iucnredlist.org/details/15967/0. Accessed 1 Aug 2017
Jayal NG (2001) Balancing political and ecological values. Environ Polit 10(1):65–88
Korf B (2004) Wars, livelihoods and vulnerability in Sri Lanka. Dev Chang 35(2):275–295
Kothari A, Suri S, Singh N (1995) People and protected areas: rethinking conservation in India. Ecologist 25:188–194
Le Billon P (2005) Corruption, reconstruction and oil governance in Iraq. Third World Q 26(4):685–703
Lele SC (1991) Sustainable development: a critical review. World Dev 19(6):607–621
Leon M (1994) Avoidance strategies and governmental rigidity: the case of the small-scale shrimp fishery in two Mexican communities. J Polit Ecol 1:67–81
Li TM (2007) Practices of assemblage and community forest management. Econ Soc 36(2):263–293
Lohmann, L (1996) Freedom to plant: Indonesia and Thailand in a globalising pulp and paper industry. In Parnwell, M. and Bryant, R. (Eds), Environmental change in South East Asia. 23–48. London: Routledge
Lynch O, Talbot K (1995) Balancing acts: community based forest management and national law in Asia and the Pacific. World Resources Institute, Washington, DC
Mabee H, Hoberg G (2006) Equal partners? Assessing co-management of forest resources in Clayoquot sound. Soc Nat Resour 19(10):875–888
Martinez-Alier J (2003) The environmentalism of the poor: a study of ecological conflicts and valuation. Edward Elgar, Cheltenham
Midgal J (1988) Strong societies and weak states: state-society relations and state capabilities in the third world. Princeton University Press, Princeton
Mollinga P (2008) Water, politics and development: framing a political sociology of water resources management. Water Altern 1(1):7–23
Moorcroft W (1820) Excerpt from a letter from Moorcroft to Mr. Metcalfe, the East India Company's resident at Delhi. India Office, London
Moore DS (1993) Contesting terrain in Zimbabwe's eastern highlands: political ecology, ethnography and peasant resource struggles. Econ Geogr 69:380–401

Moore DS (1996) Marxism, culture and political ecology: environmental struggles in Zimbabwe's eastern highlands. In: Peet R, Watts M (eds) Liberation ecologies: environment, development, social movements. Routledge, London, pp 125–147

Moore DS (2000) The crucible of cultural politics: reworking "development" in Zimbabwe's eastern highlands. Am Ethnol 26(3):654–689

Mosse D (2003) The rule of water: statecraft, ecology and collective action in South India. Oxford University Press, Oxford

Neumann RP (1992) Political ecology of wildlife conservation in the Mt Meru area of northeast Tanzania. Land Degrad Rehabil 3:85–98

Nordstrom C (2004) Shadows of war: violence, power and international profiteering in the twenty-first century. University of California Press, Berkeley

Omara-Ojungu PH (1992) Resource management in developing countries. Longman Scientific, Essex

Painter M, Durham W (eds) (1995) The social causes of environmental destruction in Latin America. University of Michigan Press, Ann Arbor

Peluso NL (1993) Coercing conservation: the politics of state resource control. In: Lipshutz R, Conca K (eds) The state and social power in global environmental politics. Columbia University Press, New York, pp 199–218

Peluso NL, Watts M (2001) Introduction. In: Peluso NL, Watts M (eds) Violent environments. Cornell University Press, Ithaca/New York, pp 1–30

Poffenberger M, Singh C (1996) Communities and the state: re-establishing the balance in Indian forest policy. In: Poffenberger M, McGean B (eds) Village voices, forest choices: joint forest management in India. Oxford University Press, New Delhi, pp 56–85

Rangan H (1995) Contested boundaries: state policies, forest classifications, and deforestation in the Garhwal Himalayas. Antipode 27(4):343–362

Rangan H (1997) Property versus control: the state and forest management in the Indian Himalayas. Dev Chang 28(1):71–94

Rangan H (2004) From Chipko to Uttaranchal: development, environment, and ocial protest in the Garhwal Himalayas, India. In: Peet R, Watts M (eds) Liberation ecologies. Routledge, London, pp 205–226

Redclift M (1987) Sustainable development: exploring the contradictions. Routledge, London

Rehman S, Jafri N (2006) Kashmiri shawl from Jamavar to paisley. Mapin Publishing House Ltd., Ahmedabad

Ribot J (1998) Theorising access: Forest profits along Senegal's charcoal commodity chain. Dev Chang 29:307–341

Robbins P (2000) The rotten institution: corruption in natural resource management. Polit Geogr 19(4):423–443

Robbins P (2004) Political ecology: a critical introduction. Blackwell, Oxford

Rocheleau D, Thomas-Slayter B, Wangari E (eds) (1996) Feminist political ecology. Routledge, London

Rose N (1999) Powers of freedom: reframing political thought. Cambridge University Press, Cambridge

Rossi B (2004) Revisiting Foucauldian approaches: power dynamics in development projects. J Dev Stud 40(6):1–29

Saberwal V, Rangarajan M (2003) Introduction. In: Saberwal V, Rangarajan M (eds) Battles over nature: science and politics of conservation. Permanent Black, New Delhi, pp 1–30

Sachs W (ed) (1993) Global ecology: a new arena of political conflict. Zed Books, London

Saigal S (2000) Beyond experimentation: emerging issues in the institutionalisation of joint Forest management. Environ Manag 2(3):269–281

Sarap K, Sanrangi TK (2009) Malfunctioning of forest institutions in Orissa. Econ Polit Wkly XLIV(37):18–22

Schaller GB (1993) Tibet's remote Chang Tang. Natl Geogr 184(2):62–87

Schonsberg BE (1853) India and Kashmir. Hurst & Blackett Publishers, London

References

Scoones I (1998) Sustainable rural livelihoods: a framework for analysis, Working Paper No. 72. Institute of Development Studies, Sussex

Scott J (1985) Weapons of the weak: everyday forms of peasant resistance. Yale University Press, New Haven

Shackleton CM, Wills T, Brown K, Polunin NVC (2010) Editorial: reflecting on the next generation of models for community-based natural resources management. Environ Conserv 37(1):1–4

Shepherd A (1998) Sustainable rural development. Macmillan, Basingstoke

Singh S (2008) Contesting moralities: the politics of wildlife trade in Laos. J Polit Ecol 15:1–20

Springate-Baginski O, Blaikie P (2007) Forests, people and power: the political ecology of reform in South Asia. Earthscan, London

Stonich S (1993) "I am destroying the land!": the political ecology of poverty and environmental destruction in Honduras. Westview Press, Bolder

Stott P, Sullivan S (eds) (2000) Political ecology: science, myth and power. Arnold, London

Sundar N (2001) Beyond the bounds? Violence at the margins of new legal geographies. In: Peluso NL, Watts M (eds) Violent environments. Cornell University Press, Ithaca, pp 328–353

Sundar N (2004) Devolution, joint forest management and the transformation of social capital. In: Bhattacharyya D, Jayal N, Mohapatra B, Pye S (eds) Interrogating social capital: the Indian experience. Sage, New Delhi, pp 203–232

Sundar N, Jeffery R, Thin N (2001) Branching out: joint forest management in India. Oxford University Press, New Delhi

Sunderland T, Campbell B (2008) Conservation and development in tropical forest landscapes: a time to face the trade-offs? Environ Conserv 34(4):276–279

Tisdell C (1999) Conditions for sustainable development: weak and strong. In: Dragum AK, Tisdell C (eds) Globalisation and the impact of trade liberalisation. Edward Elgar, Cheltenham, pp 23–36

Utting P (1993) Trees, people and power: social dimensions of deforestation and forest protection in central America. Earthscan, London

Vasan S (2006) Living with diversity. Indian Institute of Advanced Study, Shimla

Vira B (1999) Implementing joint forest management in the field: towards an understanding of the community-bureaucracy interface. In: Jeffery R, Sundar N (eds) A new moral economy for India's forests? Discourses on community and participation. Sage, New Delhi, pp 254–275

Wardell DA, Lund C (2006) Governing access to trees in northern Ghana: micropolitics and the rents of non-enforcement. World Dev 34(11):1887–1906

Watts M (2000) Political ecology. In: Barners T, Scheppard E (eds) A companion to economic geography. Blackwell, Oxford, pp 257–275

WCED (1987) Our common future. Oxford University Press, Oxford

Wiggins S, Marfo K, Anchirinah V (2004) Protecting the forest or people? Environmental policies and livelihoods in the forest margins of southern Ghana. World Dev 32(11):1939–1955

Yearley S (1996) Sociology, environmentalism, globalization. Sage, London

Younghusband FE (1909) Kashmir. Adam & Charles Black, London

Chapter 2
Jammu and Kashmir: Contextualising Conservation in Specific Sites

Abstract This chapter lays out background information on the state of J&K and a brief introduction to the field sites, methods and techniques used for collection of data on which the book is based. Section 2.1 provides an overview of the geographical features and demographic composition as well as unique political challenges witnessed by the state since independence, owing to its special status in the Indian Constitution. Section 2.2 describes methods, design, and techniques of data collection for the research. I discuss the specific challenges faced during fieldwork, and present the limitations of the study towards the end of the chapter.

Keywords Jammu and Kashmir · Conflict and conservation · Research method · Multi-sited ethnography · Qualitative research

This chapter lays out background information on the state of J&K and a brief introduction to the field sites, methods and techniques used for collection of data on which the book is based. Section 2.1 provides an overview of the geographical features and demographic composition as well as unique political challenges witnessed by the state since independence, owing to its special status in the Indian Constitution. Section 2.2 describes methods, design, and techniques of data collection for the research. I discuss the specific challenges faced during fieldwork, and present the limitations of the study towards the end of the chapter. The historical and social milieus specifically relating to the two nature conservation interventions will be discussed in detail later in the book – the origin, growth and development of the shawl industry in Kashmir in Chap. 3, and history of forest management and control (from the early colonial period to the late 1980s) in Chap. 6.

© Springer International Publishing AG 2018
S. Gupta, *Contesting Conservation*, Advances in Asian Human-Environmental Research, https://doi.org/10.1007/978-3-319-72257-3_2

2.1 Research Setting

2.1.1 Jammu and Kashmir: An Introduction

J&K occupies a special status in the Indian Union as it is the only state of India with a separate constitution. Geographically, the state is divided into three regions viz. Jammu, Kashmir and Ladakh, covering an area of 2,22,236 km^2 (GoJ&K 2001). Srinagar is the summer capital while Jammu is the winter capital of the state. A large part of the state territory is under Pakistani administration (since 1947) and also under Chinese control (since the Indo-Chinese war in 1962), resulting in a long lasting dispute between the three countries (see Map 2.1 below). Effectively, the Indian administered area of the state is estimated to be 1,01,387 km^2 (ibid.). The Jammu region is separated from the rest of the state by the Pir Panjal range of the Himalayas. The Kashmir valley, situated between Jammu and Ladakh, is surrounded by the inner Himalayas from the northeast to the northwest, the Pir Panjal range from the southeast to the southwest and Pakistan in the west. Ladakh region is characterised by a rugged topography at an average altitude of over 3000 m (Dame and Nüsser 2008). The western and central parts of the region are dominated by high mountain ranges and carved valleys whereas eastern Ladakh is characterised by the high altitude plateau of Changthang. This peripheral region is only accessible by road from the Indian lowland via Srinagar or Manali, crossing some of the highest motorable passes in the world (ibid). The sparsely populated former kingdom of

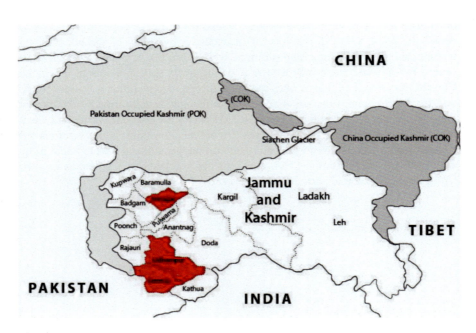

Map 2.1 Field locations in Jammu and Kashmir

2.1 Research Setting

Ladakh had been ruled by the Maharaja of J&K for more than a century before India's and Pakistan's independence in 1947.

In 2011, the population of the state was 12.54 million of which 72.62% resided in rural areas and 27.38% in urban areas (Census of India 2011). Of the total population of the state, Muslims constitute nearly 68.3% (largely concentrated in Kashmir valley), Hindus approximately 28.4% (mainly residing in the Jammu region), Sikhs 1.9% (concentrated in the Jammu region) and Buddhists 0.9% (mostly in Ladakh) (ibid.). Nearly 90% of the population is dependent on agriculture, forests, tourism and the handicraft industry for sustenance (Ramasubramanian 2004). The dependence of the local population on forest resources, particularly timber and fuelwood, is very high due to harsh winter conditions in the region. Other forest produce such as medicinal plants, herbs and fruits also provide additional incomes to the village residents, especially the women. The state has not seen large-scale industrialisation due to poor infrastructural facilities (mainly transportation) and political factors (such as militancy and terrorism), limiting the range of livelihood options available to the citizens. Owing to a rich handicraft tradition, the art and craft industry occupies a significant place in the social and economic life of the people in J&K. This industrial activity is mainly unorganised and limited to small-scale. Carpets, shawls, wood carving, papier mache and silver work are some of the prominent handicraft articles produced in J&K. Since the late 1980s, the rise in militancy has severely affected the tourism industry and, in turn, the arts and crafts industry in the state.

In Ladakh region of the state, subsistence-oriented land use has remained a dominant feature of the local economy for long but it has witnessed a decline over the last years. Nevertheless, agro-pastoral activities persist as the central pillar of local livelihoods (Dame and Nüsser 2011). Additionally pashm, the winter undercoat of pashmina goats which are bred on the high plateaux of Ladakh, Tibet and Central Asia, represents a lucrative trade commodity in the region. The trade in pashm from Ladakh to Srinagar, supplying the Kashmir shawl production, has found reference in several historic documents (Rizvi 1999). Due to the present geopolitical situation, Ladakh is the only supplier of pashmina wool for the Kashmir shawl industry (Ahmed 2004). Following this preliminary information on the state, I now turn to note some significant political developments in the state since the year of independence.

In 1947, with the end of British rule, when the princely states were given an option to join India or Pakistan, the then Maharaja of J&K (a Hindu Dogra ruler) kept delaying the decision to accede to either of the dominions. Challenged by the popular movement for democratic change within the state that had been going on since the early 1930s, he tried to find ways to retain his autocratic power (Bose 1997). However, a Pakistan sponsored attack of tribal bands in the same year left him with no option than to ask the Indian government for help (Jan 2005: 68). The Indian government laid the condition that the instrument of accession should be duly signed by him and the state should first accede to Indian Union. The Maharaja agreed to sign the instrument of accession but asked for the special status of the state in recognition of the unique conditions of accession. By the time the J&K state signed the instrument of accession in October 1947, large parts of the former princely state were already under Pakistan's control. The Indian government took the Kashmir dispute to the United Nations, whose resolution asked Pakistan to vacate the occupied areas and asked India to organise a plebiscite for determining the will of the people on the question of accession. However, the issue of plebiscite has been lingering and Pakistan has refused to vacate the occupied areas resulting in continued hostility between the two countries. The separatist sentiments within Kashmir have been prevalent since its accession and the Indian government has tried to curb them through various mechanisms such as democratic suspension, liquidation of special status and heavy deployment of armed and para-military forces.[1]

The J&K state originally ceded to the Indian Union in matters of defence, foreign affairs and communications, retaining autonomy in all other matters. However, the autonomy has significantly declined in the decades after independence mainly on the pretext of national threat (posed by Pakistan) and disintegration. While large parts of its territory continue to remain under Pakistani and Chinese administration, the state has witnessed fluctuating fortunes with regard to the democratic process as well as democratic functioning of institutions in the post-independence period (Punjabi 1990). There are, however, a few important markers in the history of democracy in J&K such as introduction of land reforms, *panchayati raj* institutions, and rights for shawl workers and artisans which illustrate the pre-dominance of pro-poor policies in the initial years after independence (Gupta 2013).

The roots of participatory management of political institutions in J&K can be dated as far back as 1944 when the All Jammu and Kashmir National Conference Party under the leadership of Sheikh Abdullah[2] formulated its agenda for radical (social, economic and political) transformation in a document titled *Naya Kashmir*. The transformatory policy spelt out in this manifesto reads: 'All the regions of the state shall have equal right(s) to participate in the political power and the same will be decentralised up to district, block and panchayat levels' (Shafi 1990: 32). Guided by this manifesto, some important policies were framed in the early 1950s (when Sheikh Abdullah assumed power in the state) to bring about political and socio-economic reforms, which are pertinent to note.

Firstly, land reforms or the 'land to the tiller' policy implemented soon after independence changed the social fabric of J&K. In 1947, following the collapse of princely rule, the first state government under Sheikh Abdullah based its development agenda along the lines of building a social and participatory democracy (Gupta 2013). The 'backwardness' of the state was largely attributed to landlordism and it

---

[1] For details on the issues of autonomy, separatism and political dynamics in J&K, see Jha (2003), Sharma (2002), Widmalm (2002), Ganjoo (1998) and Puri (1993/1999).

[2] Sheikh Abdullah (1905–1982) was the leader of the 'All Jammu and Kashmir Muslim Conference' (later renamed the 'All Jammu and Kashmir National Conference') formed in 1932 to redress the grievances of Muslims and other marginalised sections under Hindu Dogra rule in the princely state of J&K (Jan 2005: 61). He also started 'Quit Kashmir Movement' in 1946 to directly challenge the autocratic rule of Maharaja Hari Singh of J&K. Soon after the accession of the state to India in 1947, he served as the Prime Minister of the state but was dismissed in August, 1953 at the behest of the Indian government. He was immediately arrested and later jailed for 11 years on charges of conspiracy against the state. He again came to power in 1975 and served as Chief Minister until his death in 1982.

was held that to safeguard the interests of the class of tenants, there was a need for the redistribution of land ownership rights (Prakash 2000).[3] Hailed as one of the most radical and successful land reform programmes in India, the Big Landed Estates Abolition Act, 1950 of the J&K government made possible the transfer of land to the tiller without any compensation to former landlords (Sharma and Bakshi 1995: 136).

Secondly, the seeds of democratic decentralisation (establishment of *panchayats*) in the state were sown way back in 1935 with the promulgation of the J&K Village Panchayat Regulation. The striking feature of this regulation was the provision regarding the right of franchise by the people in the villages and the right to be elected as a *panch* or head of the village (Sharma 2002: 434). The regulation amended in 1941 was an improvement on the earlier one in that now the *panchayats* were also assigned to perform the development functions in the countryside (ibid.: 436). Following the guidelines of *Naya Kashmir*, more powers were devolved to *panchayats* in J&K in the early 1950s. These two reform policies (land reforms and *panchayats*) illustrate the state's commitment to promote pro-poor reforms and benefit the marginalised sections in the early 1950s. Yet, the *panchayats* in the state continued to remain dormant in the 1960s through to the 1980s, the period of decreasing autonomy of the J&K state.[4]

Thirdly, in order to systematically curtail the influence of shawl merchants, the manifesto also outlined some specific measures to be taken. It provided for the nationalisation of all key industries through the establishment of the National Industrial Council in order to abolish the monopoly of big private entrepreneurs (Jan 2005). The manifesto emphasised on improvement in the conditions of artisans and shawl workers by providing minimum wages to them, freedom to form and join trade unions, reduction of the working day to 8 h and universal social insurance (KIN 2006). Ironically, the zeal for pro-poor reforms started to wane in the following decades, mainly due to the sabotage of democracy in the state by the central government.

While regular elections were held in most parts of the country from 1952 onwards, the first State Assembly election in J&K was held only in 1962 and the first parliamentary election did not take place until 1967. However, most elections held in J&K before the mid 1970s are generally considered to have been fraudulent or rigged by New Delhi (Widmalm 2002: 56). This period saw political instability, authoritarianism, pervasive corruption and popular discontent against the state governments (Bose 1997: 32). Also, due to the constitutional order of 1954 by the President of India, there have been drastic curbs on fundamental liberties such as freedom of speech, assembly and association in the state (ibid.: 33). The separatist sentiments in Kashmir have been alive and in the absence of a strong democratic outlet along with liquidation of the special constitutional status in the years after independence, the discontent of the people in Kashmir took a violent turn. The Hindu minority in Kashmir valley came to be associated with the Indian state,

---

[3] For details on land reforms in J&K, see Dhar (2004), Misri and Bhatt (1994) and Bamzai (1962).
[4] For details on the functioning of Panchayati Raj institutions in J&K, see Sharma (2002) and Puri (1999).

against which the resentment of the separatists kept growing. In the mid 1980s, widespread anti-Hindu riots struck the southern districts of Kashmir province (Sunita 2006: 107). This violence spread to other parts of the state and was the first expression of organised armed resistance against the Indian state, very often fuelled by terrorist groups based in Pakistan. In the following years, J&K saw massive turmoil due to militancy and terrorism, counterpoised by heavy deployment of armed forces and anti-terrorist measures taken by the Indian state. Bose (1997) notes the extent and nature of violence in J&K in the 1980s and 1990s as involving recurrent massacres of civilians, widespread torture, random killings of bystanders and pedestrians, and defenceless people inside their homes. There were extended periods of blanket curfew in major towns and cities, a persistent pattern of search-operations and road-block checks frequently involving beatings, intimidation and verbal abuse (Bose 1997: 55–56). Every conceivable occasion that could set off strikes and protests rallies was suppressed with lethal force (Puri 1993: 56–57). Even peaceful demonstrations against a steep increase in the electricity tariff in 1988 resulted in numerous deaths from police firing (Bose 1997: 47; Puri 1993). All institutional channels of protest and dissent were effectively 'blocked' (Bose 1997: 49), and all protest activities came to be seen as 'anti-national'. This violent process also resulted in the migration of hundreds of thousands of Kashmiri *Pandits* from the Valley to Jammu, Delhi and other places in India, thus, ending the centuries old co-existence of the Hindu minority and the Muslim majority in the Kashmir valley (ibid.: 71).

In 1990, the State Assembly was dissolved and the President's rule was imposed in the state to retain peace and stability. In 1996, elections took place and since then, formal democracy has been re-established in the state with three rounds of assembly elections by the year 2008. It can be, thus, argued that although the J&K state started with a strong base and commitment to bring about social and participatory democracy in the form of agrarian reforms and empowerment of local bodies, it had to struggle even to maintain regular electoral processes for formal or representative democracy. I now turn to introduce the field sites (highlighted in the Map 2.1 below) where I conducted my research on the *shahtoosh* trade and forest management in J&K.

2.1.2 Field Locations

For the first case-study (ban on the *shahtoosh* trade), I conducted most of my fieldwork in Srinagar. The city inhabits a population of approximately 8,00,000 with Muslims constituting more than 95% of the population. Srinagar city is the centre of the shawl industry and accounted for 97% of all *shahtoosh* workers in the Valley prior to the ban. Within Srinagar, 85% of the *shahtoosh* workers were concentrated in the downtown or older areas of the city (IFAW and WTI 2003). These localities include Rathpora, Tawheedabad, Syedpora, Merjapora, Khanyar, Mondibal, Hamchi, Nowhatta, Khaiwan, Malpura, Nawa Kadal, Waniyar, Eidgah, Nowshera, Narwara, Khaiwan, Zoonimar, Nawab Bazar, Fateh Kadal, Malabagh, Saffa Kadal and Noor Bagh. Some of these localities are numerically dominated by the shawl workers engaged in a particular skill, for example, the majority of the weavers are concentrated in Nowshera, Khaiwan, Eidgah, Rathpora and Zoonimar, while the embroiderers mainly reside in Syedpora, Hamchi and Malabagh. Apart from Srinagar city, some *shahtoosh* workers also resided in nearby villages namely Tulamula, Ganderbal and Khanmulla. I interviewed the *shahtoosh* workers in Srinagar city as well as in these three villages. For analysing the politics of forest conservation, I chose two villages of Udhampur district in the Jammu region. In order to protect the identity of my respondents, I shall call these twin villages Navni and Chinnora. The choice of the villages was made on the recommendation of senior forest officials, who considered the NAP to be most successful in Udhampur and Nowshera forest divisions. I selected Udhampur division because of the problem of militancy in Nowshera. The choice of Udhampur district was also made considering the fact that the forest density in the district is one of the highest in the Jammu region.[5]

2.2 Research Methods

2.2.1 Ethnography of Conservation Interventions

Due to the nature of my topic, my field research was likely to be less quantitative, as it was more qualitative data that I needed to collect in my effort to analyse how nature conservation interventions permeate at various levels, from macro to micro. Since the objective of my study was to understand processes in which actors are multiply situated, I designed a research plan based on multi-sited ethnography coupled with an examination of secondary materials including archival records, governmental documents and NGO reports. Multi-sited ethnographies can be useful when the aim is to understand complex *processes*, or to look for evidence in relation to the questions relating to *how* rather than *what* (see Bryant 2005; Mosse 2005). Such ethnographies, as Marcus (1995: 110) suggests, help to understand something broadly about the system as much as to understand local subjects. Moreover, they enable us to rethink the 'relationship between places, projects, and sources of knowledge' (Des Chene 1997: 81) and open up spaces that may otherwise be invisible from the single site (Hovland 2005). By employing ethnographic approaches to the processes and relations operating within and between multiple sites, recent work on natural resources has succeeded in transforming the global-local divide (Mehta et al. 1999; see also Mosse 2005).

While conducting my study at multiple locations, I relied on various techniques of data collection including use of schedules with open ended questions, semi-

---

[5] The geographical area of Udhampur district is 4550 square kilometres (sq. km.) with a forest area of 2343 sq. km. i.e. 51.49% of the total area making it the most densely forested district after Poonch (56.81%). Due to the high incidence of militancy in Poonch, I selected Udhampur district for personal safety considerations.

structured interviews, group discussions, informal conversations and direct observation.[6] In the case of the *shahtoosh* trade, I interviewed a total of 117 respondents. The first set of interviewees included 92 *shahtoosh* workers, selected from each of the 21 categories according to their population size and a variety of issues to be investigated.[7] The census survey of *shahtoosh* workers conducted by the International Fund for Animal Welfare (IFAW) and Wildlife Trust of India (WTI) in 2002–03 (see Table 2.1 below) served as the sample frame in order to select respondents for my study.[8] The largest number of my respondents were weavers (24%), followed by embroiderers (17%), spinners (17%), separators (7%), agents of manufacturers (7%), manufacturers and traders (3%), hawkers (3%), wool dealers (2%), thread-processors (2%), warp-dressers (2%), warp-threaders (2%), clippers (2%), darners (2%), dyers (2%), designers (2%), printers (2%), washers (2%), carpenters (1%) and shuttle-makers (1%).

The second set included respondents (16 in total) from the Ministry of Environment and Forests and Ministry of Textiles (Delhi), Handloom Department, Handicrafts Department and Police Department (Srinagar), Wildlife Department (Srinagar and Jammu), and Weaver's Service Centre, Craft Development Institute, School of Designs and a private firm called Cashmere Marketing Agency (Srinagar). The third set included seven respondents from conservationist groups – WTI (Srinagar) and WTI, WWF and Traffic India (Delhi). I interviewed two leading politicians in Srinagar, one from the ruling party and the other from the opposition to understand their position and perspectives in relation to the ban on the *shahtoosh* trade.

---

[6] Selection of the interviewees in both the case-studies was made on the basis of 'purposive', 'snowball' and 'quota' sampling. 'Purposive' sampling was most helpful in identifying key informants in the initial stages for both the cases and selecting respondents from the various government departments and NGOs. The information gained through these initial interviews helped in spotting the 'plausibility probes' (Esser, 2008) and in framing the questions accordingly. Through these interviews, I was also able to gauge the possibility of conducting interviews with the selected respondents and the problems I could face. The 'purposive' sampling approach employed in the initial interviews was useful in figuring out the critical areas. Since the manufacturing of *shahtoosh* shawls requires 21 different categories of workers, I chose respondents from each of these categories following 'quota' sampling. Once contacts were made in the different categories of *shahtoosh* workers, I 'snowballed' out through introductions from one informant to another. A snowball survey, in which trust is established by interviewing 'friends of initial contacts' (Tripp, 1997: 27) is also the best method for research on activities 'outside the law' (MacGaffey and Bazenguissa-Ganga, 2000: 24), such as the illegal *shahtoosh* and timber trades for this study. The 'quota' and 'snowball' sampling were also used in the study of the NAP while choosing village respondents from different social groups based on caste, gender and wealth.

[7] Of the 92 *shahtoosh* workers, 20% were female, mainly spinners, separators and warp-threaders. 92% of the respondents were from Srinagar while the remaining 8% were from the villages of Tulamula, Khanmulla and Ganderbal in the Srinagar district.

[8] Note that the census survey does not include finishers and warp-dressers as separate categories but finishers are clubbed with washers and warp-dressers with weavers, possibly because of their overlapping tasks. However, in my study, I have treated them as separate categories. Also, I could not interview any *sazgurs* (cotton loop-makers) and fringe-makers in the field due to their non-availability.

2.2 Research Methods

Table 2.1 Occupational distribution of *shahtoosh* workers

Name of category	Srinagar	Budgam	Pulwama	Baramulla	Total	%
Spinners	9803	86	13		9902	69.28
Weavers	1808	5			1813	12.68
Thread-processors	952				952	6.66
Embroiderers	562	81		31	674	4.72
Separators	359				359	2.51
Agents	112	2			114	0.80
Manufacturers	108				108	0.76
Hawkers	74				74	0.52
Warp-threaders	57				57	0.40
Washers/finishers	56				56	0.39
Clippers	44				44	0.31
Dyers	42				42	0.29
Traders	33				33	0.23
Designers	29				29	0.20
Dealers	14				14	0.10
Darners	13				13	0.09
Sazgurs	3				3	0.02
Fringe-makers	3				3	0.02
Carpenters	2				2	0.01
Shuttle-makers	1				1	0.01
Total	14,075	174	13	31	14,293	100.00

Adapted from IFAW and WTI (2003)

In the case of joint forest management, I interviewed a total of 104 respondents. The first set of respondents (80 in total) included 73 villagers, five members of the JFMC and two *ex-sarpanch* of Navni and Chinnora.[9] I interviewed members belonging to all caste and religious groups in the villages – *Brahmins* (21%), *Thakurs* (18%), Scheduled Tribes (22%), Scheduled Castes (19%) and Muslims (20%).[10] The second set of interviewees (9 in total) included five members of forest field-staff (one forester, three forest guards and one watchman), three local timber contractors and one NGO worker. The third set of respondents (15 in total) included

---

[9] In 2006, at the time of my study, Navni had a population of around 2990 with approximately 450 households. In Navni, 43% of the population is upper caste Hindus (such as *Thakurs* and *Brahmins*), 40% Muslims, 15% Scheduled Castes (such as *Mahshe, Meg, Chamar*) and 2% Scheduled Tribes (*Gaddis*). Chinnora had a population of 2870 with around 320 households. In Chinnora, 55% of the population is *Gaddis*, 25% upper caste Hindus (*Brahmins, Thakurs and Mahajan*), 11% Muslims and 9% Scheduled Castes. Approximately 12% of the residents of Navni (mainly Scheduled Castes) and 15% of Chinnora (mainly Scheduled Tribes) were officially categorised as Below Poverty Line (BPL). This information was provided to me by *patwari*, the local revenue official.

[10] Of the total respondents, 40% were female and 60% were male. 52% of respondents were residents of Navni village while the rest 48% belonged to Chinnora. 17% of the respondents I interviewed were 'BPL'.

government officials from the Ministry of Environment and Forests (Delhi), Forest Department (Jammu and Udhampur), Department of Soil Conservation (Jammu), an official from the State Forest Corporation (Jammu) and two timber contractors from the State Forest Corporation (Udhampur).

2.2.2 Research in the Context of Violent Conflict

Working in the context of violent conflict involved several challenges in doing fieldwork. The insurgency in Srinagar caused problems with travel and communication, which made it difficult to plan interviews with respondents.[11] My Hindu identity in a situation of a generally hostile attitude towards Hindus and non-Kashmiris prevalent in Srinagar around that time brought considerable risk in conducting interviews.[12] Walton (2009) suggests that in conflict situations, interviewees are more cautious about sharing sensitive information and more suspicious about the motivations of a researcher. In response to the considerable methodological issues associated with conducting research in conflict affected regions, Barakat et al. (2002) have argued for a 'composite research approach' which draws on a variety of methods depending on the availability of information and access. A process of methodological triangulation can help to compensate for the restrictions imposed by conflict. While also relying on other techniques for data collection such as conducting interviews, group discussions and direct observation, the triangulation method was particularly helpful in understanding the perceived connections between the militancy and the *shahtoosh* trade, the illegal production of *shahtoosh* shawls in Kashmir valley and the illegal timber trade in the Jammu region.

---

[11] For example, during my stay in Srinagar, bomb blasts took place on two different occasions in the localities where most of the *shahtoosh* workers reside. Although no curfews were imposed after the blasts, I avoided going to those areas for the next couple of days for safety reasons.

[12] In the case of my respondents in Srinagar, I observed that my Hindu style of greeting invited some unwelcoming gestures in the beginning but after a few minutes of interacting with them, I felt that the issue of different religious identities did not matter much in our conversations. Yet, on other occasions, I experienced both the risk of being a Hindu working in Srinagar and the situation of embarrassment it brought for some people I interacted with. For example, one day, after conducting interviews at Hazratbal, Srinagar, I took an auto-rickshaw to return home. The auto-rickshaw driver could identify me as being a non-Kashmiri when I spoke to him in a mix of Hindi and Urdu languages. However, assuming that I am also a Muslim, he did not refrain himself from speaking against the Hindus of the state and narrating stories about his transformation from being an active member of a militant group engaged in bomb blasts in Srinagar in the 1990s to auto-rickshaw driving since the year 2000. On the way, he also narrated stories about members of his family killed in police encounters and pointed to the places in Srinagar where these encounters took place. When I was about to reach home after an hour long journey, he asked me: 'By the way, sister, are you a Muslim or Hindu?' My reply of being 'Hindu' shocked him, he looked at me, stared and apologised as I handed him the fare for the ride.

2.2.2.1 Debunking Presumed Identities

Investigating global processes such as wildlife or forest conservation is a complex task that addresses 'a great number of interactions between actors of different statuses, with varying resources and dissimilar goals' (Lewis and Mosse 2006: 1). Thus, research at multiple sites involves working in a context where 'unequal power relations loom large' (Markowitz 2001: 41). I faced difficulties in explaining my identity as a 'researcher' and the purpose of this research to a diverse set of actors ranging from senior government officials, NGO workers, timber contractors to forest field-staff, *shahtoosh* workers and village residents. While some *shahtoosh* workers associated me with the WTI, the Delhi-based organisation involved in banning the trade and refused to speak to me, others took me as an official from the State Handicrafts Department, eager to enquire about the benefit they could gain from the interaction. Some villagers in Navni and Chinnora were initially reluctant to engage in any conversations for they considered me to be associated with the Forest Department, which polices the neighbouring forest areas. In an attempt to lessen misconceptions about my work and gain access to the interviewees, I made clear the purpose of interviews right at the beginning. Research on sensitive issues may damage the position of officials in respective departments and organisations, and can even cause harm to the local residents. Therefore, some officials and interviewees wished not to disclose their identity while sharing the information with me. Their anonymity has, thus, been protected in the text either by stating this or providing pseudonyms for both individuals and villages chosen for the study.

2.2.2.2 Researching Activities Outside of the Law

The investigator's primary problem in research on illegal activities is to find ways to win informants' trust. Generally, people are wary of discussing activities that can land them in trouble with the police (MacGaffey and Bazenguissa-Ganga 2000: 25). Hence, there were significant challenges involved in getting information on the issues of 'shadow networks' in the illegal *shahtoosh* as well as timber trades, and connections between militancy and *shahtoosh*. On the issue of connections between the trade of *shahtoosh* and the militancy, I decided to deliberately stay away from any direct investigation with my respondents. However, I realised later that I had acquired some data on it by coincidence based on informal discussions with my respondents, local gossip and rumours among the interviewees. For more information, I informally discussed the issues with the respondents suggested by my research assistant or a few others with whom I had built a good rapport in the field, having made frequent visits to their houses. Some additional information on the connections between the illegal trade, militancy and *shahtoosh* were also provided to me by the local officials of the WTI in Srinagar which helped in triangulating the information on these issues obtained from different sources.

Before I delve into wildlife and forest conservation interventions in J&K in the rest of the book, it is pertinent to point out two main limitations of this research. In the case of ban on the *shahtoosh* trade, while I present primary information to suggest the loss of livelihoods of the *shahtoosh* workers due to the ban and their indulgence in the now illegal production of shawls, I could not gather primary data on the Tibetan herders who are involved in poaching *chirus* and supplying wool to the traders in India. This was due to sheer practical challenges involved in accessing high altitude areas of a disputed territory like Tibet. Therefore, for this information, I have relied on secondary sources to argue for the loss of livelihoods of the Tibetan herders. In the case of forest conservation, I wanted to do in-depth research on community forest management by selecting a village site where the forest officials considered the programme to be most successful. However, I could not assess the functioning of the JFMC formed in November, 2007 in the selected villages since no concrete activities were undertaken by the new JFMC until November, 2008 when I conducted the second round of my fieldwork. My discussion on this case, therefore, needs to be seen in the light of the functioning of the JFMC from 2003 to 2007.

References

Ahmed M (2004) The politics of pashmina: the Champas of Eastern Ladakh. Nomadic Peoples 8(2):89–106

Bamzai PNK (1962) A history of Kashmir. Metropolitan Book Co. Private Ltd, Delhi

Barakat S, Chard M, Jacoby T, Lume W (2002) The composite approach: research design in the context of war and armed conflict. Third World Q 23(5):991–1003

Bose S (1997) The challenge in Kashmir: democracy, self-determination and a just peace. Sage, New Delhi

Bryant RL (2005) Nongovernmental organisations in environmental struggles: politics *and the making of moral capital*. Yale University Press, London

Census of India (2011) Series 2, Jammu and Kashmir provisional population totals, paper 2 of 2001, rural urban distribution of population. Registrar General and Census Commissioner of India, New Delhi

Dame J, Nüsser M (2008) Development perspectives in Ladakh, India. Geographische Rundschau International Edition 4(4):20–27

Dame J, Nüsser M (2011) Food security in high mountain regions: agricultural production and the impact of food subsidies in Ladakh, Northern India. Food Security 3:179–194

Des Chene M (1997) Locating the past. In: Gupta A, Ferguson J (eds) Anthropological locations: boundaries and grounds of a field science. University of California Press, Berkeley, pp 66–85

Dhar DN (2004) Kashmir: land and its management. Kanishka Publishers, New Delhi

Esser D (2008) How local is urban governance in fragile states? Theory and practice of capital city politics in Sierra Leone and Afghanistan. PhD thesis, London School of Economics and Political Science

Ganjoo SK (1998) Kashmir: history and politics. Commonwealth Publishers, New Delhi

GoJ&K (2001) Digest of statistics 2000–01. Directorate of Economics and Statistics, Planning and Development Department, Jammu

Gupta S (2013) Democratic transition in Jammu and Kashmir: lessons from recent nature conservation interventions. In: Arora V, Jayaram N (eds) Roots and routes of democracy in the Himalayas. Routledge, New Delhi

Hovland I (2005) What do you call the heathen these days? The policy field and other matters of the heart in the Norwegian mission society. Paper presented at workshop on problems and possibilities in multi-sited ethnography, Brighton, University of Sussex, 27–28 June 2005

References

IFAW and WTI (2003) Beyond the ban: a census of shahtoosh in the Kashmir valley. International Fund for Animal Welfare and Wildlife Trust of India, New Delhi

Jan A (2005) Protest movements in J&K. Zeba Publications, Srinagar

Jha PS (2003) The origins of a dispute: Kashmir 1947. Oxford University Press, New Delhi

KIN (2006) Naya Kashmir, Legal Document No 81 (Extract), Kashmir information network. http://www.kashmir-information.com/LegalDocs/81.html. Accessed 10 Apr 2010

Lewis D, Mosse D (2006) Development brokers and translators : the ethnography of aid and agencies. Kumarian Press, Bloomfield

MacGaffey J, Bazenguissa-Ganga R (2000) Congo-Paris: transnational traders on the margins of the law. International African Institute in association with James Currey and Indiana University Press, London

Marcus G (1995) Ethnography in/of the world system: the emergence of multi-sited ethnography. Annu Rev Anthropol 24:95–117

Markowitz L (2001) Finding the field: notes on the ethnography of NGOs. Hum Organ 60:40–46

Mehta L, Leach M, Newell P, Scoones I, Sivaramakrishnan K, Way S (1999) Exploring understandings of institutions and uncertainty: new directions in natural resource management. IDS discussion paper 372, Institute of Development Studies, Sussex

Misri ML, Bhatt MS (1994) Poverty, planning and economic change in Jammu & Kashmir. Vikas Publishing House, Delhi

Mosse D (2005) Cultivating development: an ethnography of aid policy and practice. Pluto Press, London

Prakash S (2000) Political economy of Kashmir since 1947. Econ Polit Wkly 35(24):2051–2060

Punjabi R (1990) Panchayati raj in Kashmir: yesterday, today and tomorrow. In: Matthew G (ed) Panchayati Raj in Jammu & Kashmir. Concept Publishing Company, New Delhi, pp 37–49

Puri B (1993) Kashmir towards insurgency. Rekha Printers Private Limited, New Delhi

Puri B (1999) Jammu and Kashmir regional autonomy. Mehra Offset Press, New Delhi

Ramasubramanian R (2004) Can environmental security bring peace to Jammu and Kashmir. Article 1407, Institute of Peace and Conflict Studies, New Delhi

Rizvi J (1999) Trans-Himalayan caravans: merchant princes and peasant traders in Ladakh. Oxford University Press, New Delhi

Shafi M (1990) Revival of a democratic tradition. In: Matthew G (ed) Panchayati raj in Jammu & Kashmir. Concept Publishing Company, New Delhi, pp 31–36

Sharma YR (2002) Politics dynamics of Jammu & Kashmir. Radha Krishan Anand & Co., Jammu

Sharma KS, Bakshi SR (1995) Economic life of Kashmir. Anmol Publications, New Delhi

Sunita (2006) Politics of state autonomy and regional identity: Jammu and Kashmir. Kalpaz Publications, Delhi

Tripp A (1997) Changing the rules: the politics of liberalisation and the urban informal economy in Tanzania. University of California Press, Berkeley

Walton O (2009) Negotiating war and the liberal peace: National NGOs, legitimacy and the politics of peace-building in Sri Lanka. PhD thesis, University of London, UK

Widmalm S (2002) Kashmir in comparative perspective. Routledge, London

Chapter 3
Tibetan Antelope and *Shahtoosh* Shawl: A Brief History

Abstract *Shahtoosh* shawl production in Kashmir has a long history. Various historical, political and social factors have influenced the present state of the shawl industry, now on the verge of decline due to the international ban on *shahtoosh*. While scientific knowledge traces the source of *shahtoosh* as the Tibetan antelope, the skilled craftspeople in Kashmir hold diverse myths and legends regarding the source of *shahtoosh* as well as the origin of shawl industry. It is also useful to examine how successive regimes have shaped the production and trade in *shahtoosh* and the working conditions of shawl workers. Such a historical overview of the shawl industry in the pre-ban period can provide insights for analysing the politics surrounding the recent controversy over the banning of *shahtoosh*. This background chapter, thus, presents the myths and realities of the origin of *shahtoosh*, and also provides a brief history of the evolution and growth of the shawl industry in Kashmir.

Keywords Kashmir shawls · *Shahtoosh* · Pashmina · Shawl industry · Chiru · Tibetan antelope

Shahtoosh shawl production in Kashmir has a long history. Various historical, political and social factors have influenced the present state of the shawl industry, now on the verge of decline due to the international ban on *shahtoosh*. While scientific knowledge traces the source of *shahtoosh* as the Tibetan antelope, the skilled craftspeople in Kashmir hold diverse myths and legends regarding the source of *shahtoosh* as well as the origin of shawl industry. It is also useful to examine how successive regimes have shaped the production and trade in *shahtoosh* and the working conditions of shawl workers. Such a historical overview of the shawl industry in the pre-ban period can provide insights for analysing the politics surrounding the recent controversy over the banning of *shahtoosh*. This background chapter, thus, presents the myths and realities of the origin of *shahtoosh*, and also provides a brief history of the evolution and growth of the shawl industry in Kashmir.

The chapter is divided into three sections. In the first section, I discuss the general characteristics of *chiru*, the animal from which the wool is obtained. I explain how the diverse myths relating to the source of wool have persisted for centuries

---

© Springer International Publishing AG 2018
S. Gupta, *Contesting Conservation*, Advances in Asian Human-Environmental
Research, https://doi.org/10.1007/978-3-319-72257-3_3

until the connection between the massacre of *chirus* and the expensive *shahtoosh* shawls was uncovered by the scientific community in 1988. In Sects. 3.1.1 and 3.1.2, I introduce the various stages of the production process and explain how the wool and the product pass from one hand to another. I also provide the population estimates of shawl workers in Kashmir. In Sects. 3.1.1, 3.1.2 and 3.2.1, I present a brief description of the growth and development of the shawl industry since the medieval era, highlighting the suffering of poor shawl workers in the Valley who experienced exploitation at the hands of both rulers and shawl merchants. Following this, I examine the conditions of the shawl workers after independence, the initiatives taken by the new state, and the strategies adopted by the manufacturers to increase their profits and maintain their privileged position in the shawl industry.

In this chapter, I refer to the works of wildlife biologists to describe the characteristic features of *chiru* and its habitat. I also use the accounts of travellers and historians in order to understand the legends relating to the *shahtoosh* wool, the origin of the shawl production, the growth of the shawl industry and the conditions of the shawl workers. Due to the absence of any authentic historical accounts specifically on *shahtoosh*, I have described the picture of the shawl industry and the artisans in general with incidental references to *shahtoosh*.

3.1 Tibetan Antelope

3.1.1 Chiru and Its Habitat

The Tibetan antelope or *chiru*, as it is commonly known, is endemic to the high plains of the Tibetan plateau which extends from Lhasa in Tibet to Ladakh in India (Kumar and Wright 1997: 5).[1] In the summer, the male *chirus* appear reddish fawn in colour, with light grey and brown tones grading to white on their stomach while in the winter, the body colour is predominantly light grey and tan with a white undercoat running from the chin to the belly (Schaller 1998: 43). Females and young *chirus* are fawn coloured, almost pinkish with a rust brown nape that fades into white on the underside (ibid.). *Chirus* grow thick coats as insulation to conserve body heat. This coat has two types of hair: first, long, rough guard hair and second, underwool, which is composed of fine woolly hair. This densely packed underwool grows close to the skin and remains covered by the guard hair. These antelopes shed their thick guard hair when the temperature rises with the onset of the summer (Chundawat and Talwar 1995). However, it is the underwool of *chirus* (obtained by killing the animal) which is known for its fine quality and is used for making highly expensive woollen shawls and scarves.

---

[1] The male *chiru* is 80–85 cm high at the shoulder, weighs 35–40 kgs. and has slender, slightly curving black horns, 50–60 cm in length. The female *chiru* is 75 cm tall, weighs 25–30 kgs. and is hornless (Schaller 1998: 42).

Chirus primarily inhabit the Tibet Autonomous Region (mainly Chang Tang Reserve), Qinghai Province and the Xinjiang Autonomous Region (mainly Arjin Shan Reserve) of the People's Republic of China (Schaller 1993). Chirus prefer flat to rolling terrain, usually above 4000 m, although their habitat ranges from 3250 m altitude to as high as 5500 m (Schaller 1998: 44). In the summer months from May to August, a small population of 200–220 chirus migrate to Ladakh in India (Kumar and Wright 1997: 6).[2]

There are no clear estimates with regard to the population size of the *chirus*. In 1905, C.G. Rawling, a colonial officer, described herds of *chirus* migrating across the Tibetan plateau in the following words: 'Almost from my feet away to the north and east, as far as the eye could reach, were thousands upon thousands of doe antelope with their young [...] there could not have been less than 15,000 or 20,000 visible at one time' (cited in Schaller 1998: 41). In 1990, the total *chiru* population for the entire Tibetan plateau had fallen considerably and was estimated to be between 65,000 and 72,500 which declined to about 45,000 in 1998 (ibid.: 59). Since the ban on *shahtoosh*, there are reports suggesting rise in the population size of *chirus*, with varied range of population numbers. For example, a Chinese news agency states that the population of the endangered Tibetan antelopes has expanded to about 2,00,000 by 2011 with an average annual growth rate of 7% in the last one decade (Xinhua 2011). Leslie and Schaller (2008) suggest the number has increased and may have reached up to 100,000, Feng (1999) estimates the number between 100,000 and 120,000, Xi and Wang (2004) estimate it 150,000 and Liu (2009) states that the population has doubled since the mid-1990s, reaching 150,000 approximately. We can note here that while the population of *chirus* has increased since the ban, it is difficult to ascertain the exact population size due to the lack of any systematic census surveys.

3.1.2 The Shahtoosh Wool: Myths and Realities

The name 'shahtoosh' is derived from two Persian words: the word *shah* meaning emperor and *toosh* meaning nature. Therefore, *shahtoosh* wool means the 'king of wools' or the 'wool that is fit for the emperors' (Traffic India n.d.). The wool is also referred to as *asli-tus* in Kashmir, meaning genuine or real wool. This wool is used to manufacture the high quality *shahtoosh* shawls that are so fine that they can pass through the ring of a finger and are thus known as 'ring shawls'. In the following discussion, I present the various myths and realities surrounding the origin of *shahtoosh* wool, the poaching of *chirus* to obtain wool, the trade routes of *shahtoosh* and the various international, national and state legislations to protect *chirus*.

Animal rights organisations claim that for centuries, the shawl industry has hidden the unsavoury origin of the *shahtoosh* wool by perpetuating the myth that wool

---

[2]The range encompasses the upper Chang Chenmo valley and its neighbouring Lingti Tsiang plains, which are opposite the Daulat Baig Oldi in the extreme north eastern part of Ladakh (Kumar and Wright 1997).

is derived from the wild goats, Ibex, or even 'Siberian geese' in the Himalayas (IFAW and WTI 2001: 19). William Moorcroft (1820), a British traveller and an employee of the East India Company, on the origin of *shahtoosh* notes that it is derived from Ibex, a mountain goat. Ahad (1987: 18) writes, '*Kyl-phamb* or *asli-tus* is grown by Tibetan Ibex [...] this animal is called Ra-ba in Ladakhi'. In the historical accounts, while there is uncertainty regarding the species from which the wool is obtained, the source of wool has been mainly described as Tibet (see for example, Bamzai 1980; Irwin 1973). There are also diverse myths regarding the collection of *shahtoosh* wool. For example, Ganhar (1979: 34) notes the process of the shedding of wool in the following words: 'The poor animal is pestered by a fly, the bot fly, which buries deep into the skin and lays eggs there. When the eggs hatch, the wriggling of the puppas makes the animal mad with discomfort. The pestered animal runs here and there and rolls over in sand to scratch his body and relieve his pain, thus, shedding some of the hair of his winter coat down in the sand'. Also, Wilson (1841: 347) writes, 'the fleece is cut once a year and wool, coarsely picked, is sold by the importers to the merchants who sell it to Kashmir'.

The origin and collection of *shahtoosh* has remained obscured for centuries, due to the fact that the entire wool trade was controlled by only a few influential Kashmiri merchants who possibly did not want to share its true origin with the local people for fear of losing their exclusive control over trade. The merchants, with their tightly knit kin groups, had been purchasing the raw material in large quantities from Chang Tang, Ladakh, Yarkand and a few other centres in Chinese territories for centuries (see Rizvi 1999). The control of Kashmiri merchants over the raw wool trade facilitated the Treaty of Tingmosgang (1683 A.D.), under which the Tibetans and the Mughals agreed that the fine wool of goats of western Tibet would be sold only to the Kashmiri merchants (Ahad 1987: 19). Wilson (1841: 347) states that the export of wool from Ladakh and western Tibet was exclusively confined to Kashmir and all attempts to convey it to other countries were punished by confiscation (see also Rizvi 1999).

During my fieldwork, I found that the *shahtoosh* workers in Kashmir still hold many myths regarding the origin of the wool. The majority of the shawl workers claimed that the nomadic shepherds collect the shed wool of the animal from the bushes and rocks it brushes up against while grazing. They maintained that this process takes place in the summer as the animal sheds its winter coat with the rise in temperature. Others believed that the wool is derived from the moulted breast feathers of a bird that eats red sand. Besides these myths, *kasturi* or musk deer, snake-eating Ibex, rabbit and sheep are also believed to be the sources of *shahtoosh* wool.[3] It can be argued that the Kashmiri merchants and Tibetan traders might have hidden the realities and perpetuated false stories which then became accepted as fact by the local people in Kashmir. The fact that the name of the product (*shahtoosh* shawls) is different from the name of the animal source (*chiru*) may represent a deliberate strategy on behalf of the traders to disguise the link between the *shahtoosh* shawls and the Tibetan antelope. Several animal rights organisations maintain that the pro-

---

[3] More discussion on the persistence of myths regarding the source of wool follows in Chap. 5 of this book.

tected status of the *chiru* and the scarcity of *shahtoosh* may also have motivated the manufacturers and exporters to even commercially rename a sub-species of Capra Ibex as the '*shahtoosh* goat' (see IFAW and WTI 2001). They argue that by using the pictures of an Ibex in *shahtoosh* promotional brochures, it is also possible for manufacturers to divert suspicion away from its original source (ibid.).

The truth about the connection between the massacre of the Tibetan antelope, and the highly prized *shahtoosh* shawls of Kashmir became known to the conservationist groups only in 1988 when wildlife biologist George Schaller spent several months working in the habitat of the Tibetan antelope. The reason for the rapid decline of *chirus* was revealed to Schaller during his visit to a small town called Gerze in 1988. He saw herdsmen plucking the *shahtoosh* wool from the hides of the antelope to sell it to local dealers (Schaller 1993). Schaller notes, 'in the courtyard of such a dealer were sacks of wool ready for smuggling into western Nepal and from there to Kashmir, where the wool is woven into scarves and shawls' (ibid.: 81). In 1992, Schaller visited a camp of hunters in Chang Tang and stated that hides of *chirus* were '[...] stacked in their tent, the frozen bodies outside; they had also saved the heads of males because the horns are widely used in traditional medicine' (ibid.: 82). This truth about the massacre of *chirus* alarmed wildlife conservationists and governments, leading to a number of surveys to investigate the poaching of *chirus*.

The surveys conducted by wildlife conservation groups such as IFAW, WTI, WWF provided evidence to challenge the existing myths on *shahtoosh*. First, it was revealed that the Tibetan antelope does moult its wool slowly over a period of time in spring, but on the icy wind-swept plains of Tibet, there are no rocks and bushes to which wool could possibly cling and subsequently be collected (IFAW and WTI 2001: 20). Second, it was argued that the small quantity of shed wool would not be sufficient to meet the 3000 kilograms of raw *shahtoosh* wool processed per annum in Kashmir (Kumar and Wright 1997: 35). Therefore, it was maintained that the supply of wool in Kashmir necessarily involves the poaching of this wild species and the extraction of underwool from its hide. Since 150 grams of *shahtoosh* wool is produced by one animal, it has been calculated that approximately 20,000 Tibetan antelopes were required to meet the market demand around the late 1990s (ibid.).[4] It was claimed that the poaching of the antelope has resulted in their decline from one million in the 1950s and 1960s to less than 75,000 in the year 2000 (IFAW and WTI 2001). Third, the hair samples taken from *shahtoosh* shawls were tested under the microscope, and confirmed to have patterns that are unique to the Tibetan antelope (ibid.). In 1998, the China Exploration and Research Society discovered the first known cases of the poaching of calving females in Xinjiang (Chundawat and Talwar 1995). However, no evidence has been found of the killing of Tibetan antelope within the Indian territories.

Historically, the Tibetan herdsmen have hunted *chirus* primarily for subsistence needs. Huber (2003) suggests that the *chirus* have been harvested in a sustainable manner for more than a millennium by indigenous hunters living adjacent to major

---

[4] It is estimated that three *chirus* are needed to make one shawl for women where as five *chirus* are killed to produce one shawl for men.

antelope populations in Tibet. It is claimed that while the trade in *shahtoosh* has been going for centuries, owing to the growing fame of *shahtoosh* shawls in the West and the Middle East by the 1970s, the killing of *chirus* acquired a commercial dimension (ibid.). The commercial hunting, Schaller (1993) notes, has often provoked the poachers to use various barbaric methods. Hunters began to shoot into herds of *chirus* from moving vehicles, killing as many as 500 animals in a hunt (Kumar and Wright 1997: 8). After poaching, the wool is bought by the traders and smuggled through various border crossings into Kashmir in exchange for other wildlife products.[5] The wool is also sold by the poachers in exchange for money, who make significant profits out of this trade (ibid.). Referring to the highly lucrative trade of wool, Bamzai (1980) quotes a famous Kashmiri saying:

Yus gav Las suh zah na av.
Av ai tas nah zah wav
He who went to Lhasa (Tibet), never returned
If he came back, he was a rich man forever.

The *shahtoosh* wool is transported from Tibet through a number of border crossings into Nepal and India. The city of Lhasa and the town of Burang are the major centres for wildlife trade in Tibet. The *shahtoosh* wool and other goods are said to be brought by *khampas*, the Tibetan herdsmen, to the Indo-Nepal border. Kathmandu serves as another important hub for *shahtoosh* wool trade. Near the Indo-Nepal border, the *khampas* hand over the wool to *bhotias,* the pastoral nomads of the Himalayan region who have largely taken up farming in Tibet and the border regions of Nepal and India. The *bhotias* then bring the *shahtoosh* south to Dharchula and Pithoragarh in Uttaranchal where it is sold to Kashmiri traders and then carried to Srinagar to be made into expensive shawls (Kumar and Wright 1997: 15).[6] In some cases, the raw wool is brought to Delhi via Kathmandu and to Srinagar via Ladakh. I further discuss the illegal trade of *shahtoosh* wool in the next chapter.

As mentioned in chapter one, the trade in *shahtoosh* wool and its products is prohibited by a number of international and national treaties and legislations. The Tibetan antelope is listed in Appendix I of the 'Convention on International Trade in Endangered Species of Wild Fauna and Flora', making all international trade in *chiru* body parts and its derivatives illegal (CITES 1999). Since China, Nepal and India are signatories to this convention, any cross-border movement of raw *shahtoosh* wool into India through its northern border with Tibet or Nepal is illegal. The species is also listed as 'near threatened' in the 2016 International Union for the Conservation of Nature and Natural Resources' 'Red List of Threatened Animals' (IUCN 2016). The *chiru* is a

---

[5] These mainly include tiger and leopard bones and skins, musk pods and bear gall bladders (Kumar and Wright 1997: 15).

[6] In the 1980s, the price of raw *shahtoosh* wool was approximately Rs. 6000 per kg. This rose to nearly Rs. 16,000 in the early 1990s and to Rs. 30,000 by 1998. This sudden rise in the price of wool can be attributed to the growing global concern for the protection of *chirus*, which limited the supply of the wool in the market. Moreover, the growing controversy over the ban on the *shahtoosh* trade in Kashmir also evoked public curiosity and increased the desire to possess it.

'protected' species in the United States of America and the European Union. The wildlife protection laws of China, Nepal and India also prohibit any harm to the Tibetan antelope and trade in its body parts including wool. The *chiru* is listed as an 'endangered' species under schedule I of Nepal's National Parks and Wildlife Conservation Act, 1973 (Kumar and Wright 1997: 29). Although enforcement has been greatly enhanced in the areas administered by the Nepalese Department of National Parks, enforcement outside the park is the responsibility of the Forest Department and police, who have taken little interest in controlling the illegal wildlife trade (ibid.: 31). The weak implementation of laws has made Nepal one of the largest illegal trade centres, with a well-connected network linking Nepal with China and India.

In India, national law protects the species through its Wildlife Protection Act of 1972. Five years after its enactment, the *chiru* was placed in Schedule I of the Act, providing 'regulated' protection, which permitted trade under 'licence' from the government. An amendment of the same Act in 1986 completely banned hunting of and trading in derivatives of Schedule I species, thus giving *chiru* complete protection in the country. Therefore, the trade in *shahtoosh* has been 'illegal' in India since 1986. J&K is the only state in India which has its own Wildlife Protection Act. Under the J&K Wildlife Protection Act of 1978, *chiru* was listed in Part I of Schedule II under the 'Special Game' category. This legislation permitted hunting and regulated trade in its derivatives under the 'licence' of the state. However, the state wildlife officials report that they have never issued any licences for *shahtoosh* wool trade (ibid.: 38). As such, the production and trade of shawls continued in the state until 2002 (when the ban was imposed) without any interference from the enforcement agencies within J&K.

3.2 *Shahtoosh* Shawl

3.2.1 *From Raw Wool to Finished Shawl: The Production Process*

Before I provide an overview of the growth and development of the shawl industry since medieval times, and the conditions of shawl workers, it is pertinent to understand the complex processes involved in the shawl production and the roles of the various categories of shawl workers.[7] By explaining the complex division of labour and the tasks performed by each category, I demonstrate that the shawl workers who perform the most strenuous jobs have experienced marginalisation within the *shahtoosh* worker community. In this section, I have explained the production process prevalent at the time of my fieldwork.

---

[7]The information presented in this section is based on my interviews with various categories of *shahtoosh* workers and observation during the fieldwork in Srinagar. Apart from this, my understanding of the Kashmiri shawl manufacturing process has also benefitted from Rehman and Jafri (2006), Ahuja (2006) and Ahad (1987).

The entire production process involves five broad functions: raw wool supplying, spinning, weaving, finishing and trading. Out of 21 categories of shawl workers involved in the process, 13 are directly involved in the shawl production, six in its trading and the remaining two (carpenters and shuttle-makers) in the production of hardware products required in the manufacturing process.[8] The diagram below (Fig. 3.1) shows how the raw material and product passes from one worker to another, and highlights the relatively powerful actors in the *shahtoosh* shawl production and trade. The production process begins with *bhotias* procuring raw *shahtoosh* wool from Tibet.

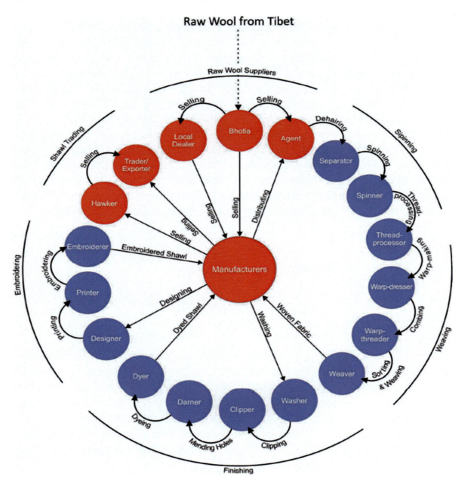

Fig. 3.1 Production process of the *shahtoosh* shawl

---

[8]The carpenter makes spinning wheels and handlooms from deodar wood. The shuttle-makers make shuttles or *muks* for weaving and clips used for brushing the unwanted threads of the fabric. With the increasing use of machines for spinning and weaving, the demand for their products is declining rapidly.

3.2.1.1 Bhotias and Local Dealers

As mentioned above, *bhotias,* the pastoral nomads of the western Himalayan region, bring the unprocessed wool from Tibet and Nepal to Uttaranchal and Delhi in order to sell it to the Kashmiri dealers. Due to strict enforcement of the ban on *shahtoosh* outside J&K state from the mid 1990s onwards, the *bhotias* started bringing the wool directly to Srinagar via Ladakh. Some of them have members of their families residing in Srinagar who act as local dealers. The *bhotias* also sell the wool to the manufacturers or their agents in Kashmir.

3.2.1.2 Manufacturers or Vastas

Shawl manufacturers are the most powerful actors in the shawl industry not only because they accrue profits and control the production process but also because they determine the relationship of the industry with the outside world. They are involved in every stage of production from purchasing the wool to the sale of finished shawls. Some of the big manufacturers buy the wool directly from the Tibetan and Nepalese traders and bypass the local Kashmiri dealers. They buy large quantities of wool for the whole year and control the production process. The manufacturers are the owners of the shawl and pay the subordinate workers for their labour. Previously, the manufacturers provided these shawls to the hawkers for selling, but since the 1980s, many manufacturers have also taken up the trading functions. Besides supplying the shawls to the hawkers on credit, they now also sell the shawls directly to the exporters or traders outside the state.[9]

3.2.1.3 Agents or Poiywans

The *poiywans* act as middlemen or agents of the manufacturers. They either get the wool from the manufacturers or purchase the wool directly from the *bhotias* or local dealers in Srinagar. These agents distribute the wool to the separators and spinners for cleaning and spinning, and the yarn to the weavers for weaving. They also distribute the fabric to the skilled workers for finishing and embroidering. Being the middlemen between the manufacturer and spinners/weavers, these agents make their earnings by purchasing the yarn at lower prices from the spinners and the weavers, and selling it at slightly higher prices to the manufacturers. I shall discuss this point in detail in chapter five of the book.

---

[9]The manufacturers also have direct contact outside the state with the customers who give them special orders for producing desired shawl patterns and designs.

3.2.1.4 Separators or Charun Wajens

The separators are exclusively women and their job is to separate the inner soft fleece from the outer rough guard hair. This process of separation by hand is known as *charun*. They return the separated fleece to the manufacturers or their agents who weigh both the waste and the inner fleece to ensure that the quantity of raw wool provided has remained the same.

3.2.1.5 Spinners or Katun Wajens

The next job is that of the spinners, mainly female. They generally learn this skill from their mothers, and begin to spin as early as the age of ten. Spinning is one of the most skilled tasks in the production process.[10] The spinning starts with the spinner running the spinning wheel with her right hand and simultaneously lifting the fleece with her left hand (see Fig. 3.2). The movement of the wheel must be in perfect harmony with the lifting up and down of the fleece in order to produce yarn. The yarn is then made thick by combining two threads on a wooden cone known as the *pritz*. The *pritz* is put on the spinning wheel to twist the two threads into one.

Fig. 3.2 Spinning yarn with a traditional hand-driven wheel

---

[10] The spinners first clean the wool by mixing it with wet rice flour or *kharott*. This mixture lubricates the wool, absorbs its impurities, and also acts as a bleaching agent. After being treated for about an hour, the wool is cleared from the flour and is then combed on an instrument called a *kangyn*. The soft combed hairs are piled up into soft tufts known as *temb*, which are kept in an earthen pot to keep the wool away from dust and impurities.

3.2 *Shahtoosh* Shawl

This process of twisting two threads is known as *absaween*. Finally, knots or *gund* are made with the help of a wooden instrument known as a *thinjour*, which has a wooden base with two sticks. The yarn is wrapped around these sticks and a knot is tied after every nine threads. On average, out of ten grams of wool, the spinners make 120–130 knots every 3 days.

The agents of manufacturers then distribute the spun threads to the weavers for weaving the shawls. Weaving involves various sub-stages such as strengthening the thread, preparing warp, weft-making and combing the threads. These pre-weaving tasks are generally carried out by workers employed by the weavers. In some cases, the weaver himself engages in these pre-weaving processes for an additional income. The family members of the weaver, including his wife and children, usually help him in this tedious process.

3.2.1.6 Thread-Processors or Perkumgors

The first pre-weaving stage is strengthening the spun thread. The threads are first cleaned by washing with herbal soap or *reetha*. These threads are then rolled on *pritz*. Upon drying, the threads are untied and dipped in a starch or *maya* made of rice flour. These wet threads are again rolled and untied many times so that the starch does not form lumps and get fixed to the yarn. This processing of thread is called *phirayee* and the workers performing this task are called thread-processors or *perkumgors,* who are generally male. The threads are then returned to the weaver who pays them for their labour. If the threads are to be dyed, they are sent to the dyers, otherwise they are woven in their natural shades. The weaver then sorts the threads on the basis of their thickness and strength. The better quality stronger threads are used to make warp and those which are somewhat thicker are kept for the weft.

3.2.1.7 Warp-Dressers or Penkumgors

The second stage is warp-making or *yen yaren* which is either done by the weavers themselves or the warp-dressers employed by them. The number of threads to be used depends on the desired width of a shawl. The strong threads are stretched by means of four to six iron rods dug into the ground at regular intervals. Usually two people, normally male, walk along the length of these rods by carrying a *pritz,* on one hand and a *yarunwej,* a long stick with a hook, on the other. While walking, they transfer the thread from the *pritz* by using the *yarunwej* and tie it around the iron rods, crossing at each end. The iron rods are then replaced at the crossings by the wooden sticks. The threads are then lifted up and spread evenly, after which the wooden sticks are again replaced by thick threads to keep the crossing at its place and avoid any tangling. The warp is then wrapped on a wooden roller or *navrada*.

The third pre-weaving stage is weft-making or *watchekein wallen,* which is done by the weavers themselves. The threads are rolled on a spindle, after which they are taken out and fitted into a special metallic reel to create the weft.

3.2.1.8 Warp-Threaders or Barangors

The fourth pre-weaving stage is to pass the warp wrapped on a *navrada* through a wooden comb. This is done by warp-threaders or *barangors* who are generally women. They perform this job in two steps; first they pass each thread of warp through thicker cotton threads fixed on an instrument known as a *saaz* and second, these threads are passed through a wooden comb known as a *kangi*. Each warp consists of 1000–1500 threads.

3.2.1.9 Weavers or Vovers

The weavers now fix the warps on the looms for weaving and the wefts are gently inserted into them with the help of a shuttle or *muk* (see Fig. 3.3). Weaving is done exclusively by men who start learning this skill at around the age of ten. A weaver gets approximately 190 grams of yarn for a single shawl. It may require a minimum of 7 days to weave a *shahtoosh* shawl, which is then returned either to the agent of the manufacturer or directly to the manufacturer. The weavers are paid by the manufacturers and, in turn, the weavers pay the wages to the thread-processors, warp-dressers and warp-threaders for their labour.

Fig. 3.3 A weaver weaving a pashmina shawl in a karkhana in Srinagar

3.2 *Shahtoosh* Shawl

3.2.1.10 Washers or Dhobis

The manufacturers, after assessing the quality of the woven fabric, send the shawls to washers or *dhobis* for their first wash to remove the mixtures such as soap and starch used in the making of threads and fabric. The *dhobis* wash them with herbal soap or *reetha* at the banks of the Jhelum river by using its water which is well known for its softening properties. The shawls are then taken by the agents to the clippers.

3.2.1.11 Clippers or Puruzgars

The clippers scrub the surface of the shawl on a frame which is made of two wooden beams set at a distance of 1 m apart. The fabric is placed on this frame where the uneven threads, knots and any other flaws are removed by a long clip known as a *wouch*. The fabric is also rubbed with a husked corn cob known as a *kashir* to make the surface smooth and even.

3.2.1.12 Darners or Raffugars

The darners have a unique skill of mending the holes in the fabric and their services may be required at any stage between the weaving and finishing of the shawls. The holes may appear when the weavers lose any thread while weaving or when the clipper brushes the fabric. To repair these holes, the darners draw out the threads from within the fabric and interlock new threads with the old by using a needle.

3.2.1.13 Dyers or Rangrez

Before the second wash, the fabrics are taken to the dyers or *rangrez*. The dyers are proficient in dyeing them with brilliant colours. However, the natural vegetable dyes such as *kirmizy* or *badamy* have been replaced these days by chemical dyes. The dyed shawls are then taken by the agents again to the manufacturer to assess the quality of finishing, after which the manufacturer sends the shawls for designing and embroidery.

3.2.1.14 Designers or Naqqashs

To embroider the shawls, the designers create patterns or designs. They draw patterns on the paper to be printed repeatedly on the fabric. Although the designers are very few in number, their job is indispensable for producing new patterns to satisfy the diverse tastes of the customers.

3.2.1.15 Printers or Chappawals

The *chappawals* print the designs on the fabric by using wooden printing blocks. These blocks are made of walnut wood because it is soft and takes the shape of the designs easily. The *chappawals* have a large collection of different sized blocks carved with various designs. They heat a mixture of black and white carbon powder with rice starch to use as a paste to print the design on the shawls.

3.2.1.16 Embroiderers or Ragbars

The shawls are then embroidered by using silk threads or *resham* dyed in different colours. Depending upon the pattern or design, embroiderers can work on a shawl from several months to years to finish embroidering it. The process can cost anything from a few hundreds to thousands of rupees. After the embroidering, the shawls are sent to the washers for a final wash, and finishers for steam ironing and then returned to the manufacturers.

3.2.1.17 Hawkers or Pheriwalas

The manufacturers supply the shawls on credit to the trusted hawkers for sale outside the state. The hawkers travel to the big cities such as Delhi, Kolkata and Mumbai and stay there for 5–6 months in a year. They contact the exporters or traders for the sale of their products and also their old customers. The profits are shared between the manufacturers and hawkers.

3.2.1.18 Traders and Exporters

The traders make significant profits by selling the *shahtoosh* shawls in the big cities of India. They also export the shawls in large numbers to fashion outlets worldwide including France, the UK, Belgium, Switzerland, the USA, Hong Kong, and those of the Middle and Far East.

In the above discussion, we can observe that categories of workers such as the spinners, weavers and embroiderers contribute the most to the production of shawls, by performing the most arduous jobs and adding value to the finished products. Yet, economically these have remained the most exploited and marginalised categories, a point with which I engage in detail in the next section. Manufacturers, traders and *poiywans* who contribute the least in the actual production process are economically the most powerful actors who are able to make profits and control the entire manufacturing and trading of shawls. This topic is further detailed in the analysis of the differential impact of the ban on the livelihoods and wages of different categories of *shahtoosh* workers in Chap. 5.

3.2.2 Shahtoosh Workers: Population and Distribution

The Government of J&K has no records on the population or distribution of the *shahtoosh* workers in the state. However, on the basis of some historical accounts on shawl workers in Kashmir, some approximate figures can be estimated. For example, Ames (1986: 20) states that during the Mughal period (1586–1753), 40,000 looms were in operation in Kashmir, with each loom employing three people. Notably, the Mughal period was the one of relative peace and stability in the region in comparison to the successive Afghan and Sikh regimes. In times of Afghan rule (1753–1819), there was a decline in the trade with the number of looms reducing to 16,000 (Ahad 1987: 49). However, at the turn of the eighteenth century, 24,000 looms operated in the Valley, giving employment to 72,000 weavers (ibid.: 49). During the Dogra rule (1846–1877), there were 27,000 weavers working at 11,000 looms (Ames 1986: 32, see also Younghusband 1909: 212). However, these accounts do not specifically indicate the total number of *shahtoosh* workers in the Valley but provide a general estimate of shawl workers.

The state's last published census information on Kashmir in 1981 does not differentiate *shahtoosh* workers from the other textile workers. The first census to determine the total number of *shahtoosh* workers and their families was conducted in 2002–03 by the IFAW and WTI. The total size of the population surveyed was 45,405 covering the Srinagar, Budgam, Baramula and Pulwama districts. Out of this figure, 14,293 individuals were directly or indirectly involved in the production of *shahtoosh* shawls. Of the remainder, 26,044 worked in other professions, and 5068 individuals were classified as dependents (IFAW and WTI 2003: 7). There is also a clear division of work by gender and females outnumber males in this industry. On average, each manufacturer keeps 75 separators and around 300 spinners on its payroll (ibid.: 7). These two categories form the largest workforce involved in the manufacturing of *shahtoosh* products. It is estimated that females constitute 74% of the population of *shahtoosh* workers in the Valley (ibid.: 7). The Wildlife Protection Society of India (WPSI) claims that this industry is largely in the hands of about 20 powerful families who, in turn, employ thousands of shawl workers in small household units (see Kumar and Wright 1997: 22). The IFAW and WTI estimate the number of manufacturers to have been 120 in the 1990s, when the industry was at its peak (IFAW and WTI 2001: 56).

3.3 Kashmir Shawl Industry

3.3.1 Origin and Development of the Shawl Industry

The word 'shawl' is commonly accepted to be derived from a Persian word *shal* which originally denotes a class of woven fabric rather than a particular article of dress (Irwin 1973: 1). It has also been argued that the word *shawl* owes its origin to the word *sha* of the Dardic group of languages to which Kashmiri belongs (Ahad 1987: 5). The *shahtoosh* shawls are well known for their lightness, warmth and softness, and are typically embroidered with beautiful floral, faunal and geometrical designs.

The origin of shawl weaving in Kashmir is obscure. The local people associate the origin of this art with Shah Hamdani (1314–1384), a Persian Sufi saint and a popular figure of medieval times in Kashmir. They believe that besides preaching Islam, he brought the weavers from Persia to teach the skill of shawl weaving to the people of Kashmir for their economic prosperity. He introduced the indoor economic activities to help them secure their livelihoods in winter conditions. Another popular belief regarding the origin of the shawl industry among Kashmiris is that Shah Hamdani was given a pair of socks made of *pashmina* wool by the local people of Ladakh on his way to Turkistan (historically, the region of Central Asia). Observing its warmth, he presented some *pashmina* wool to the Sultan Qutub-ud-din (1374–89 A.D.) in Kashmir, who had it manufactured into a shawl and, thus, the shawl became a popular item of manufacture among the Kashmiri craftsmen (Hassnain 1992: 169).

Scholars of Kashmir history such as Riazuddin (1988), Ames (1986), Strauss (1986) and Irwin (1973) hold the view that the shawls of Kashmir originated during Sultan Zain-ul-Abidin's reign (1420–1470). Zain-ul-Abidin brought the weavers from Turkistan to the Valley in order to train the people of Kashmir. This consensus has been challenged by writers such as Ahad (1987) and Jayakar (1959) who argue that the art of shawl weaving was known to the Kashmiri craftsmen since ancient times. Historians such as Rehman and Jafri (2006) and Ahad (1987) have maintained that the art of shawl weaving was merely a subsidiary occupation pursued by agriculturalists in Kashmir before the conquest of the Sultans in the fourteenth century. The shawl was woven on a rough loom with the help of 'family labour, primitive know-how and little of the division of labour that characterises current production' (Ahad 1987: 9). However, under the reign of the Turkish Sultans, shawl weaving began its slow transition from a subsistence base to a factory system based on the complex division of labour. The industry flourished in Mughal times and was a significant source of revenue to the subsequent Afghan, Sikh and Dogra regimes. Below, I discuss the development of the shawl industry in Kashmir during different periods.

3.3.1.1 Kashmir Sultans (1320–1569): Introduction of Persian Designs to Craftsmen in Kashmir

The Sultans of Kashmir showed great enthusiasm for developing Kashmiri handicrafts. As mentioned above, the art of shawl weaving developed during the reign of Sultan Zain-ul-Abidin (1423–1474), who took several systematic measures to organise the shawl industry. Besides improving the traditional looms, it is claimed that he invited experts from Turkistan and provided them with facilities in order to train the people of Kashmir in weaving the kinds of designs prevailing in Persia at the

3.3 Kashmir Shawl Industry

time (ibid.: 10). The art of weaving in Kashmir, thus, came to be largely influenced by the Central Asian experts during this time. Furthermore, due to the geographical proximity of Kashmir to Central Asia and its linkage with the old 'silk route', Kashmir became an important transit point in the Indo-Central Asian trade during this period (Warikoo 1989: 55).

3.3.1.2 Mughal Period (1586–1753): *Shahtoosh* Shawls as Symbolic Capital

Although Sultan Zain-ul-Abidin is credited as the first monarch to have developed the shawl weaving art in Kashmir, the Mughal emperor, Akbar (1542–1605) is venerated for making a quantum leap in its manufacturing. Abul Fazl, the court's chronicler, describes in *Ain-i-Akbari* how the emperor Akbar was famously known to hold the *shahtoosh* shawl in high regard (cited in Rehman and Jafri 2006: 35). He referred to these shawls as *paramnaram,* meaning the softest fabric (Ahad 1987: 12). Abul Fazl describes the king as having had such a great passion for the *shahtoosh* shawls that his empire legislated against the use of *shahtoosh* shawls outside the court (cited in Rehman and Jafri 2006: 36). References to the Kashmiri shawls are also found in *Tuzk-i-Jehangiri*, in which Jehangir (1569–1627), the son of Akbar, describes the shawl as one of his favourite items of dress. His love for naturalistic forms inspired the shawl designs with flowering patterns (ibid.: 57). Until Shah Jahan's (1592–1666) reign, the *toos* shawls were reserved for the exclusive use of the royal household (ibid.: 45). They were also given as tribute to the foreign kings of Rome, Persia, Egypt, Golcanda and Bijapur (Ahad 1987: 11). The other shawls available for the courtiers and noblemen were *pashmina*. This illustrates that the *shahtoosh* shawls were not considered just as commodities, but goods imbued with symbolic capital to be possessed only by the ruling class.

The state owned factories (known as *karkhanas*), developed rapidly under the Mughal rule, becoming a vital source of prosperity for Kashmir (ibid.: 45). Ahad (1987: 45) describes *karkhanas* as 'large halls seen in many places for the artisans who are busily employed and superintended by a master'. The officers called *Khans* or *Maliks* exercised control through the master craftsmen over subordinate workers (ibid.). They were in charge of the workshops, collecting the materials and distributing the wages. The master craftsmen, known as *vostas*, were employed to supervise the work done by the artisans in the production of shawls. The state, however, discouraged private entrepreneurs from gaining a foothold in the shawl production and sale. A few entrepreneurs who were permitted by the state to employ weavers and embroiderers established looms in ordinary dwelling rooms of their small houses (Schonsberg 1853: 131–133). They were subordinate to and dependent upon the government, which exercised absolute authority over their small workshops through *Dag-i-shawl*, an early department of excise (Ahad 1987: 46).

54

3.3.1.3   Afghan Period (1753–1819): Expansion of Markets and Emergence of Independent Karkhanas

The state factory system became a major source of revenue during the period of Afghan rule. There was a time in Kashmir in the early nineteenth century, when the revenue given by the shawl industry was more than the land revenue (Dhar 2001: 31).[11] The most notable development of this period was the extensive trade of shawls with Europe. The Kashmiri shawl was largely adopted as a fashion article at weddings, especially during the reign of Queen Victoria (Ahad 1987: 12). The merchants from Europe came to Kashmir to purchase the shawls to export them to various European countries including England, Russia, France and Italy. This resulted in the expansion of shawl markets, which led some artisans to start their own *karkhanas*. However, the number of weavers employed by private individuals was very small and state officials controlled the *karkhanas*, both state owned and private (Schonsberg 1853: 106). Ames (1986: 26) suggests, 'a fine piece of shawl was sold at the rate of forty rupees by the merchant during this period', although the weavers were paid only in kind. Moorcroft (1820) argues that the main profit makers of this industry were not the loom owners but the shawl brokers or *mohkuns*, who were intermediaries between the producers and foreign merchants. By this time, Kashmir had not only evolved a system of factory and complex division of labour but also a fully-functioning brokerage and marketing system.

3.3.1.4   Sikh Period (1819–1846): Emigration of Shawl Workers to Punjab

During the Sikh rule, the manufacture of shawls increased manifold. Maharaja Ranjit Singh (1780–1839) made it a convention to present these shawls to Europeans who visited the court (Ahad 1987: 13). These shawls were often used to bribe the officials of the British East India Company by the princes to seek personal favours (Irwin 1973: 10). The industry continued to be a significant source of revenue to the government. The working conditions of the shawl workers, however, were very poor during this period (Rehman and Jafri 2006: 88). The *karkhanadars* (factory owners) completely controlled the weavers, who were bound to work only for the employer who trained them. The practice of luring away a weaver by others was punishable by law (Ames 1986: 32). The workmen were kept under perpetual bondage by their *karkhanadars* and their wages were extremely low. The last Sikh governor of Kashmir, Sheikh Imam-ud-din gave the shawl weavers some relief in 1846, by setting them free from the bondage of *karkhanadars* (ibid.: 32).

During this period, Kashmir's trade links with Central Asia and Europe were firmly established. Merchants from Persia, Turkey, British India and Europe regu-

---

[11] The available archival material does not contain any specific figures on the comparison between land and shawl revenue or revenues coming from pashmina and *shahtoosh* respectively in precolonial times. Therefore, it is not possible to take the description beyond the sketchy view of the growth of shawl industry drawn from works of historians and travellers referred to in this section.

larly came to Kashmir to purchase its fabrics (Ahad 1987: 13). The British also tried to bring the shawl workers to Punjab, apparently to feed the supply to England (Ames 1986: 40). The oppression of the rulers and *karkhanadars*, and the great famine in 1834 in Kashmir resulted in the emigration of a large number of workers to Amritsar and other towns in Punjab. Ames notes that 'since about 1810, Loudhiana has been the site of a new industry employing over a thousand specialised workers and 400 looms' (ibid.). This was also the time when the industry was experiencing an economic tug-of-war between England and France, who both wanted to establish their own separate shawl industries, independent from Srinagar (ibid.).

3.3.1.5 Dogra Period (1846–1877): The Rising and Falling Demand of *Shahtoosh* Shawls

A new situation developed when the colonial interests of the Sikhs in Kashmir clashed with the British commercial interests. The British defeated the Sikhs led by Maharaja Ranjit Singh and took control over all of their territories. However, in order to meet their commercial interests and reduce administration costs, the British government sold the kingdom of Kashmir to Maharaja Gulab Singh (1792–1857), the Dogra ruler of Jammu, after signing the treaty of Lahore in 1846. This treaty gave shape to the princely state of J&K including Ladakh. The treaty contained an article which said, 'Maharaja Gulab Singh acknowledges the supremacy of the British Government and will, in token of the supremacy, present annually to the British Government, one horse, twelve perfect shawl goats of approved breed (six male and six female) and three pairs of Kashmir shawls' (Dhar 2001: 41). The treaty demonstrates the importance of the Kashmiri shawl industry and by extension the trade in *pashm* and *shahtoosh* wool (Rizvi 1999: 50). It also appears that the British were contemplating rearing the shawl goats from the mountains of western Tibet. However, the experiments of rearing these goats in England failed as the new habitat did not suit the species (see Dhar 2001; Schonsberg 1853).

In the Dogra period, European taste, especially English and French, completely dominated the Kashmiri shawl design. The early accounts show that weavers themselves resented this foreign interference. In the initial years, much difficulty was experienced in persuading the native designers to alter or amend their patterns, but eventually they started adopting hints (Irwin 1973: 17). The French and English traders attempted an imitation of the art of *shahtoosh* weaving and embroidering. Norwich is regarded as the pioneering imitation shawl centre in Europe, started as early as 1784, followed by Edinburgh and some other Scottish towns (ibid.: 20). Competition to the Kashmir shawl manufacturing was also posed by the flourishing industry in Lyons in 1800. It is notable that when the British brought Kashmir under their control, they stopped Kashmiri craftsmen from having direct contact with the French agents in order to establish their monopoly. The British also blocked the Kashmir shawl trade with Russia by controlling the frontiers at Yarkand and Kashgar (Dhar 2001). It is reported that the popularity of these shawls declined in Europe with the Franco-Prussian War of 1870 and a sudden change in European fashion

(Ames 1986: 48; see also Younghusband 1909). In short, the commercial rivalry of the colonial powers and the popularity of shawls in European markets contributed among other things to the rise and fall of the shawl industry of Kashmir.

3.3.2 Marginalisation and Exploitation of the Shawl Workers: Pre-independence

The available literature provides very limited information on the conditions of *shahtoosh* workers in the Kashmir valley. However, we find some references to it in the accounts of travellers such as Moorcroft and Trebeck (1841), Schonsberg (1853) and Younghusband (1909). In the discussion below, I highlight the poor living and working conditions of the shawl workers and the harsh system of taxation under different regimes.

Describing the living conditions of the artisans in the mid nineteenth century, Ahad writes: 'the artisans lived in old dilapidated houses situated in the places of inconceivable filth' (Ahad 1987: 68). The dress of artisans too reflected their acute poverty. The majority of them did not possess sufficient clothes to protect themselves from the cold and moved around almost bare-foot (ibid.: 67). Moorcroft (1820) states that the weavers were the most oppressed section of the industry, the majority of them ill-clothed, under-nourished and permanently indebted to the state tax-collectors and *karkhanadars*. Without the supplementary earnings of his wife and children, it was impossible for an average weaver to support his family (ibid.). The artisans worked under strict regimentation as bond-slaves or day labourers from morning to evening (Moorcroft and Trebeck 1841). Wakefield (1879) observes that the shawl workers spent their childhood, youth and old age with unending poverty, disease and suffering.

The poor conditions of artisans described in historical accounts reflect that their wages in the Mughal period must have been meagre (see Verma 1994; Ahad 1987; Lawrence 1895). In the time of the Afghans, the system of remuneration was both in kind and cash. The food grains were sold to the shawl weavers in exchange for their labour at the rate of 50% above their market price (Ahad 1987: 54). When paid in cash, the wages were very low for shawl workers but the merchants and shawl brokers earned profits and lived a prosperous life (ibid.). During the Sikh period, the economic exploitation of the artisan class and mass of peasantry had reached new heights. The weavers earned seven to eight rupees per month, but were forced to pay five rupees per month as tax (Dhar 2001: 31). They had to undergo heavy loans, which ultimately left their families starving. For spinners, the wages did not exceed three and a half rupees per month. Out of this amount, they had to spend two rupees on the purchase of wool (Ahad 1987: 55). Treating them as bonded slaves, the government coerced the workers not only to pay the tax but also to work very hard. The dreadful cruelties of the rulers can be observed when, for the smallest offence, the government punished the poor artisans with the loss of their noses and ears (Ames

3.3 Kashmir Shawl Industry

1986: 30). The weavers are said to have cut off their thumbs to escape the tyrannies of the rulers and *karkhanadars* or proprietors (ibid.: 32). The Sikh government levied heavy taxes on the wool trade. The import of raw wool was subjected to innumerable duties at various places between Ladakh and Tibet. On each horse-load, 95 rupees was to be paid to the government as duty. On its arrival in Kashmir, the wool was sold by the government at 20% profit to the merchants. During the Sikh period, there was 25% ad valorem duty on each shawl and 25% more as private exaction of the corrupt officers (Ahad 1987: 105).

After Kashmir was handed over by the British to Maharaja Gulab Singh, conditions of the weavers deteriorated even further. The Maharaja levied a poll-tax of 47–48 rupees per annum on each shawl weaver, which caused their living conditions to deteriorate considerably (Irwin 1973: 9). Heavy duties were imposed on the import of wool, and the production and export of the fabric greatly hindered the flow of the trade, discouraging the purchaser (Ahad 1987). This accelerated the downfall of the industry. During this period, the shawl workers were forced to pursue their craft in order to create profits for their masters even though many were half-blind or incapacitated (ibid.). If any shawl worker gave up his employment or succeeded in escaping to Punjab, the *darogah* of *Dag-i-shawl* would send a sepoy to the house of the fugitive to drag his wife, mother or father to the court, where they were fined a heavy sum or imprisoned for a few days (ibid.: 104). However, the shawl merchants gained huge profits from the sale of shawls to European markets during this period. For example, Zutshi (2004: 85) notes that in 1850s, shawls were sold at a 500% profit on its original cost of production in Kashmir.

We can observe that the formation of each new state attempted to control the shawl industry and the shawl workers by way of different strategies. For example, during the Mughal period, the *shahtoosh* shawls were reserved for the exclusive use of the royal household or presented as a tribute to the foreign kings, thus making shawls a symbolic capital. In the eighteenth and nineteenth century, with the expansion of the shawl industry and the extension of trade links with foreign markets, we saw the emergence of private *karkhanas* and a fully-functioning brokerage system. Yet, there were several restrictions imposed by the Mughal and Afghan rulers on private entrepreneurs in order to control them and protect state revenues. Apart from this, the state forced the workers to work for long hours in return for very low wages and, at times, controlled labour by paying them only in the form of food grains. During the Sikh period, there were checks placed on the mobility or freedom of *shahtoosh* workers to work with other *karkhanadars* and if the shawl workers fled the Valley, they were punished brutally. We can observe that the harsh taxation system under the different regimes imposed on the shawl workers contributed to the increase in the state revenue but, at the same time, it also led to the deteriorating living conditions of the artisans. We also witnessed the attempts by British colonialists to re-territorialise the shawl trade by closing off older trans-regional networks with Russia, France and Central Asia for their own commercial interests by way of monopolising the trade.

3.3.3 Shahtoosh Workers and the New State: Post-independence

In 1948, soon after independence, the shawl industry received a severe blow in the wake of partition of the country when many artisans migrated to Pakistan, where they tried to earn a living by adopting other professions (Shah 1992). The control over the production and trade of shawls now passed from the feudal rulers and merchants to the manufacturers of that time. The *shahtoosh* workers now experienced exploitation in the industry at the hands of manufacturers and their agents, and continued working in an unorganised sector without any fixed contracts and wages (ibid.). Ahad (1987), however, describes that after independence, the working conditions of shawl workers improved slightly after the creation of cooperative production centres and the formation of trade unions. Although few in number, government production centres and embroidery units run on cooperative basis had well-ventilated halls, with proper heating and cooling arrangements (ibid.: 56). Since the shawl industry could be one of the important sources of revenue for the state government, the government started making efforts to foster its growth by providing subsidies to exporters, credit facilities to manufacturers and starting training schools for artisans (see Shah 1992).

According to the shawl workers, the period of Sheikh Abdullah's government represents a glorious period in the history of artisans in the Valley. In the early 1950s, Abdullah took some measures to curtail the influence of shawl manufacturers and traders, and to improve the conditions of artisans by providing them the minimum wages. He also encouraged the shawl workers to form trade unions in order to check the exploitative practices of their employers. As mentioned in Chap. 2, in 1950, Abdullah introduced land reforms in the state and distributed the land to the tiller. The tenants were now entitled to three-quarters of the produce of the land. These changes resulted in the improvement of the economic conditions of artisans who partly depended on agriculture for meeting their survival needs. During this period, the government sponsored Kashmir Industrial Arts Emporium set up outlets in Amritsar, Delhi, Mumbai and Shimla with a view to find a market for the Kashmiri handicrafts and to eliminate the middlemen altogether. Cooperative societies started organising exhibitions for Kashmir arts and crafts to strengthen the industry. According to the shawl workers, Sheikh Abdullah also prohibited the trade of any unfinished products in order to protect the incomes and livelihoods of local people against machine-made products outside the state. In 1975, when Abdullah came to power again, Debt Conciliation Boards were formed in order to help the artisans who depended on the manufacturers for advances and loans. These boards helped generate amicable settlements between debtors and creditors and dismissed all claims in which one and a half times the principal had been paid (Sharma 2004: 83). Yet, the manufacturers found ways to maximise their profits and retain control over the production process, a point that I explain below.

Until the 1970s, each job in the production process was mutually exclusive, with *poiywans* (agents of manufacturers) purchasing the wool from the *bhotias* or local dealers and supplying it to the manufacturers. The manufacturers, in turn, employed the skilled artisans for processing, and supplied the finished shawls to hawkers to be sold within and outside the state. Since the 1980s, in order to derive profits from the raw wool trade, the manufacturers started bypassing the *poiywans* and the local Kashmiri dealers in procuring the raw wool.[12] They started buying the wool directly from the *bhotias* or from the Nepalese traders by making cash payments. The *bhotias* preferred to sell the wool to the manufacturers directly than to the *poiywans* as the latter were only able to pay the amount in instalments unlike the manufacturers who had ready cash available. The manufacturers now started employing the *poiywans* to distribute the raw material to the separators, spinners and weavers, and also to supervise the whole process of shawl production. Furthermore, owing to the rising demand for *shahtoosh* shawls in foreign countries, the big manufacturers, bypassing the hawkers, started directly engaging in the marketing of shawls outside the state. As such, the conventional jobs of the manufacturers, *poiywans* and hawkers, most often, now came to be performed by the same person.

Another important shift occurred in the manufacturing process itself. We saw a decline of big *karkhanas* in the post-independence period, possibly, due to new labour regulations regarding minimum wages and working conditions. The production process got de-concentrated into small *karkhanas* where typically weavers would employ subordinate workers on their residential premises. Spinners, separators and embroiderers also started working from their own houses, and *poiywans* acted as supervisors of the entire production process on behalf of manufacturers. Although the *karkhanas* owned by manufacturers declined in number in the post-independence period, the manufacturers, through their agents, were still able to control the entire production process. The de-concentration of the manufacturing site could be understood as a strategy on the part of manufacturers to discourage unionisation of workers and reduce production costs. Since there was a significant difference in the costs incurred in the production of the *shahtoosh* shawls and their selling price, some weavers also felt motivated to engage in small-scale manufacturing. Their initiatives, however, were discouraged by the established manufacturers and they devised different tactics to keep these weavers at a disadvantage. For example, when the weavers approached the manufacturers to purchase *shahtoosh* wool, the manufacturers charged them very high rates for the wool. The weavers had to purchase the wool from the manufacturers on credit as, most of the time, they did not have the cash readily available. This indebtedness further led to the vulnerability of the weavers. As the weavers did not have easy access to the market, they were forced to sell their products to the manufacturers at a low cost in order to pay for the raw material they had borrowed. Owing to this, very few weavers were able to succeed in becoming manufacturers in their own right and others simply returned to the status of subordinate workers of big manufacturers. Some weavers also took over the roles of hawkers to sell their products in the market outside the state. But the manufacturers again adopted various ways to convince the customers and exporters

---

[12] The discussion presented is made on the basis of my interviews with senior *shahtoosh* workers, mainly embroiderers and weavers during my fieldwork in Srinagar.

outside J&K that the products brought by these weavers were adulterated and did not represent the pure hand-made *shahtoosh* shawls. Thus, due to the lack of capital, the high cost of wool, and the absence of a market for their products, the weavers were forced to accept the domination and exploitation by the established manufacturers, who managed to remain the most powerful actors in the shawl production process. I shall return to this point in Chap. 5 of the book.

3.4 Conclusion

In this chapter, I have discussed the myths and realities concerning the source of *shahtoosh* wool and showed that the origin of *shahtoosh* is obscure in the historical accounts as well as amongst the shawl workers. Historically, the Kashmiri shawl merchants have enjoyed monopoly over the wool, both *shahtoosh* and *pashmina*, originating from western Tibet and Ladakh. The fact that the local shawl workers still hold the traditional beliefs and legends concerning the source of *shahtoosh* possibly point to the deliberate strategy adopted by the merchants and traders over the centuries to secure their interests and maintain control over the trade. Despite the fact that the wool is obtained after the killing of *chirus* and the meagre quantity of shed wool is not sufficient to meet the requirements of the shawl industry, the manufacturers and exporters have tried to conceal the true origin of *shahtoosh* by perpetuating myths relating to the collection of wool as well as using the pictures of the mountain goat Ibex in the *shahtoosh* promotional brochures. The frontier location and the special status of J&K made it possible for the trade in *shahtoosh* to continue even after the trade involving the antelope and its body parts was made illegal by the Indian government in the mid 1980s.

We have observed that *shahtoosh* shawl production involves a complex division of labour. The powerful actors such as *bhotias*, local dealers, manufacturers, agents, hawkers and exporters, who contribute very little to the actual production process, accrue profits by exploiting the skilled workers engaged in the most laborious tasks involved in the production of *shahtoosh* shawls. The lack of capital and access to markets has traditionally kept these artisans in a subordinate position. We also witnessed that the shawl industry had undergone important changes during its growth under different regimes. The production of the *shahtoosh* shawl for exclusive use by the ruling classes gave way to the production for mass consumption with the decline of the Mughal Empire. While the state owned *karkhanas* remained dominant, some privately owned *karkhanas* also emerged by the eighteenth century, although under the control of the state officials. We see an increase in the volume of long distance trade in shawls especially with European countries around this time. The British succeeded in gaining a monopoly over the shawl trade by restricting the access of French and Russian agents. Historically, the shawl trade has remained an important source of revenue for the various regimes in Kashmir. While the merchants and shawl brokers amassed wealth as the industry flourished, the conditions of the artisans, as I explained, did not improve much. On the contrary, harsh taxes imposed on the shawl workers and their poor working conditions have forced them to live in misery in the period before independence.

In the post-independence period, I argued that the manufacturers and agents replaced feudal rulers and shawl merchants in controlling the shawl industry and production process. Although the new state took some initiatives to help the poor artisans by granting loans, cancelling debts and abolishing the taxes imposed on them, it could not eliminate the exploitation of the artisans at the hands of manufacturers and their agents. I explained that the manufacturers started taking up wool and shawl marketing jobs, bypassing *poiywans* and hawkers, in order to gain exclusive control over the production process as well as marketing. Rather than being associated with their previous job of supplying wool to the manufacturers, *poiywans* now came to be employed by the manufacturers to distribute wool and supervise the production process. The established manufacturers, as noted, also discouraged the shawl weavers from becoming small-scale manufacturers by charging a higher price for the raw wool and purchasing their fabric at lower rates. The weavers, due to the lack of capital and easy access to the market, often ended up working under the big manufacturers. Furthermore, the manufacturers have been able to reduce the production costs and prevent the unionisation of artisans by fragmenting the site of production but, at the same time, maintaining control over their subordinate workers through their agents, *poiywans*. I further elaborate on the complex relations of domination and subordination in the post-ban period as well as examine the roles and agendas of the different stakeholders in the banning of *shahtoosh* trade in the next two chapters.

References

Ahad A (1987) Kashmir to Frankfurt: a study of arts and crafts. Rima Publishing House, New Delhi

Ahuja B (2006) The unique and eloquent legacy of Kashmir handmade pashmina. Wildlife Trust of India, New Delhi

Ames F (1986) The Kashmir shawl. Antique Collectors Club, Suffolk

Bamzai PNK (1980) Kashmir and Central Asia. Light and Life Publishers, New Delhi

Chundawat RS, Talwar R (1995) A report on the survey of Tibetan antelope (Pantholops Hodgsoni), in the Ladakh region of Jammu and Kashmir. Ministry of Environment and Forests, New Delhi

CITES (1999) Conservation of and control of trade in the Tibetan antelope. www.cites.org/eng/res/11/11-08R13.shtml. Accessed 1 Feb 2006

Dhar DN (2001) Dynamics of political change in Kashmir. Kanishka Publishers, New Delhi

Feng Z (1999) Status and conservation of Tibetan antelope in China. In: Zhen RD. (ed) The future of Tibetan antelope. proceedings of an international workshop on conservation and control of trade in Tibetan antelope, October 12–13, 1999, Xining, Qinghai, Beijing, China, pp 27–28

Ganhar JN (1979) The wildlife of Ladakh. Srinagar Press, Srinagar

Hassnain FM (1992) The beautiful Kashmir valley. Rima Publishing House, New Delhi

Huber T (2003) The chase and the dharma: the legal protection of wild animals in pre-modern Tibet. In: Knight J (ed) Wildlife in Asia: cultural perspectives. Routledge, London, pp 36–55

IFAW, WTI (2001) Wrap up the trade: an international campaign to save the endangered Tibetan antelope. International Fund for Animal Welfare and Wildlife Trust of India, New Delhi

IFAW, WTI (2003) Beyond the ban: a census of shahtoosh in the Kashmir valley. International Fund for Animal Welfare and Wildlife Trust of India, New Delhi

Irwin J (1973) The Kashmir shawl. Her Majesty's Stationary Office, London

IUCN (2016) IUCN red list of threatened species. www.iucnredlist.org/details/15967/0. Accessed 1 Aug 2017

Jayakar P (1959) Cotton Jamdanis of Tanda and Benaras. Lalit Kala, Delhi

Kumar A, Wright B (1997) Fashioned for extinction: an expose of the shahtoosh trade. Wildlife Protection Society of India, New Delhi

Lawrence W (1895) The valley of Kashmir. Henry Frowde, London

Leslie D Jr, Schaller G (2008) Pantholops hodgsonii. *Mammal species* 817:1–13

Liu W (2009) *Tibetan Antelope*. China Forestry Publishing House, Beijing (in Chinese)

Moorcroft W (1820) Excerpt from a letter from Moorcroft to Mr. Metcalfe, the East India Company's resident at Delhi. India Office, London

Moorcroft W, Trebeck G (1841) Travels in the Himalayan provinces of Hindustan and the Punjab in Ladakh and Kashmir. John Murray, London

Rawling C (1905) The great plateau. Edward Arnold, London

Rehman S, Jafri N (2006) Kashmiri shawl from Jamavar to paisley. Mapin Publishing House Ltd., Ahmedabad

Riazuddin A (1988) History of handicrafts, Pakistan-India. National Hijra Council, Islamabad

Rizvi J (1999) Trans-Himalayan caravans: merchant princes and peasant traders in Ladakh. Oxford University Press, Delhi

Schaller GB (1993) Tibet's remote Chang tang. Natl Geogr 184(2):62–87

Schaller GB (1998) Wildlife of the Tibetan steppe. The University of Chicago Press, Chicago

Schonsberg BE (1853) India and Kashmir. Hurst & Blackett Publishers, London

Shah MA (1992) Export marketing of Kashmir handicrafts. Ashish Publishing House, New Delhi

Sharma SP (2004) Kashmir's political leadership: sheikh Mohammed Abdullah and his legacy. RBSA Publishers, Jaipur

Strauss L (1986) The romance of the cashmere shawl. Mapin Publishing Pvt. Ltd., Ahmedabad

Traffic India (n.d.) Shawls of shame. Traffic India, New Delhi

Verma T (1994) Karkhanas under the Mughals: from Akbar to Aurangzeb. Pragati Publications, Delhi

Wakefield W (1879) The happy valley. Searle and Rivington, London

Warikoo K (1989) Central Asia and Kashmir: a study in the context of Anglo-Russian rivalry. Gyan Publications, Delhi

Wilson HH (1841) Travels in Himalayan provinces of Hindustan and the Panjab from 1819 to 1825. Himalayan Gazetteer, Calcutta

Xi Z, Wang L (2004) Tracking down Tibetan antelopes. Foreign languages press, Beijing

Xinhua (2011) Tibetan antelope population hits 200,000. 21 January, 2011. http://news.xinhuanet.com/english2010/china/2011-01/21/c_13701284.htm. Accessed 3 Feb 2011

Younghusband FE (1909) Kashmir. Adam & Charles Black, London

Zutshi C (2004) Languages of belonging: Islam, regional identity and the making of Kashmir. Hurst, London

Chapter 4
The Ban on *Shahtoosh*: Sustainability for Whom?

There's restlessness in every heart
But no one dare speak out —
Afraid that with their free expression
Freedom may be annoyed.

— Ghulam Ahmad Mahjoor (A selection from the poem
"Aazaadee" by Ghulam Ahmad Mahjoor, a Kashmiri poet. The
English translation of the poem can be found in 'An Anthology
of Modern Kashmiri Verse (1930–1960)' by Trilokinath Raina
(1972))

Abstract This chapter examines the politics that surround the process of banning the *shahtoosh* trade in J&K and analyses how power determines access to and control over *local* resources which are now seen as *global* commons. The following questions are addressed in this chapter: (a) What were the agendas and interests of a diverse set of actors (international, national, state and local) associated with the ban on the *shahtoosh* trade? (b) How can the state be understood in its dual role, acting as an agency for imposing the ban and at the same time, allowing illegal production and trade of *shahtoosh* to continue? (c) How did *shahtoosh* workers respond to the ban and how were the interests of both powerful actors and poor artisans served? (d) What were the linkages, if any, between militancy in the state and the now illegal trading of *shahtoosh*? Attending to these questions, I explain how the ban on the *shahtoosh* trade was understood, accepted, negotiated and resisted at different levels from global to local.

Keywords *Shahtoosh* · Chiru farming · Split Role of State · Illegality · Corruption · Militancy · Environmental politics · Macropolitics · Jammu and Kashmir

© Springer International Publishing AG 2018
S. Gupta, *Contesting Conservation*, Advances in Asian Human-Environmental
Research, https://doi.org/10.1007/978-3-319-72257-3_4

63

The debates on nature conservation and livelihoods are far from conclusion and have tended to follow two entrenched and opposed lines of argument. The mainstream environmental conservation policies aimed at achieving goals of environmental sustainability have, thus far, largely ignored the politics and power relations crucial in the access, control and distribution of resources. The alternative viewpoints have mainly considered nature conservation interventions as coercive and detrimental to the interests of the local communities dependent upon natural resources for sustenance. Before taking sides, it is important to examine *how* nature conservation policies permeate different layers of politics and are further shaped and reshaped according to the interests of the various stakeholders involved in the process of implementation.

This chapter examines the politics that surround the process of banning the *shahtoosh* trade in J&K and analyses how power determines access to and control over *local* resources which are now seen as *global* commons. The following questions are addressed in this chapter: (a) What were the agendas and interests of a diverse set of actors (international, national, state and local) associated with the ban on the *shahtoosh* trade? (b) How can the state be understood in its dual role, acting as an agency for imposing the ban and at the same time, allowing illegal production and trade of *shahtoosh* to continue? (c) How did *shahtoosh* workers respond to the ban and how were the interests of both powerful actors and poor artisans served? (d) What were the linkages, if any, between militancy in the state and the now illegal trading of *shahtoosh*? Overall, I explain how the ban on the *shahtoosh* trade was understood, accepted, negotiated and resisted at different levels from global to local.

The structure of this chapter is as follows: Sect. 4.1 explains the chronology of events that led to the imposition of the ban on *shahtoosh* in 2002, and presents the arguments of various actors with regard to the prospect of *chiru* breeding. I identify three forms of politics associated with law, science and the constitutional status of the state and argue that the ban on the *shahtoosh* trade was addressed largely along conservation lines without adequate attention paid to the livelihoods of the poor *shahtoosh* workers. Section 4.2 examines the response of the state and *shahtoosh* workers to the ban and unravels embedded local political complexities. I analyse the 'split role' of the enforcement officials and limited avenues of protest for the poor *shahtoosh* workers. I argue that the powerful actors within the *shahtoosh* workers' community are able to secure their interests while the relatively powerless and poorer workers are marginalised. Following this, in Sect. 4.3, I examine shadow networks of *shahtoosh* and argue that the trade still continues, albeit illegally. Towards the end of this section, I also discuss the various arguments presented by the *shahtoosh* workers about the connection between militancy and the *shahtoosh* trade. My principal argument is that the imposition of a blanket ban often proves to be detrimental for poorer groups that are dependent upon natural resources for sustenance. At the same time, it can potentially open up new opportunities for rent seeking (for enforcement agencies), profit making (for traders and manufacturers), and the legitimisation of a conservation agenda along the lines of science and sustainability (to the benefit of international conservation community).

4.1 Ban on *Shahtoosh*

4.1.1 A Chronology of Events

I have discussed the legal status of the *chiru* in detail in Chap. 3. Here, I explain the chronology of events that led to the change of the *chiru*'s status from a protected to a banned species.

In 1988, with the revelation of the truth about the connection between the massacre of the Tibetan antelope and the highly prized *shahtoosh* shawls of Kashmir, the conservation of the *chirus* became a global concern. CITES began to create awareness of the issue among policy makers through international workshops and directing the various governments to coordinate, collaborate and enforce CITES resolution to protect the *chirus* from extinction. Supporting this agenda, wildlife conservation NGOs launched various campaigns worldwide to educate the masses about the unsavoury origin of the *shahtoosh* shawl. In India, NGOs such as Traffic India, WTI, IFAW and WWF played a major role in generating awareness of the issue. The IFAW and WTI organised various programmes, supported by the fashion fraternity, such as 'Say No to *Shahtoosh*', 'Save *Chiru*' and 'Wrap Up the Trade' to educate the people about the origin of *shahtoosh* wool. Similarly, the Traffic India launched another campaign 'Don't Buy Trouble' with the aim to get the message across to consumers and discourage the demand for *shahtoosh* shawls in the market. Oddly, these awareness building programmes were confined to metropolitan cities and elite consumers. Apart from a small-scale campaign organised by the WWF in Srinagar, there were no programmes organised to generate awareness among the local people in Kashmir.

In 1995, CITES accused the Indian MoEF of failing to stop the *shahtoosh* production and sale and ordered the government to take strict measures in this regard. In response, the MoEF appointed an expert committee[1] to undertake a survey in Ladakh to determine the conservation status of the *chiru* in India and the trade in its wool. The first objective of the survey was to identify the areas in India where the Tibetan antelope is found, and to give a rough estimate of its population. The second task was to verify the area from where the shed *shahtoosh* wool (if any) is collected as claimed by local herders and traders. The third aim was to verify the existence of a breeding farm for the species in Ladakh, as claimed by the traders. The fourth objective was to identify the border areas through which the barter trade of *shahtoosh* wool is carried through Ladakh (Chundawat and Talwar 1995).

The survey confirmed the presence of Tibetan antelope in the upper Changchenmo valley and Aksai Chin region of Ladakh. The *chiru* habitat that lies within the Indian territory was estimated to be around 2500 km^2. The traders were unable to provide any evidence of shed wool of the *chirus* or any breeding farms. The team also

---

[1] The survey team was comprised of the Chief Wildlife Warden of J&K, the Regional Wildlife Warden of Leh, a scientist from the Wildlife Institute of India, a representative of Traffic India and a few traders from Kashmir.

reported that no evidence of killing of the Tibetan antelope was found within Indian territory (ibid.). According to the survey team, the figure for the consumption of *shahtoosh* was initially stated by the traders as about 3000 kg per year. Subsequently, on being asked to confirm this, the traders stated that accurate figures of consumption were not available. Considering this initial figure, the expert group estimated that since 150 g of *shahtoosh* wool is produced by one antelope, approximately 20,000 Tibetan antelopes are required to meet the current market demand (ibid.). The experts argued that since a very small population of 200–220 *chirus* enter into the Indian territory, commercial collection of wool was not possible to sustain the *shahtoosh* trade in J&K. While not rejecting the possibility of captive breeding of the *chirus* in Ladakh in India, the experts stated that it is economically not viable considering the high costs involved in the project.[2]

4.1.1.1 WPSI vs. J&K State: Petition and the Response

A complex situation emerged in the mid 1990s when the trade of *shahtoosh* shawls was banned in most parts of the world, yet due to the J&K's special constitutional status, the production and sale of *shahtoosh* shawls continued in the state. A careful study of constitutional provisions and legal precedents was then undertaken by the Supreme Court of India. The court indicated that there were two ways of ensuring that the J&K government banned the trade. First, at the time of accession of the state to the Indian government, it was agreed that all international treaties that are ratified by the central government would apply to J&K and that when a state law differed from a central law, the latter would prevail. Since the *chiru* is a protected animal under the provision of CITES, the import and export of its parts from China and Nepal signalled its illegality and violation of an international law. The second argument was that since no regulation of the *shahtoosh* trade has taken place in the state under 'licence', the *shahtoosh* trade was already in violation of the Wildlife Protection Law of 1978 (IFAW and WTI 2001).

In 1997, the WPSI issued a notice to the Government of J&K to include all endangered species in the State Wildlife Protection Act along the lines of the Central Wildlife Protection Law (Kumar and Wright 1997: 39). It stated that the trade in *shahtoosh* was being carried out without permission from the relevant authorities and without the issuance of licences. It directed the state to control the illegal trade of raw wool. However, due to the inaction of the state, in 1998 the WPSI filed public

---

[2]The then Wildlife Warden of Leh and a member of the survey team informed me that captive breeding of the *chirus* is possible although it requires high investments and co-ordination with the Indian army as antelopes need to be airlifted in helicopters from Changchenmo valley to the closure in Leh, which is the only suitable place (owing to its altitude, transportation facilities, availability of food etc.) in Ladakh for establishing closures. Also in 1995, there were no tranquilising experts in the state and very little funds were available with the Wildlife Department in J&K to materialise this project. Militancy was at its peak and the entire government machinery and funds were directed to tackle it. Therefore, the plan to explore *chiru* breeding was dropped (Interview in Jammu, 22 March, 2007).

4.1 Ban on *Shahtoosh*

interest litigation in the J&K High Court on the grounds that there had been several seizures of *shahtoosh* wool and shawls from 1992 to 1998 and that India accounts for most of the seizures (IFAW and WTI 2001). It stated that there is a paradoxical contrast between the Central Wildlife Act and State Wildlife Act with regard to the status of the *chiru*, and, therefore, it appealed to order the state government to ban the hunting and trade of all species listed in Schedules I to IV (ibid.).

It can be observed that the law set the first stage for politics among various actors[3] involved in the controversy over the ban on *shahtoosh*. I understand this relationship between law and politics in two ways: First, as soon as the two propositions were made by the Supreme Court (mentioned above), it expanded the power and arena of actions of various players associated with the *shahtoosh* controversy. Second, I observe that laws, if unclear, can create uncertainties and disputes and hence can be used by the various parties to their advantage. Such a controversy found its way through the ambiguities in the meanings of 'wool' and 'hair' of the *chiru* in the Wildlife Protection Act of J&K. In the following discussion, I illustrate these two points by presenting the arguments made by the WPSI and the central MoEF in relation to the urgency of the ban and the competing claims made by the J&K state and the other interest groups such as the Handicraft Traders Association and the *Shahtoosh* Weavers Association.

In the public interest litigation, the WPSI stated that the *chiru* falls under the definition of 'animal article' and 'trophy' in the State Wildlife Act of 1978.[4] The definition of 'animal article' reads: 'an article made from any captive animal or wild animal, other than vermin, and includes an article or object in which the whole or any part of such animal has been used'.[5] Further, the definition of 'trophy' states: 'trophy means the whole or part of any captive animal other than vermin, which has been kept or preserved by any means whether artificial or natural and includes a) rugs, skins and specimens of such animal mounted in whole or in part through a process of taxidermy and b) antler horn, rhinoceros horn, hair, feather, nail, tooth, musk, eggs and nests'.[6] Thus, the WPSI held that since the *chiru* is included under 'animal article' and its hair falls in the category of 'trophy', the trade in its wool or products was illegal.

The MoEF supported the arguments of the WPSI and added that the state of J&K cannot maintain its own laws if they conflict with those of the central government. Therefore, it asked the State government to bring the State Wildlife Act in line with the Central Wildlife Act. The Ministry claimed to have taken steps to train and sensitise the enforcement agencies such as the Indo-Tibetan Border Police, the Border

---

[3] Here, various actors include the WPSI, the Union of India represented by the MoEF, the Government of India, the CITES Secretariat, the Supreme Court of India, the J&K High Court, the State of J&K represented by the Secretary, Department of Forests, the Chief Wildlife Warden, J&K, the Presidents of Handicraft Traders Association and the *Shahtoosh* Weavers Association.

[4] High Court of Jammu and Kashmir (2000). Public Interest Litigation PIL (CWP) No. 293/98. This PIL was filed in 1998 and the decision came out in May, 2000.

[5] Ibid.

[6] Ibid.

Security Force, postal guards, customs officers and police to prevent any illegal import or export of the *shahtoosh* and other animal articles. However, the Ministry maintained that enforcement in the state is a subject that falls within the domain of state government and therefore it was the state government's responsibility to take corrective measures.[7]

In 1998, the state government rejected the WPSI directive, stating that J&K has a separate constitutional status and is not obliged to comply with central wildlife regulations. Moreover, the state government argued that there has been no evidence of the poaching of *chirus* in the state and maintained that shawls are made from the wool shed by the animal and collected by the local herdsmen, which are then sold in the market against which there is no prohibition. It was further added that the *chiru* figures in Schedule II of the State Wildlife Act and that hunting is not allowed without permission from the Chief Wildlife Warden of the State. Therefore, even if the animal is not moved to Schedule I, hunting of the animal is not possible without the knowledge of the wildlife authorities. It is, therefore, immaterial whether the *chiru* figures in Schedule I or Schedule II of the State Wildlife Act.[8]

On account of the provisions based on the special status of the state, the J&K Handicraft Traders Association also held that the state can maintain its own laws and take appropriate measures to protect the species rather than banning the trade. The association resisted the WPSI petition on two grounds: first, it argued that the *shahtoosh* shawl is made of wool (and not hair), which is not covered under 'animal article' or 'trophy' as defined in the State Wildlife Act, and which therefore requires no licence. Second, it held the belief that since this wool is shed by the animal which the local herdsmen collect and sell in the market, no killing is therefore involved.[9] Supporting this position, the *Shahtoosh* Weavers Association added that 'the prohibition of *shahtoosh* trade in the state will help those who are competing with the shawl industry of Kashmir. Foreign countries, especially those in the Middle East, are overtaking the Indian markets in such local products'.[10] The association emphasised that the state, which is suffering from ongoing militancy and which is primarily dependent on the handicrafts and tourism industry, would be hit hard. The total ban would cause irreparable damage to the art, artisans and culture of the state and, therefore, the trade should be regulated and not prohibited.

As we observe in the narratives above, the ambiguity over the question of 'hair' and 'wool' became a controversial subject for the players involved. Below, I present the arguments made by different parties involved in the petition hearing in the J&K High Court from 1998 to 2000.

The counsel for the WPSI stressed that '"animal article" as defined in the State Wildlife Act includes shawls made from the hair of Tibetan antelope, and that the definition of "trophy" also includes the hair of the animal and that wool is nothing

---

[7] Ibid.

[8] Ibid.

[9] Ibid.

[10] Ibid.

4.1 Ban on *Shahtoosh*

but hair'.[11] He added that under the State Wildlife Act, wild animals, including 'animal articles', shall be the property of the government and the person who obtains them by any means must report it to the enforcement agencies or police officials. Therefore, according to him, no person can be a manufacturer or dealer of any of the animal products. He further argued that since the *chiru* is a protected animal under the provisions of CITES, import and export of the animal article is not permissible except in accordance with the law. Moreover, since no licences have ever been issued by the state authorities, the manufacture, trade and possession of the *shahtoosh* shawl is illegal.

In response to the above arguments, the counsel for the J&K state argued that the state is acting strictly in accordance with the provisions of the State Wildlife Act. Since the Central Wildlife Act is not applicable to the state, he argued that the *chiru* is not a protected animal and there is no need to regulate the trade.[12] The counsel for the Handicraft Traders Association added that the petition should not be maintained because it involves no public interest. He reiterated the argument that it is not the hair of the *chiru* but the underwool shed by the animal from which the *shahtoosh* shawls are manufactured.[13] Further, the counsel for the private respondents argued that unless poaching of the animal is proved to be for the purpose of trade, the manufacture of *shahtoosh* shawls cannot be prohibited. He stressed that if killing is necessary for wool to be collected, the trade would not have survived for centuries. He also maintained that it is the underwool shed by the animal that is collected by the local people of Tibet.

In response, the counsel for the WPSI stated that the justification provided by the Handicraft Traders Association that *shahtoosh* is not 'hair' but 'wool' can be questioned on the grounds that wool is defined as 'fine wavy hair from the fleece of sheep, goats etc.' and 'any of the fine thread like strands growing from the skin of mammals'.[14] He further added that the claim of *shahtoosh* not being obtained from killing *chirus* is misleading. He stated that 'the yield from a dead animal is only between 125 to 150 grams of the *shahtoosh* and when it is shed, the quantity may be much less, and hence it cannot meet the market demand'.[15] He added that the expert committee appointed by the MoEF in 1995 also found no evidence for the collection of shed wool of any kind and argued that the Union government as well as the State government are under legal obligation to enforce the provision of CITES. The WPSI maintained that since the J&K state has acceded to the central government in foreign affairs, the union government alone is competent to enter into trade agreements with foreign countries and has the obligation to enforce it throughout India, including in the state of Jammu and Kashmir.

The above discussion suggests that the law set the ground for the interplay among various actors by enhancing their power to address concerns and opening up new

---

[11] Ibid.

[12] Ibid.

[13] Ibid.

[14] Ibid.

[15] Ibid.

possibilities of negotiations because of the ambiguities in the policy or in the law itself. We also observed that ambiguities around the origin of the *shahtoosh* wool (discussed in Chap. 3) became useful in the controversy for various parties.

The J&K High Court asked the *shahtoosh* traders for evidence to prove that the *shahtoosh* wool is shed by antelope, which they failed to provide. On the other hand, the plaintiff (WPSI) provided irrefutable evidence of the slaughter of the Tibetan antelope and the smuggling of raw wool from Tibet to India (IFAW and WTI 2001: 28). Moreover, the hair samples taken from *shahtoosh* shawls were tested and confirmed to have patterns that are unique to Tibetan antelope (ibid.: 20). As a result, the J&K High Court upheld the view that the claim made by the *shahtoosh* traders about the origin of *shahtoosh* wool is unacceptable due to lack of evidence.

In May 2000, the J&K High Court directed the state government to strict *regulation* of the *shahtoosh* trade according to its Wildlife Protection Act, 1978.[16] The J&K state, however, did not take any effective measures to control or regulate the *shahtoosh* trade for fear of losing the revenue and to protect the age old shawl industry. It maintained that the special status of J&K (under Article 370 of Indian Constitution) provides autonomy to act in its own interests. The then Chief Minister of J&K, Farooq Abdullah stated:

> As long as I am the Chief Minister, *shahtoosh* will be sold in Kashmir. The campaign to ban the trade maligns the people of the state [...] There was no evidence of Tibetan antelope being reduced in number or their being shot to acquire wool for *shahtoosh*.[17]

Here, we see that the controversy on the ban on *shahtoosh* became more complex due to the special status of J&K state. The conflict around the status of the *chiru* in the State Wildlife and Central Wildlife Acts provided a podium for both the parties to justify their claims. The political instability encouraged the representatives of the state to plead for the ban not to be imposed and to argue that complete prohibition would further damage the already fragile economy of J&K.

In the same year, the WPSI filed an appeal to the Division Bench of the J&K High Court against the previous judgement, calling for a ban or *complete prohibition* of the trade in products made from the Tibetan antelope and parts thereof. The central government was concerned about its 'international image' and persuaded the J&K government to provide complete protection to the *chirus* and to bring its status within the J&K Wildlife Act in line with the Central Wildlife Protection Act, 1972.[18] It reminded the state government that all international treaties to which India is a signatory, including CITES, are also enforceable in J&K. Following this, in September, 2002, the J&K legislature ended the long drawn legal battle by passing

---

[16] High Court of Jammu and Kashmir (2000). Public Interest Litigation PIL (CWP) No. 293/98.

[17] Cited in the article by Peter Popham (1998), 'These animals are dying out. And all because the lady loves *shahtoosh*', The Independent, 20 June 1998. This statement by Farooq Abdullah also needs to be seen in the light of tense relationship between the J&K state and the centre historically. It is also likely that Abdullah was using this language to consolidate his political position in the state. I discuss these points in detail in Sect. 4.2 of this chapter.

[18] Letter written by S.P. Prabhu, the Minister of Environment and Forests, Government of India to the J&K Chief Minister, Farooq Abdullah (MoEF n.d.).

4.1 Ban on *Shahtoosh*

the J&K Wildlife Protection (Amendment) Act and moved the Tibetan antelope from Schedule II to Schedule I, thus giving it complete protection in the state. As a consequence of the amendment, any trade in *shahtoosh* was declared *illegal* and all persons with stocks of *shahtoosh* wool and shawls were required under law to declare the stocks and obtain ownership certificates from the Chief Wildlife Warden.

Interestingly, in 2003 the Division Bench of the J&K High Court disposed of the writ petition filed by the WPSI in 2000 pleading complete prohibition. The court noted that in light of the amendments to the J&K Wildlife Act in 2002, which changed the status of the *chiru*, no further direction was required to be given to the J&K state in this matter.[19] It is the responsibility of the state to abide by the international treaties (such as CITES) and cooperate with the central government to enforce the ban. However, the court observed that 'those who have been deprived of their livelihoods, both the artisans as well as nomadic tribes, be adequately compensated'.[20] It is notable that apart from this observation, the court did not put the onus of responsibility for compensation and welfare of affected artisans on either the state/central governments or the conservation community. The conservationists, thus, eventually succeeded in making the trade in *shahtoosh* 'illegal'. While the trade still continues illegally (as explained later in the chapter), this legislative change, nevertheless had significant implications for the livelihoods of shawl workers.[21]

On the basis of the above discussion, it can be argued that the global environmental concerns can act as potential grounds for conflict between nature conservation interests and the livelihoods of the local communities. Despite the initial resistance to the ban by the J&K state on the pretext of its special status, and the observation made by the High Court regarding compensation to artisans, we can see that the issue of *shahtoosh* was addressed largely along conservation lines. To a certain extent, this supports the claims of scholars who are of the view that the global conservation agenda is, inherently, oppressive in nature. I will examine the arguments of some of these scholars below, and validate their claims of the disciplinary and coercive nature of conservation policies. I will also indicate (and demonstrate later

---

[19] High Court of Jammu and Kashmir (2003). Verdict of the Division Bench on the writ petition filed by the WPSI.

[20] Ibid.

[21] Different claims about the number of people associated with the *shahtoosh* trade are made by the local workers' associations and political parties in Kashmir. For example, the former Chief Minister of J&K Farooq Abdullah declared the number to be 20,000 at a public meeting in Srinagar (IFAW and WTI 2001). The President of the People's Democratic Party puts the figure between 40,000 and 50,000 (Interview with Mehbooba Mufti, Srinagar, 25 November, 2006). The Kashmir Valley Weavers and Manufacturers Association claims the figure to be around 25,000–30,000 and the Kashmir Handicraft Traders and Weavers Association at over 100,000. The J&K Department of Handlooms has no data on the total population engaged in *shahtoosh* production or trade and refers to it as an unorganised sector. As mentioned in the previous chapter, according to a survey by the IFAW and WTI in 2002–2003, 14,293 individuals were directly involved in the production and marketing of *shahtoosh* (IFAW and WTI 2003). Even the modest estimate from the survey indicates that banning of the *shahtoosh* trade has rendered a large population unemployed in the Valley.

in Sects. 4.2 and 4.3) that the enforcement of such policies is influenced by competing interest groups and governed by local political realities.

Scholars such as Peluso (1993) argue that the conservation agenda, which is described as being in the common interests of the entire global community, is an excuse for external intervention in what had previously been the sole affairs of states. Similarly, Gupta (1998) suggests that increasing environmental concerns are resulting in a new form of 'governmentality' to discipline third world societies. Neumann (2001) argues that the conservation agenda has a disciplinary effect by limiting the access of local people to their natural resources. This has also been characterised by a neglect of human needs and interests to the extent that Dietz labels it 'ecototalitarian' (Dietz 1996: 13) and Goldman terms this new hegemonic regime as 'eco-governmentality' (Goldman 1996: 33). There has also been considerable debate within urban contexts on the role of PIL in advancing the interests of middle classes on environmental issues (often described as bourgeois environmentalism) at the cost of interests of the poor urban residents (see Mawdsley 2004; Baviskar 2002; Chaplin 1999). Randeria notes that the new conservation regime, aided by international NGOs, has brought more and more territories and communities on the peripheries under the control of the postcolonial state in the name of 'ecological development' (Randeria 2007: 13). These claims can be supported by my findings which suggest the primacy of conservation over local interests. We observed that although the J&K state resisted the imposition of ban for a long time, due to pressures by the international conservation community and the Indian government, it was forced to comply with their demands of supporting the conservation interests even against the will of its population, who remained almost voiceless in the legal battle.

In the discussion on legal proceedings, I have also illustrated that science and the faith in its objective nature can be used to serve and legitimise conservation interests. This point draws us closer to the arguments of scholars (such as Saberwal and Rangarajan 2003; Greenough and Tsing 2003; Feyerabend 1993; Pickering 1992) who have demonstrated how the authoritarian nature of science and its ability to claim expert knowledge can undermine indigenous or local understandings. It has also been argued that for conservation management, the local is a site for instruction, implementation and control with specific scientific objectives in mind (Blaikie 2006: 1951). Some other scholars have even questioned the objective nature of science and hence its appropriateness in resolving conservation related problems. For example, Agrawal (1995: 11) states that the problem with science arises only when it claims to be necessarily better by basing itself on notions of objectivity and neutrality, and hence supporting the scientific community to claim primacy in decision-making (see also Saberwal 2003). In the discussion below, I further examine the ways in which science and scientists are enrolled within the argument for a total ban, and against captive breeding.

4.1 Ban on *Shahtoosh* 73

4.1.2 The Prospects of chiru *Farming: Observations of the 'Expert Group'*

In 2004, the Prime Minister of India visited the J&K state. During his visit, some politicians of the ruling party (People's Democratic Party), the President of the Chamber of Commerce, and the *shahtoosh* workers appealed for the ban to be lifted. Soon after, an expert committee was formed by the central government to investigate the prospects of *chiru* farming so as to protect the shawl industry in J&K. The expert committee comprised of representatives of the Union Ministry of Textiles, the MoEF, the President of the Chamber of Commerce, J&K, the Director of Animal Husbandry and senior members of wildlife organisations such as the WTI and the WPSI. From the outset, the committee split into two groups on the basis of different perspectives held on the prospects of *chiru* breeding. The first group, comprising the MoEF and conservationist NGOs, believed that *chiru* farming was not a viable alternative. The other group, comprising the Chamber of Commerce and the Animal Husbandry Department of J&K, contended that the breeding is possible as it has been done in the case of other protected or endangered species such as the mink in Estonia, eider-duck in Iceland, and vicuna and alpaca in South America. Below, I present some of their arguments in relation to possibilities of *chiru* farming in Ladakh.

The various representatives of the J&K state maintained that *chiru* is not killed in the Indian territory and that it is the shed wool which is collected by the herdsmen. They argued that serious efforts to farm *chirus* should be made to sustain the art and livelihoods of the people involved in the production of *shahtoosh* shawls. The Director of Animal Husbandry suggested that while poaching may be a major factor in the decline of the *chiru* population, other factors such as disturbances in the *chiru* habitat due to the construction of roads and railways, mining and tourism are equally noteworthy.[22] He contended that the vicuna in Latin American countries was also on the verge of extinction but that initiatives by the government authorities to involve local communities in its farming resulted in positive outcomes.[23] Not only

---

[22] Minutes of the meeting of the Expert Group set up to look into the issues relating to *shahtoosh* (Ministry of Textiles 2005a).

[23] The management of vicuna in the Andes is one of the success stories of international wildlife conservation. The vicuna population recovered from 10,000 to about 2,50,000 in the period between 1965–2005 (McNeill et al. 2009). This was made possible in two ways: first, in 1969, the five countries- Argentina, Bolivia, Chile, Ecuador and Peru entered into the Convention for the Conservation of the vicuna to make a joint effort to strictly protect the species in the region (ibid.). Second, the government authorities realised that the armed park-guard model was inadequate for providing protection from poaching of the species. In the 1980s, there was a paradigm shift from state-centred control towards involvement of the local people for its breeding and conservation (Wells and Brandon 1992). Furthermore, considering the fact that the remote regions were resource poor and with very few economic alternatives for local people, giving the ownership of wool to the local communities was considered to be an important incentive for them to participate in its breeding. The local communities, who earned less than £300 in a year, had great economic interest in farming this species because 1 kg of its fleece costs £200–300 on the European market (McNeill et al. 2009).

have these measures resulted in an increase in the vicuna population but the local people have also been able to generate income by selling vicuna fibre. The Director of Animal Husbandry, therefore, suggested that if a turnaround in the vicuna population was possible, it could also be done in the case of *chirus*. He stressed that the issue of conservation would not be successful unless local populations living in and around the *chiru* habitat were involved for the protection of wildlife with a vested interest in such a way that the nomadic tribes would find it more beneficial to protect the *chirus* rather than simply being involved in their killing. He proposed that this would require the involvement of the Chinese authorities, since 90–95% of the *chiru* habitat falls in the Tibetan territory (Ministry of Textiles 2005a).

Supporting the above argument, the President of the J&K Chamber of Commerce added that financial incentives provided to the local herdsmen to conserve and farm *chirus* could help to increase its population (ibid.). He presented the example of eider-ducks[24] in Iceland which were also on the verge of extinction due to relentless hunting. To avert this, eider-ducks were made the property of farmers on whose land they bred. As a result, the farmers felt motivated to protect these ducks from natural predators and poachers. This provided a new business opportunity for local communities. The President of the J&K Chamber of Commerce argued that if such initiatives were put in place in the case of *chirus*, the art and the livelihoods of the people could be preserved. The Wildlife Department of J&K maintained that captive breeding of *chirus* is possible. However, it requires heavy investment costs and coordination with the Indian Army and para-military forces, not just for transportation purposes but also in checking the solid waste pollution created by the army which is posing a major threat to the local ecology and *chiru* population in the Indian territory.[25]

The proposal of *chiru* breeding, however, was not supported by conservationist groups such as the WTI and WPSI on the basis that it seemed impractical.[26] These groups argued that it would require high investment to set up a breeding farm in a vast and inaccessible area of Tibetan highland. Furthermore, they emphasised that no parallel can be drawn between the rearing of vicuna in Latin America and *chiru* in the Changthang region for the reason that vicuna is amenable to be shorn and *chiru* is not. It was stated that no in-depth study regarding the physiology or the growth of *shahtoosh* fibre has been conducted to date which adds uncertainty to the prospects of *chiru* farming. Belinda Wright, a representative of the WPSI reiterated

---

[24] The eider-duck is a protected species in Iceland. Eider down is used in expensive luxury items such as duvets, sleeping bags and padded winter clothing. To conserve this species, the government involved the local farmers in its breeding and granted the right to gather and export eider down. The breeding of these ducks resulted in the increase of populations along with providing sustainable income generation opportunities for the local farmers (Chaulk et al. 2005). Similarly, in Estonia, serious efforts were made to conserve European mink through captive breeding. Although the history of maintaining and breeding these populations dates back to 1983, regular breeding was only achieved in the mid 1990s. For more details, see Maran (2003).

[25] Interim Survey Report of Tibetan Antelope in Daulat Beg Oldi (D.B.O) & Changchenmo valley prepared by the Wildlife Warden Leh (Wildlife Department, J&K 2005).

[26] Draft report of the Expert Group on the issues related to *shahtoosh*, Ministry of Textiles (2005b).

4.1 Ban on *Shahtoosh*

that it is not possible to breed *chirus* in captivity. She argued that breeding prospects were possible for vicuna in South America because the species could be herded and kept in a semi-captive state unlike the *chiru*.[27] She stated that the only example of a captive *chiru* was recorded in *Natural History of the Mammalia of India and Ceylon* (Sterndale 1884) wherein it is observed that 'the captive *chiru* was reduced to a living specimen'.

George Schaller, an expert on Tibetan wildlife, was contacted by the WTI to give his opinion on the prospects of *chiru* breeding. In response, Schaller stated that captive breeding of *chirus* in India is a poor idea.[28] India has only a very small population which migrates from Tibet during the summer. Historically, *chirus* have never been kept in captivity and the problems of maintaining them are still unknown. Besides this, he reported that nothing is known about the nutritional requirements, veterinary care or social interactions in confinement of these sensitive and nervous animals. Capturing and transporting them is likely to cause deaths, stress and injury to the animal. He added that to maintain such a farm requires a facility that needs high initial costs, besides recurrent costs including staff salaries, food for animals, vehicle maintenance etc. To initiate a captive breeding program under present conditions is, therefore, morally and economically indefensible. He suggested that any available funds would be better spent on protecting the *chirus* and providing alternative jobs to people in the *shahtoosh* business.

After discussing the various issues and possibilities of *chiru* breeding, the Expert Group felt that the matter should be discussed with the Chinese authorities. In this regard, the WPSI contacted the CITES Management Authority of China. In response to CITES Management Authority of India, it was stated that the decline in population of the *chirus* in Tibet was due to demand for their wool in J&K.[29] Further, it was held that the animal has not been successfully bred in captivity until now and hence, the proposal to breed *chirus* did not seem to be a feasible solution. However, the Chinese authorities stated that they were willing to cooperate with the Indian government to crack down on any illicit trade in endangered species in the future (ibid.). The WTI, in 2005, arranged a meeting between the representatives of the Chinese government and the Ministry of Textiles to discuss the possibilities of *chiru* farming. In 2005, the Ministry proposed that if the Chinese government agreed to provide 1000 *chirus*, India would take the initiative to farm them.[30] However, this proposal from the Ministry of Textiles was rejected by the Chinese government. The reasons were: first, technically, it would not be possible to domesticate or breed the animal in captivity as the animal lives in open steppes at an altitude of above 5000 metres and cannot survive at lower altitudes. Second and more importantly, it was

---

[27] Letter written by Belinda Wright, WPSI to the Director, Ministry of Textiles, Government of India. 10 October, 2005 (WPSI 2005).

[28] Letter written by George Schaller to the WTI. 15 November, 2003 (Schaller 2003).

[29] Letter written by CITES Management Authority of China to CITES Management Authority of India. 15 November, 2004 (CITES China 2004).

[30] Letter written by Vivek Menon, WTI to Shashi Bhushan, Director, Ministry of Textiles, Government of India (n.d.).

stressed that the Tibetan antelope is listed in Appendix I of CITES to which 167 countries are signatories including India and China. Captive breeding for commercial trade of Appendix I species can only be permitted if a resolution to that effect is adopted by vote at the CITES Conference of Parties (COP). Before such a resolution can be brought to the COP, it must be placed before the Animals Committee of CITES for technical appraisal. Therefore, the representatives of the Chinese government argued that the power or authority to lend the species for farming does not depend upon the mere fact that the *chiru* habitat lies within their territorial boundaries. Further, it is a matter which does not simply involve India and China but requires international deliberations.

It is important to note here how the international conservation community can re-territorialise the region in a very different way than was explained in chapter three. There I noted the efforts made by British colonialists to de-territorialise the shawl trade by discouraging the association of agents in Kashmir with non-British merchants for their own *commercial* interests. Here the argument put forward by the Chinese government with regard to prior approval by the international community or signatories to CITES can be understood as an obligation for sovereign nation-states to maintain the primacy of *conservation* interests. It signals the considerable power and authority of a new form of conservation regime, which can re-territorialise a region according to its priorities and interests. Owing to the transnational nature of the problem, CITES and other international conservation agencies could have played a vital role in *chiru* breeding (by collaborating efforts of Indian and Chinese governments) in order to secure the livelihoods of nomadic tribes along with curbing illegal poaching.

Despite the claims made by the international conservation community about the increasing reconciliation between conservation goals and local interests, the discussion above suggests the opposite. The approach pursued by the conservation community in the name of 'environmental sustainability' can be questioned on two grounds: first, it has been observed that marginal significance was attached to the local livelihoods (both herders and artisans) in the entire controversy. The proposals for community based management through participation of Tibetan herders were put forward by various representatives of the J&K state. Different forms of incentives to these herders were proposed to both protect the species and save the shawl industry in Kashmir. Yet, apart from meetings and discussions of the Expert Group, no practical efforts were made by the international agencies to experiment with *chiru* breeding. It can, thus, be argued that undertaking conservation in poverty ridden areas is not tenable if it disregards the needs of local people and further contributes to impoverishment (see Campese et al. 2007; Cernea and Schmidt-Soltau 2006; Arjunan et al. 2006). I shall return to this point while discussing the socio-economic conditions of Tibetan herdsmen later in this chapter.

Second, little efforts were made to conduct experiments of *chiru* breeding owing to the possible complexities involved such as huge investment costs, lack of studies in *chiru* physiology and wintering conditions. It is pertinent to refer to the arguments of scholars such as Chhatre and Saberwal (2006), Guha (2003), Norgaard (1994) and Lele (1993), who have noted the limitations of the scientific and conser-

4.1 Ban on *Shahtoosh*

vation community in operationalising 'sustainability', due to a lack of self-reflection, cultural sensitivity and poor understandings of social structures. It can be suggested that before dismissing the possibility of farming *chirus* by involving local herders, there is a strong need to use science in the service of local communities, and practically assess the possibility of *chiru* breeding before outright denial. This argument can be supported by other experiments as we see in the case of endangered species such as vicuna, eider-duck and mink, which succeeded in spite of specific challenges involved in each of these interventions.

On the basis of the above two arguments, I maintain that the ban on *shahtoosh* production and trade in J&K state was not based on the principle of 'sustainability'. Various political ecologists have expressed scepticism about the 'sustainable development' approach as a basis for change (see Hulme and Murphree 2001; Escobar 1996; Middleton et al. 1993; Sachs 1993) . It is suggested that the mainstream 'reformist' approach has hit an impasse both intellectually and practically in its efforts to address intensifying environmental problems (Bryant and Bailey 1997). As well as failing to address the problems of poverty and environmental degradation, this approach lacks faith in local participation which has, in turn, led to inadequacies and contradictions in policy-making (Lele 1991). Therefore, and as Yearley (1996) argues, the SD approach or its offshoot 'biodiversity conservation' must recognise the necessity of a flexible approach that involves listening to a variety of voices if the conservation goals are to be realised at the local level.

To conclude, in this section, I have explained how law, science and the special constitutional status of the J&K state set the stages for politics in the controversy regarding imposition and resistance to the ban. I have also discussed diverse viewpoints regarding the prospects of *chiru* breeding. It is noted that the issue of *shahtoosh* was largely dealt with along *conservation* lines and that the agenda of the international conservation community ultimately prevailed in making the trade 'illegal'. Based on these broad findings, I have highlighted the limitations of the mainstream nature conservation policies, and have extended my support to scholars who emphasise the coercive and violent nature of the global conservation regime, which is inimical to the interests of the people dependent on natural resources for sustenance. In doing this, however, I maintain that the 'coercive conservation' viewpoint too is inadequate in explaining and understanding the ways in which global conservation policies are understood and experienced at the local level. In the rest of the chapter, I aim to unravel the many embedded local political complexities that shape conservation policies during the process of implementation. Below, I analyse the 'split role' of the state and local forms of protest and politics to examine the response of the state and the *shahtoosh* workers against the ban.

4.2 State and Shawl Workers

4.2.1 Weak Enforcement and Split Role of the State

As discussed earlier, the J&K state tried to resist the ban and delay its imposition but finally succumbed to the pressures of the international agencies and central government. Here, I focus on the role played by the J&K state as an agency for enforcing the ban and, at the same time, acting as an important actor in allowing the 'illegal' trade to continue. I refer to this dual function as the *split role* of the state. Before proceeding to examine the split role, I present the conflicting positions of the two chief political parties in J&K regarding the ban, in order to understand the new form of party politics that generated in the state.

The National Conference (NC) Party that was in power in 2002 followed the verdict of the court regarding the obligation of the J&K state to assist the central government in enforcing CITES regulations concerning the Tibetan antelope. Mustafa Kamal, the then Minister of Industries, informed me that the government at that time had no choice but to comply with the decision of the court and the will of the central government.[31] Subscribing to the myths sustained by *shahtoosh* workers in the Valley, he argued that the ban on the *shahtoosh* trade was not based on any valid reason since the wool is shed by *chirus* and collected by the herdsmen.[32] According to him, pressurising the state to ban the production of *shahtoosh* certainly involves subjugation of the interests of J&K state by the central government.[33] He blamed the Indian government for not taking a firm stand against international conservation groups and argued that international agencies have always worked for their own interests without considering the costs incurred by the poorer countries and affected communities.[34] As a solution to the problem, he proposed that *chiru* farming should be promoted in the same way as that of the musk deer in China. He added that the central government would have taken a different stand had this issue involved any other state than J&K, and that the central government has never trusted the people or popular governments of J&K.

---

[31] Interview with Mustafa Kamal, Srinagar (30 November, 2006).

[32] Interestingly, the Director of the Handlooms Department in my interview (20 November, 2006), while describing the origin of *shahtoosh* wool expressed a similar opinion: '*chiru* is not slaughtered but it sheds wool which gets stuck to the bushes. The raw wool is stained with blood, at times, because the hands of the herdsmen get scratched by the thorns of the bushes while collecting wool [...] Sometimes, even the animal bleeds while rushing through the bushes'. He further added that it has long been believed that if the wool is from a dead *chiru*, the fibre cuts frequently and cannot be spun. According to him, the reason that it is possible to spin this wool validates the fact that it is shed by the living *chiru*.

[33] Kamal blamed Maneka Gandhi, the then Indian Minister of State for Social Justice and Empowerment and a well known advocate of animal rights, to put pressure on the J&K government for the imposition of the ban.

[34] The NC lost elections in 2003 and was the main opposition party in 2006 at the time of my fieldwork.

4.2 State and Shawl Workers

The main opposition party at the time of the imposition of the ban was the People's Democratic Party (PDP). Drawing support from the *shahtoosh* traders' and weavers' associations, the party started campaigning for removing the ban to preserve the shawl industry of Kashmir and pointed to the resulting massive unemployment in the state, loss of cultural heritage and state revenue. It proposed that the vast Changthang pasture, where the antelope is found, be used to farm them in such numbers so that the issue of livelihoods could be taken care of along with *chiru* conservation.

In 2003, the PDP came to power. Mehbooba Mufti, the President of the party, argued that the ban on the *shahtoosh* trade could not be justified.[35] She blamed the NC government for surrendering the interests of the state to those of international agencies and the Indian government. She alleged that China had recently started manufacturing cheap shawls and carpets for the international market, imitating Kashmiri designs and weaving patterns, and was thus also involved in banning *shahtoosh* so as to deal a blow to the Kashmir shawl industry which remains its biggest competitor. She stated that the PDP government was trying to discuss the possibility of lifting the ban with the central government and was also trying to explore the prospects of farming *chirus* in Ladakh and Siachin. However, she held that once the law was passed, it would be difficult to overturn it. It is pertinent to note here that the PDP government did not take any rehabilitative measures for the poor *shahtoosh* workers in the form of providing any compensation or alternative job opportunities. Rather than providing compensation to poorer workers, the state allowed the weak enforcement of the ban, creating room for manoeuvre for the powerful shawl manufacturers and exporters.

During my fieldwork, I found that although the production of *shahtoosh* shawls has been banned in the state, the trade has not stopped entirely. Due to the high rate of corruption in J&K, informal networks of *shahtoosh* trade within, as well as outside the state, remained. While the manufacturers were reluctant to share information on the now illegal *shahtoosh* trade, the poor skilled workers were open to such questions. Out of 92 *shahtoosh* workers interviewed, 24 were still engaged in *shahtoosh* processing. A few of them, especially women spinners, were not aware of the ban and others were working on it simply because the raw wool and products were not owned by them but manufacturers. It is important to mention that the majority of interviewees could not specify the year in which the ban was initiated. Their responses varied from as early as 1996 to 2002. This suggests severe shortcomings on the part of conservationist groups and the state enforcement agencies to disseminate information about the ban to the affected communities.

The poor hawkers described several cases where the local police officials harassed and grabbed money from them by confiscating their *pashmina* shawls, claiming that the shawls were made of mixed *shahtoosh* and *pashmina* wool and hence 'illegal'.[36] A few *poiywans* (agents) described how enforcement officials sometimes even confiscated raw *pashmina* wool and refused to admit that it was not

---

[35] Interview with Mehbooba Mufti, Srinagar (25 November, 2006).

[36] Interviews with various hawkers, Srinagar (October–November, 2006).

shahtoosh.[37] However, shawl workers remarked that the officials did this to obtain bribes and that after making illicit payments to the officials, the wool and shawls would be released.

The poor shawl workers explained that although there have been several restrictions placed on them, the products stored and sold by rich manufacturers and traders have never been seized. They argued that the manufacturers in Srinagar have strong links with the politicians, wildlife and police officials and that it is this support from the enforcement agencies that facilitates the illegal production and trade of *shahtoosh* shawls in the state. For example, a poor hawker, wishing anonymity, remarked:

> Why are the police targeting us? Why are not they raiding the houses of politicians, bureaucrats and businessmen? We've sold *shahtoosh* shawls to most of them.[38]

This information is further supported by my observation during the fieldwork that although the shops in Srinagar did not openly display *shahtoosh* shawls for sale, the shawls were available on the demands of the customers and could be easily purchased. It points to the strong links between traders and enforcement officials who share the profits of this *illegal* trade.[39] This illustrates that in spite of legal restrictions, the trade still persists in the Valley, although in a 'shadow' form. The ban on *shahtoosh* thus remains de jure but not de facto.

It is also interesting to note that although *shahtoosh* shawls are produced exclusively in Kashmir, there have been no seizures in the Valley to date. Contrary to the information provided by hawkers and weavers, a senior police official informed me that the staff lack any training in identifying *shahtoosh* and are unable to distinguish it from other varieties of wool.[40] He stated that another reason for the lack of seizures in the state is due to the fact that the police machinery is primarily preoccupied with the problem of terrorism and insurgency. Nonetheless, some embroiderers stated that state officials themselves indulge in the illegal trade of *shahtoosh*. For example, a senior embroiderer, on the condition of anonymity, stated that a wildlife official was found carrying one quintal of *shahtoosh* wool in the official conveyance on his way back from Ladakh in 2004. No police action was taken against him. He added that there are some manufacturers and traders in Srinagar who have more than 20 quintals of undeclared *shahtoosh* in stock, and that no police action has been taken against them because they are affluent and well-connected.[41]

The above discussion signals the split role of the state. Neither has the ban been implemented effectively nor is there a complete disregard to the law and the regulations surrounding it. Weak states, as Midgal (1988) observes, are characterized by

---

[37] Interviews with various *poiywans*, Srinagar (October–November, 2006).

[38] Interview in Srinagar (6 November, 2006).

[39] During my fieldwork, a trader in Srinagar intimated that he could sell a fine quality *shahtoosh* shawl. Also, a senior hawker and well-to-do embroiderer who was on good terms with the manufacturer of his locality, stated that he could provide me with a pure *shahtoosh* shawl at the lowest rate possible.

[40] Interview with the Deputy Inspector General of Police (DIG), Srinagar (23 November, 2006).

[41] Interview in Srinagar (29 October, 2006).

4.2 State and Shawl Workers 81

high capabilities of penetration into societies while being markedly weaker when it comes to regulation and appropriation. It can also be argued that a set of perverse incentives have sprouted from the ban and have created rent seeking opportunities for local officials. This resonates with the findings of Wardell and Lund (2006: 1900), in the context of Ghana that 'a tacit understanding between public servants and local elites prevails such that the access to resources is not prevented but kept illegal in order for various rents to be extracted'. The local elites in a community are able to build a nexus with the local state and bureaucrats, becoming willing participants in corruption networks; what Veron et al. (2006), in Indian context, refer to as 'decentralised corruption'. Likewise, Bloomer (2009: 49) adopts the term 'extra-legal' to refer to criminalised networks in Lesotho between formal political actors and those engaged directly in the illegal trade. In the case of the *shahtoosh* trade, we see the emergence of such complex networks between rich traders, manufacturers, enforcement officials and politicians, which have helped to serve the interests of these powerful players, but have shrunk the space of protest available to poorer workers.

4.2.2 *Shawl Workers Response to the Ban: Protest and Politics*

Before proceeding further, it is important to mention that symptoms of resentment among the shawl workers against harsh taxation and exploitation can be observed under different regimes. During the Afghan period, the people of Kashmir had rallied against Sayyid nobles to protest against the harsh taxation system (Panikar 1953: 139). Shawl workers had also risen in revolt in 1833 against the exploitation of the Sikh rulers (Dhar 2001). In 1847, the weavers demanded that the Dogra ruler give them permission to emigrate to Punjab or change working conditions in Srinagar but these demands did not bring any substantial change in their actual conditions. In 1865, the shawl workers rebelled against the *Dag-i-shawl* department but this rebellion was crushed mercilessly (Zutshi 2004: 85). Although the weavers showed resistance to the growing exploitation of the rulers, this marginalised class was not able to lead an organised political movement. The sporadic incidents mentioned above were largely unsuccessful in altering the economic conditions of the poor weavers, embroiderers and spinners. Also, as explained in chapter two, after independence, all popular struggles (with the exception of the separatist movement) have suffered tremendously at the hands of insurgency and counter-insurgency in J&K, limiting the space for protest. Below, I explain this point in the context of the banning of *shahtoosh*.

Before the initiation of legal battle around banning the *shahtoosh* trade in the mid 1990s, some shawl workers' associations such as the Kashmir Shawl Manufacturers and Weavers Association, the Garib Mazdoor Association, the Kashmir Shawl Workers Association, and the New Farosh Pashmina Shawl Association already existed in the state. Their primary demand in the pre-ban period had been the increase in the wages of shawl workers. In the 1990s, these associations also demanded a ban on the use of machines for dehairing and spinning which was proving to be detrimental to their livelihoods and incomes. However, neither the government nor the manufacturers responded to their demands at that time.[42] It is notable that until this time, the interests of manufacturers and shawl workers did not match with regard to ban on the use of machines, increase in wages but when the ban on the *shahtoosh* trade was being enforced on the state, both manufacturers and the shawl workers faced threat to their livelihoods and incomes. This common threat resulted in manufacturers and their agents joining associations. Not only they joined, they soon took control of the associations' activities and tried to increase the membership by motivating a large number of weavers to join them and participate in the protest demonstrations against imposition of the ban. In return, they assured the artisans that they would campaign to stop the use of machine-spun *pashmina* wool and increase their wages.

The manufacturers appointed senior workers in each locality to disseminate information about meetings, demonstrations and any other protest activities against the ban. These workers then passed the information to all their subordinate employees and motivated them to assemble for such meetings. Poor weavers' main motivation for participating in the meetings was to ensure that their wages would be increased once the issue of the ban was settled. The manufacturers financed the meetings as well as protest activities but largely refrained from attending demonstrations and instead sent their agents. In hindsight, the weavers could see that the manufacturers were merely using the poverty of shawl workers as a tool to protest against the *shahtoosh* ban, and were not motivated by an interest in their welfare.[43]

The *shahtoosh* workers started demonstrations with pamphlets, posters and banners against the ban. In 1997, when the WPSI first sent a notice to the J&K government regarding the 'illegality' of the *shahtoosh* trade (as per CITES regulations), thousands of weavers and spinners assembled at Sher-e-Kashmir Park in Srinagar and raised slogans against the prospective ban on *shahtoosh*. These protests were suppressed by the police through the use of violence; as a result of which many shawl workers were injured.[44] With a history of militant separatism, the state condemned these public protests as an 'anti-national' activity maintaining *bahari siyasat ki badboo aati hai* (these protests are being supported by a 'foreign power'). In 1998, around 15,000 shawl workers gathered at the shrine of Shah Hamdani (a Sufi saint who is believed to have introduced the skill of shawl weaving in Kashmir) demanding protection of the *shahtoosh* trade. Interestingly, the manufacturers had also presented spinning wheels to female spinners as a gesture of solidarity with the '*shahtoosh* community' during the demonstration.

---

[42] As informed by the *shahtoosh* weavers and spinners during interviews conducted in Srinagar (October–November, 2006). Several of my respondents suggested that the banning of *shahtoosh* has damaged the shawl industry and that the use of machine-spun *pashmina* wool by manufacturers and *poiywans* is further undermining the age old craft of hand-made shawls in Kashmir and rendering thousands of workers unemployed.

[43] I draw this information on the basis of my interviews with shawl workers in Srinagar (October–November, 2006).

[44] Note that democracy was suspended in J&K at that time and the state did not refrain from using brutal force to tackle all popular protests.

4.2 State and Shawl Workers

In June 2000 (in response to the decision of the J&K High Court that the regulations of CITES were applicable throughout India including J&K), the shawl workers launched the *'Ban Hatao Andolan'* (or 'Remove the Ban Movement'). The poor workers were interviewed by the media who discussed their grievances but nothing changed as a result. On the contrary, several workers were injured in police action in these agitations. The manufacturers soon realised that it would be impossible to avert the ban and began dialogues with the politicians and government officials to delay the implementation of the ban in order to allow them to sell *shahtoosh* shawls, which were being processed at that time.[45]

Eventually the poor skilled workers understood that the manufacturers and *poiywans* were neither taking the issue of wage increase seriously, nor were they stopping the use of machines for dehairing and spinning. Moreover, disputes emerged within the associations between the weavers and the manufacturers. For example, previously shawl workers in the Kashmir Pashmina Tarr Farosh Association had demanded that there should be a ban on the use of dehairing machines in the state and the government had agreed to it in principle. Some of its members, on the condition of anonymity, informed me that after the ban on *shahtoosh* was imposed, the manufacturers and well-to-do *poiywans* had changed their agenda by supporting the use of machines.[46] This resulted in a dispute within the association which finally split into two camps: one dominated by manufacturers and well-to-do *poiywans*, and the other led by poorer shawl workers. However, at the time of my fieldwork, I observed that this and other associations had become inactive after the imposition of the ban.

Although the weavers and spinners demonstrated against the growing unemployment and loss of their incomes, we can conclude on the basis of the discussion so far that this marginalised category was not able to initiate an organised protest movement. This can be attributed to two factors: First, the shawl industry has remained a disorganised sector lacking in leadership and effective mobilisation from within the exploited classes to fight for their rights. Apart from the common agenda of protesting against the ban, the *poiywans*, the manufacturers and the traders have always acted against the interests of the poor shawl workers. The weavers, spinners and embroiderers found themselves powerless to fight the legal battle initiated by international and national wildlife conservation agencies. What is more, they were completely at the mercy of manufacturers and traders who had exploited them for centuries.

Second, and as suggested by my respondents, there is limited space for protest in Kashmir. Any group that comes out for demonstration in the Valley is immediately labelled as 'anti-national' by the state authorities. This restrains them from actively engaging in such protests. For example, a hawker (on the condition of anonymity) explained to me:

---

[45] This information is based on my interviews with several *shahtoosh* workers in Srinagar (October–November, 2006).

[46] Some of my respondents even claimed that the manufacturers bribed the Minister of Industries to allow the use of machines to continue.

We are harassed by the police. We pay several thousand rupees at different check posts until we reach Delhi. Many a time, they keep the money as well as our shawls. The Delhi police calls us notorious militants and anti-India people [...] You can imagine what will happen to us after protests and agitations.[47]

In the brief analysis of the protest by shawl workers presented above, I have shown that the J&K state used its violent power to crush demonstrations by the local people against the imposition of the ban on *shahtoosh*. This corroborates the argument presented by Peluso (1993: 47) that on the pretext of resource control, the state uses 'legitimate violence' to control people, especially 'recalcitrant and marginal groups' who challenge the state's authority. It also supports the argument made by other scholars that the state uses its violent capacities to suppress the local resistance in the name of conservation (see Baviskar 2001; Neumann 2001; Ghimire 1994). Thus, conservation initiatives, as Bryant and Bailey (1997) argue, can provide a means for states to assert their authority over peoples and environments thereby strengthening the position of the state actors. We observed that even within the affected communities, the interests of poor *shahtoosh* workers were subdued by powerful manufacturers and traders. Ramirez (1999: 107) suggests that weak, disenfranchised stakeholders stand to lose much from negotiations where power differences are too acute to enable collaboration.[48] In the case of *shahtoosh*, we noted that the initial collaboration between weavers and manufacturers could not be sustained after the implementation of the ban became inevitable. When affected communities fail to dismiss interventions, recognising their own position within power hierarchies, Rossi (2004: 24) argues that the various actors bargain to retain those components that matter most to them. Taking stock of the situation, the manufacturers and rich traders, after failing to resist the ban, began to build networks with the politicians, police and wildlife officials to secure their interests. On the other hand, the poor shawl workers, after understanding the motive of manufacturers and traders, stopped participating in the movement and started weaving *pashmina* to sustain themselves.

To conclude, this section has outlined local features of protest and politics to examine how various actors resisted the ban on *shahtoosh* in order to serve their respective interests. We observed that a powerful stream of actors such as rich manufacturers and traders largely determined the role and response of the poor *shahtoosh* workers in the protest activities, and also succeeded in delaying the imposition of the ban. Moreover, the banning of the trade, I argued, has provided opportunities for extortion by the state officials. I demonstrated the split role of the state in terms of the enforcement of the ban and argued that either there is no enforcement or if there is some, it is directed more against those who are easiest to control rather than those who commit the most serious violations.

---

[47] Interview in Srinagar (2 November, 2006).

[48] Similar arguments have been made by Njogu (2005) and Edmunds and Wollenberg (2003).

4.3 Illegality and Conflict

In the previous section, I briefly analysed the ongoing illegal production of *shahtoosh* shawls in Kashmir, and I shall explain this further while presenting the narratives of local *shahtoosh* workers in the next chapter. In this section, I focus on the illegal poaching of the *chirus* and the smuggling of *shahtoosh* wool into Srinagar. I also discuss the connections between the ban on the *shahtoosh* trade and the insurgency in J&K.

4.3.1 The Trade Continues: Illegality and Shadow Networks of Shahtoosh

Despite the regulations enforced by the international conservation agencies to stop the poaching of the *chirus*, its illegal killing still continues.[49] It has been reported that the biggest seizure of *chiru* hides, estimated to be 1600 in number, was from Kekexili Nature Reserve in China in 1994 (IFAW and WTI 2001). There have been many other confiscations of hides, wool and its products in the following years. The poaching of *chirus* can be attributed to a number of factors. It has been noted that the expansion of livestock herding into remote and previously unused areas has resulted in human-wildlife conflicts (Fox and Bardsen 2005). Besides this, the building of roads in Tibet has facilitated illegal hunting, since the contractors engaged in road construction have become attracted to the lucrative *shahtoosh* wool trade (IUCN 2010). The fencing of pastures on the Tibetan plateau, and the construction of the Beijing-Lhasa railway have posed another threat to the *chirus* by dividing its habitat (ibid.). Further, gold mining in the Kekexili region of Qinghai province has motivated thousands of miners to enter into the network of the illegal poaching and wool trade (Schwabach and Qinghua 2007; IFAW and WTI 2001). While the potentially large profits made from the illegal trade have been a major motivation for the complacency of enforcement agencies in China, the poor economic conditions of Tibetan herdsmen and lack of alternative livelihoods available to them are also important reasons behind the continued killing of the *chirus*. For example, it has been estimated that in the Chang Tang region, the annual income of a nomadic herding household is approximately Rs. 25,000 per year (WWF 2006). When these hunters are able to sell Tibetan antelope skins to *shahtoosh* traders for Rs 7000–10,000 per animal, there is a high economic incentive to hunt the antelope (ibid.). Thus, desperately poor herders may resort to hunting antelope not to become rich, but merely to subsist. While the blame of poaching tends to be placed on the

---

[49] Various news reports earlier in 2014 confirm the continuity of the now illegal *shahtoosh* trade. For instance, the UK's leading news paper The Guardian reports the growing popularity of *shahtoosh* shawls, illegally exported from Kashmir, in the high end fashion circles of Pakistan (see Khan 2014).

A reporter of Indian trade magazine Business Standard also confirms the availability of *shahtoosh* shawls for sale in the market in Srinagar (see Krishna 2014).

Tibetan herdsmen, it has been reported that the Chinese hunters are also involved in poaching.[50]

According to IFAW investigations in 2001, there remain tremendous inadequacies in the existing regulatory structure of China. Different nature reserves fall under the management of different central government departments, creating departmental frictions and making protection measures very difficult (IFAW and WTI 2001: 35). Some nature reserves did not have sufficient policing power to enforce the Wildlife Protection Law, and staff shortages can be seen as an important factor responsible for the weak enforcement in the region (ibid.). For example, the total number of staff in the Tibet Autonomous Region is reported to be 163, the lowest among all provinces in China.[51] Moreover, the anti-poaching forces are short of equipment and have serious funding problems (Tibet Information 2006). In the 1990s, several high profile poaching cases involved the armed police and government officials in the Tibet Autonomous Region (IFAW and WTI 2001: 34; WWF 2006). This supports my previous argument relating to the split role of state officials, leading to the weak enforcement of the ban on the *shahtoosh* trade.

As mentioned in chapter three, the porous borders of Nepal, Tibet and India have various potential centers for illegal transactions. Lhasa, Burang and Shigatse in Tibet, Kathmandu in Nepal, and Delhi, Pittoragarh and Dharchula in India are the main hubs for the illegal *shahtoosh* trade.[52] The *shahtoosh* wool has been reported to travel with *pashmina* or sheep wool stuffed inside Chinese light foam mattresses, pillows, bedrolls, blankets and jackets via the Zangmu-Kodari border (IFAW and WTI 2001: 21). It has also been found concealed in trucks transporting sheep wool that move southward from the Qinghai-Tibet plateau. The wool is sent through individual couriers by bus, truck or by air into India where the main centre is Delhi (ibid.). Some traders also carry the wool by foot across the border to Pithoragarh and then by bus to Delhi (ibid.).

The IFAW and WTI (2001) report that up to the mid 1990s, big manufacturers and *poiywans* collected the consignments of raw wool from Delhi and carried them to Srinagar. However, after 1998, as enforcement outside J&K started getting stricter, they stopped going to Delhi and instead started buying raw wool from sup-

---

[50] The Central Tibetan Administration (CTA 2007), in a report titled 'Wildlife poaching still rampant in Tibet' claims that in upper Gertse, the Chinese run most of the shops and restaurants and Tibetans are economically marginalised. These Chinese, while running their shops and restaurants, also carry out poaching and indulge in the animal skin trade. It is reported that in their attempt to make themselves look Tibetan, the hunters have the Tibetan holy mantra *Om Mani Padme Hum* written on their vehicles.

[51] Ibid.

[52] The wool has been reported to be collected from several points on the stretch between Gar and Amdo and brought to Shigatse. *Shahtoosh* also comes from Qinghai and Xinjiang via the highway connecting to Qinghai (IFAW and WTI 2001: 21). Porters carry the wool to Simikot from where it is transported to Kathmandu or Nepalgunj in Nepal. The wool is also transported from Dongba to Mustang and Manang in Nepal which is then taken to Delhi. The porters also move the wool from Taglakot in Tibet to Dharchula in India. From Dharchula, it moves via Pittoragarh to Tanakpur and onwards to Delhi. From here a different set of suppliers and couriers take over, taking it further to Srinagar (ibid.: 21–22).

pliers who had formed new routes bypassing Delhi (ibid.). Some local *shahtoosh* workers stated that Tibetan, Ladakhi and Nepali traders have now started using the old border routes (unpaved) again, to Srinagar via Ladakh where they pass raw wool to their kin and agents settled in Kashmir.[53]

The finished *shahtoosh* shawls are smuggled out of Kashmir in a number of ways. Since they are light and compact, they are sent by post or by courier mail. They have also been found hidden in shipments of other shawls. They are exported by traders through their trusted agents to markets in France, Italy, Germany, the United Kingdom, the United States of America, Canada, Mexico, Spain, South Africa, Hong Kong, Singapore, Thailand, Japan and the United Arab Emirates (Kumar and Wright 1997: 23). Shawls are also sold illegally by the hawkers or well-to-do vendors who come from Kashmir to Delhi to sell them by making visits to houses or contacting their old customers by telephone (ibid.).

It can be argued that the geographical location of the state is such that it forms an important route for cross-border illegal transactions. Although most of the respondents were reluctant to share any information regarding illegal *shahtoosh* networks, it can be observed that regional networks and trade (explained in chapter three) still operate, although now only in 'shadow form' because of the 'illegality' involved. Nordstrom (2004: 184) examines such 'shadow economies' in Angola, Mozambique, Sri Lanka and South Africa, and argues that shadow economies 'are not simply monetary matters, but socio-political power houses'. I have illustrated the involvement of powerful actors such as enforcement officials in the illegal trade in this chapter. I shall further discuss strong networks of power and control involving *bhotias* (raw wool suppliers) and manufacturers in the next chapter.

It is suggested that the states confronting civil violence and economic crises are more prone to illegal economies and unrecorded activities, what MacGaffey (1991) refer to as the 'second economy'. Owing to the fact that the J&K state has been affected by militancy for the last three decades, I now turn to explore the links between civil violence and illegal trading.[54]

4.3.2 Militancy and Shahtoosh: Exploring the Connections

It is alleged that militants have set up various commercial bases in Kathmandu where they raise funds for insurgent activities partly through *shahtoosh* and other wildlife-related trades (Kumar and Wright, 1997). In 1994, three *shahtoosh* traders,

---

[53] Interviews with various shawl workers in Srinagar (October–November, 2006).

[54] I may mention that to explore the illegal *shahtoosh* trade and its links with militancy in the state was not the intended objective of my research. The information presented on this issue is based on the informal conversations and interviews with the state officials and *shahtoosh* workers where tangential references to illegality and possible connections with militancy were made. The discussion presented here, therefore, remains synoptic. Due to safety considerations of my respondents, I have withheld their names.

who were reported to be also working as informers for Indian intelligence units, were arrested in Ladakh. They confessed to bartering tiger bones and skins for raw *shahtoosh* and using the profits to buy arms for Kashmiri militants (ibid.). This incident signals the need to explore the links between the militancy and a scarce natural resource such as *shahtoosh*.

Some scholars have suggested that unrest and violence can be the manifestations of deeper environmental conflict over the scarcity of natural resources. For example, Peluso and Watts (2001) argue that environmental violence reflects or masks other forms of social struggles. Homer-Dixon (1999) maintains that environmental scarcity can lead to civil violence through 'resource capture' (generally by elites) and ecological marginalisation of vulnerable or disenfranchised people (see also Le Billon 2005). MacGaffey and Bazenguissa-Ganga (2000) points out that due to the loss of access to resources, the affected poor may turn to unofficial activities, and society and economy, in turn, may become more violent and chaotic. Although it is difficult to draw a direct causal relationship between environmental scarcity and insurgent activities in Kashmir, we do find connections between the two. Below, I present four sets of arguments shared by my respondents in Kashmir, which shed light on the possible linkages between the insurgency in Kashmir and the ban on *shahtoosh*.

The first argument was that there is no relationship between the militancy and the *shahtoosh* trade in Kashmir. A weaver in Srinagar remarked:

> It is difficult for a worker to feed his own family, and so there can be no question of *shahtoosh* workers feeding militancy in the state [...] I tell you these are Kashmiri Pandits who want to take revenge from the Muslims of Kashmir by spreading this rumor in order to ban the *shahtoosh* trade and ruin Kashmir economy.[55]

The second argument was that since the ban on *shahtoosh* has resulted in the unemployment of a large number of people in Kashmir, the unemployed youth are joining the militant groups. For example, a hawker pointed out:

> I believe that the growing unemployment in the state due to the ban on the *shahtoosh* trade will only feed the militancy in the times to come. I am aware of a few instances in which the youth in Srinagar town, after a steep decline in their incomes, have taken up insurgent activities and have been involved in the recent few blasts in Kashmir for a small amount of money to feed their families.[56]

Third, it was also argued that foreign money generated out of the *shahtoosh* trade helped to promote the separatist movement. A senior embroiderer explained:

> There are some well-connected hawkers in Srinagar who go to sell *shahtoosh* shawls in the UAE and the Middle East, and who are provided with huge amounts of money by the militant groups there to promote and sponsor insurgent activities in Kashmir until recently [...] This black money is shared by the manufacturers, hawkers and the insurgent groups in Kashmir.[57]

---

[55] Interview in Srinagar (11 November, 2006).

[56] Interview in Srinagar (26 November, 2006).

[57] Interview in Srinagar (10 November, 2006).

A fourth argument was suggested by the President of the J&K Chamber of Commerce who argued that militancy, in fact, has contributed to the boom in the *shahtoosh* trade in Kashmir. He stated:

> The shawl industry had its glorious period at the time of militancy starting from the late 1980s. Since shawl production is an indoor activity, at the time of militancy, when shops were kept closed, people had plenty of time to weave shawls inside their houses, resulting in more production of shawls than what was seen earlier. In the 1990s, when terrorism was at its peak, big manufacturers started moving out of Srinagar to Delhi, Mumbai and Calcutta, and established their businesses there. Using their links, they were able to increase export of *shahtoosh* shawls to France, Italy, Germany, UK, UAE and Saudi Arabia. This, in turn, multiplied the demand of *shahtoosh* shawls and their production in Srinagar.[58]

The above point corroborates the argument made by Korf (2004: 275) that 'civilians in conflict situations are not all victims, and conflict can be a threat and an opportunity, often at the same time'. As I have shown, although militancy created new opportunities for big manufacturers, traders, exporters and hawkers, it also limited the avenues of protest for poor shawl workers and posed a direct threat to their livelihoods. Following the diverse viewpoints on the question of *shahtoosh* and militancy, I emphasise that although there is no direct or linear relationship between environmental struggles and insurgency, conservation interventions in conflict regions can certainly aggravate the miseries of already marginalised communities.

4.4 Conclusion

From this chapter, I draw three main conclusions. First, we observe that CITES and international activist NGOs played a central role in imposing the ban on the *shahtoosh* trade. The Government of J&K delayed its imposition owing to its special status within the Indian Union, but ultimately surrendered to the international pressures for *chiru* conservation against the will of its population. This demonstrates that the transnational actors are increasingly influential and possess coercive powers to re-territorialise a region and control *local* natural resources or *global* commons in the name of environmental sustainability. Through my description of the process of the implementation of the ban on *shahtoosh* in J&K, I have, however, demonstrated that this power does not go unchallenged but permeates various layers of politics when global conservation policies meet local realities.

I explained three kinds of politics generated by law, science and the special status of J&K. Law set the initial stage by expanding the arena of power and actions of the different players involved. This was further seen in the negotiations based on the uncertainties in the meaning and application of the term 'hair' and 'wool', as also in the ambiguities concerning the origin of *shahtoosh*. Similarly, I have shown how science and scientists were enrolled in making a case for the banning of *shahtoosh*

---

[58] Interview with the President of the Chamber of Commerce and owner of the Cashmere Marketing Agency in Srinagar (22 October, 2006).

trade and foreclosing the possibilities of *chiru* farming. Further, I discussed how the discrepancy in the status of *chiru* in the State and Central Wildlife Protection Acts provided a new basis for the contending parties to justify their respective claims. Although the ban could not be averted, these processes led to significant delays in its imposition.

Second, the 'illegal' production of *shahtoosh* shawls and the high rate of corruption in bureaucratic and political circles reveal the *split role* of the state. On the one hand, the state imposed new regulations to ban the trade, and on the other, it has hardly taken any measures to put a check on the now illegal activities of powerful players (manufacturers and traders) involved in the *shahtoosh* trade. The ban was mainly enforced against relatively powerless and poor *shahtoosh* workers and has, at the same time, opened up new opportunities for rent seeking and corruption by state officials. We also observe that the wider separatist movement in J&K has enabled the state to limit the space even for non-violent protest movements. The state, in fact, used its violent force to control protests against the ban on the *shahtoosh* trade. I described the diverse linkages between the insurgency against the state and the *shahtoosh* trade, and argued that although there is no definite and causal link between resource scarcity and violence, conservation interventions in conflict areas can further marginalise the poor as observed in the case of *shahtoosh* workers in J&K.

Third, and most importantly, we observed that power relations within the community of *shahtoosh* workers determined the conforming and resisting behaviours of poorer workers. The manufacturers and traders played a decisive role in setting the agenda and form of protest. They tried to co-opt the poor workers in protest activities, and used the issue of a loss of livelihoods and the resulting unemployment in their negotiations with politicians and state officials. However, the original demands of poor workers for a rise in wages and a ban on machines for dehairing and spinning were abandoned once the ban on *shahtoosh* became inevitable. While the manufacturers and rich traders were able to secure their own interests by forging a nexus with the officials and politicians, the interests and concerns of the poorer workers were left unaddressed.

Based on the multifaceted understanding of the politics seen in the process of imposition and enforcement of the ban on *shahtoosh*, I challenge the simplistic viewpoints of scholars who argue that conservation interventions are based on the principle of sustainability, as well as the arguments of those who maintain that a new form of eco-governance secures fixed and determined outcomes by using its coercive powers to discipline the populations according to its own agenda and interests. What can be inferred from the discussion so far is that the power of international community, although very influential in defining the relationship between local communities and natural resources, does not go uncontested, but is resisted and negotiated by various actors depending upon their power and interests. I have shown that the process of banning and its enforcement is shaped and governed by various forms of politics from macro- to microlevel. This results in differential impact of the ban on various different actors within the affected community. I examine this issue in more detail in the next chapter.

References

Agrawal A (1995) Dismantling the divide between indigenous and scientific knowledge. Dev Chang 26:413–439

Arjunan M, Holmes C, Puyravaud JP, Davidar P (2006) Do development initiative influence local attitudes towards conservation? A case-study from the Kalakad- Mundanthurai Tiger Reserve, India. J Environ Manag 79:188–197

Baviskar A (2001) Written on the body, written on the land: violence and environmemtal struggles in Central India. In: Peluso NL, Watts M (eds) Violent environment. Cornell University Press, Ithaca, pp 354–379

Baviskar A (2002) The politics of the city. Seminar 516:41–47

Blaikie P (2006) Is small really beautiful? Community based natural resource management in Malawi and Botswana. World Dev 34(11):1942–1957

Bloomer J (2009) Using a political ecology framework to examine extra-legal livelihood strategies: a Lesotho-based case study of cultivation of and trade in cannabis. J Polit Ecol 16:49–69

Bryant R, Bailey S (1997) Third world political ecology. Routledge, London

Campese J, Borrini-Feyeraband G, de Cordova M, Guigner A, Oviedo G (2007) Just conservation? What can human rights do for conservation and vice-versa?! Policy Matter 15:6–9

Cernea MM, Schmidt-Soltau K (2006) Poverty risks and national parks: policy issues on conservation and development. World Dev 34(10):1868–1830

Chaplin S (1999) Cities, sewers and poverty: India's politics of sanitation. Environ Urban 11(1):145–158

Chaulk KG, Robertson GJ, Collins BT, Montevecchi WA, Turner B (2005) Evidence of recent population increases in common eiders breeding in Labrador. J Wildl Manag 69:805–809

Chhatre A, Saberwal V (2006) Democratising nature: politics, conservation and development in India. Oxford University Press, New Delhi

Chundawat RS, Talwar R (1995) A report on the survey of Tibetan antelope (Pantholops Hodgsoni), in the Ladakh region of Jammu and Kashmir. Ministry of Environment and Forests, New Delhi

CITES China (2004) Letter by CITES Management Authority of China to the CITES Management Authority of India. 15 November 2004

CTA 2007 Wildlife poaching still rampant in Tibet. Central Tibetan Administration. 12 July 2007

Dhar DN (2001) Dynamics of political change in Kashmir. Kanishka Publishers, New Delhi

Dietz AJ (1996) Entitlements to natural resources: contours of political environmental geography. International Books, Utrecht

Edmunds D, Wollenberg E (2003) Local forest management: the impacts of devolution policies. Earthscan, London

Escobar A (1996) Constructing nature: elements of a post-structuralist political ecology. In: Peet R, Watts M (eds) Liberation ecologies. Routledge, London, pp 46–68

Feyerabend P (1993) Against method. Verso, London

Fox JL, Bardsen BJ (2005) Density of Tibetan antelope, Tibetan wild ass and Tibetan gazelle in relation to human presence across the Chang Tang Nature Reserve of Tibet, China. Acta Zool Sin 51:586–597

Ghimire KB (1994) Parks and people: livelihood issues in national parks management in Thailand and Madagascar. Dev Chang 25:195–229

Goldman M (1996) Eco-governmentality and other transnational practices of a "green" World Bank. In: Peet R, Watts M (eds) Liberation ecologies. Routledge, New York, pp 166–192

Greenough P, Tsing AL (2003) Introduction. In: Greenough P, Tsing AL (eds) Nature in the global south. Duke University Press, Durham, pp 1–28

Guha RC (2003) The authoritarian biologist and the arrogance of anti-humanism: wildlife conservation in the third world. In: Saberwal V, Rangarajan M (eds) Battles over nature: science and politics of conservation. Permanent Black, New Delhi, pp 139–157

Gupta A (1998) Postcolonial developments in the making of modern India. Duke University Press, Durham

High Court of Jammu and Kashmir (2000) Proceedings of the Public Interest Litigation PIL (CWP) No. 293/98. May 2000

High Court of Jammu and Kashmir (2003) Verdict of the Division Bench of the J&K High Court on the Writ Petition filed by the Wildlife Protection Society of India. 10 February 2003

Homer-Dixon T (1999) Environment scarcity and violence. Princeton University Press, Princeton

Hulme D, Murphree M (2001) African wildlife and livelihoods: the promise and performance of community conservation. James Currey, London

IFAW and WTI (2001) Wrap up the trade: an international campaign to save the endangered Tibetan antelope. International Fund for Animal Welfare and Wildlife Trust of India, New Delhi

IFAW and WTI (2003) Beyond the ban: a census of shahtoosh in the Kashmir valley. International Fund for Animal Welfare and Wildlife Trust of India, New Delhi

IUCN (2010) IUCN red list of threatened species. www.iucnredlist.org/apps/redlist/details/15967/0. Accessed 20 Aug 2010

Kumar A, Wright B (1997) Fashioned for extinction: an expose of the shahtoosh trade. Wildlife Protection Society of India, New Delhi

Khan RS (2014) Pakistan's demand for shahtoosh shawls threatens rare Tibetan Antelope. http://www.theguardian.com/environment/2014/jan/17/shahtoosh-shawl-rate-tibetan-antelope-kashmir. Accessed 25 July 2014

Korf B (2004) Wars, livelihoods and vulnerability in Sri Lanka. Dev Chang 35(2):275–295

Krishna G (2014) The shahtoosh conundrum. http://www.business-standard.com/article/opinion/geetanjali-krishna-the-shahtoosh-conundrum-114071801445_1.html. Accessed 25 July 2014

Le Billon P (2005) Corruption, reconstruction and oil governance in Iraq. Third World Q 26(4):685–703

Lele SC (1991) Sustainable development: a critical review. World Dev 19(6):607–621

Lele SC (1993) Sustainability: a plural, multidimensional approach. Working paper. Pacific Institute for Studies in Development, Environment and Security, Oakland

MacGaffey J (1991) The real economy of Zaire: the contribution of smuggling and other unofficial activities to national wealth. James Currey, London

MacGaffey J, Bazenguissa R (2000) Congo-Paris: transnational traders on the margins of the law. International African Institute in association with James Currey and Indiana University Press, London

Maran T (2003) European mink. Setting of goal for conservation and the Estonian case-study. Galemys 15: 11. www.secem.es/galemys/pdf

Mawdsley E (2004) India's middle classes and the environment. Dev Chang 35(1):79–103

McNeill D, Lichtenstein G, Renaudeau M (2009) International policies and national legislation concerning vicuna conservation and exploitation. In: Gordon I (ed) The vicuña: the theory and practice of community based wildlife management. Springer, New York, pp 63–79

Menon V (n.d.) Letter by Vivek Menon, Wildlife Trust of India to Shashi Bhushan, Director, Ministry of Textiles, Government of India. New Delhi

Middleton N, O'Keefe P, Moyo S (1993) The tears of the crocodile: from Rio to reality in the developing world. Pluto, London

Midgal J (1988) Strong societies and weak states: state-society relations and state capabilities in the Third World. Princeton University Press, Princeton

Ministry of Textiles (2005a) Minutes of the meeting of the expert group set up to look into the issues relating to shahtoosh. Udhyog Bhavan, New Delhi. 17 Nov 2005

Ministry of Textiles (2005b) Draft report of the expert group on the issues related to shahtoosh. Udhyog Bhavan, New Delhi

MoEF (n.d.) Letter by S.P. Prabhu, the Minister of Environment and Forests, Government of India to the J&K Chief Minister, Farooq Abdullah

Neumann R (2001) Disciplining peasants in Tanzania: from state violence to self-surveillance in wildlife conservation. In: Peluso NL, Watts M (eds) Violent Environments. Cornell University Press, Ithaca, pp 305–327

Njogu JG (2005) Beyond rhetoric: policy and institutional arrangements for partnership in community based forest biodiversity management and conservation in Kenya. In: Ros-Tonen, Dietz

References

T (eds) African forests between nature and livelihood resources: interdisciplinary studies in conservation and forest management. the Edwin in Mellen Press, New Yrko, pp 285–316

Nordstrom C (2004) Shadows of war: violence, power and international profiteering in the twenty-first century. University of California Press, Berkeley

Norgaard RB (1994) Development betrayed: the end of progress and a co-evolutionary revisioning of the future. Routledge, London

Panikar KN (1953) Gulab Singh: foundation of Kashmir state. Sage, New Delhi

Peluso NL (1993) Coercing conservation: the politics of state resource control. In: Lipshutz R, Conca K (eds) The state and social power in global environmental politics. Columbia University Press, New York, pp 199–218

Peluso NL, Watts M (2001) Introduction. In: Peluso NL, Watts M (eds) Violent environments. Cornell University Press, Ithaca, pp 1–30

Pickering A (1992) Science as practice and culture. Chicago University Press, Chicago

Popham P (1998) These animals are dying out. And all because the lady loves shahtoosh. The Independent. 20 June 1998. www.independent.co.uk/news/these-animals-are-dying-out-and-all-because-the-lady-loves-shahtoosh-1166083.html. Accessed 8 June 2010

Raina T (1972) An anthology of modern Kashmiri verse 1930–1960. Sangam Press, Poona

Ramirez R (1999) Stakeholder analysis and conflict management. In: Buckles D (ed) Cultivating peace: conflict and collaboration in natural resource management. World Bank Institute, Ottawa, pp 101–126

Randeria S (2007) Global designs and local lifeworlds: colonial legacies of conservation, disenfranchisement and environmental governance in post-colonial India. Intervent Int J of Postcolonial Stud 9(1):12–30

Rossi B (2004) Revisiting Foucauldian approaches: power dynamics in development projects. J Dev Stud 40(6):1–29

Saberwal V (2003) Conservation by state fiat. In: Saberwal V, Rangarajan M (eds) Battles over nature: science and politics of conservation. Permanent Black, New Delhi, pp 240–266

Saberwal V, Rangarajan M (2003) Introduction. In: Saberwal V, Rangarajan M (eds) Battles over nature: science and politics of conservation. Permanent Black, New Delhi, pp 1–30

Sachs W (ed) (1993) Global ecology: a new arena of political conflict. Zed Books, London

Schaller GB (2003) Letter by George Schaller to the Wildlife Trust of India, New Delhi. 15 November 2003

Schwabach A and Qinghua L (2007) Measures to protect the Tibetan antelope under the CITES framework. Thomas Jefferson Law Rev 29. www.papers.ssrn.com/sol3/papers.cfm

Sterndale (1884) Natural history of the mammalia of India and Ceylon. Thacker, Calcutta

Tibet Information. (2006) Guarding the homes of Tibetan antelopes. http://www.tibetinfor.com/tibetzt-en/antelope/menu.htm. Accessed 10 Jan 2009

Veron R, Williams G, Corbridge S, Srivastava M (2006) Decentralised corruption or corrupt decentralisation? Community monitoring of poverty alleviation schemes in Eastern India. World Dev 34(11):1922–1941

Wardell DA, Lund C (2006) Governing access to trees in Northern Ghana: micropolitics and the rents of non-enforcement. World Dev 34(11):1887–1906

Wells M, Brandon K (1992) People and parks: linking protected area management with local communities. World Bank, Washington, DC

Wildlife Department, J&K (2005) Interim Survey Report of Tibetan Antelope in D.B.O. (Karakuram Nubra Wildlife Sanctuary) & Changchenmo valley (Changthang Cold Desert Wildlife Sanctuary). August–September 2005

WPSI (2005) Letter by Belinda Wright, Director, Wildlife Protection Society of India to the Director, Ministry of Textiles, Government of India. New Delhi. 10 October 2005

WWF (2006) The slaughter of Tibetan antelope in Tibet's Chang Tang National Nature Reserve continues. In: WWF China-Tibet program. WWF, Beijing

Yearley S (1996) Sociology, environmentalism, globalization. Sage, London

Zutshi C (2004) Languages of belonging: Islam, regional identity and the making of Kashmir. Hurst, London

Chapter 5
The Micropolitics of the Ban on *Shahtoosh*: Costs and Reparations

Abstract I have addressed the local forms of politics embedded both in the protest movement and split role of the state in the previous chapter. This chapter extends the discussion further to other forms of micropolitics that have emerged with the imposition of the ban on *shahtoosh*. The following questions are addressed in this chapter: (a) In what different ways did the ban on *shahtoosh* affect the various categories of shawl workers and their livelihoods? (b) How have local power relations within the *shahtoosh* worker community determined the control over shawl production and trade after the imposition of the ban? (c) How did the various categories of shawl workers deal with the new scenario? (d) What initiatives have been taken by the state and non-state agencies to rehabilitate or compensate the losses of affected shawl workers? Dealing with these questions, I assess the differential impact of the ban and differential abilities of various categories of shawl workers to secure their respective interests after the imposition of ban.

Keywords *Shahtoosh* · Pashmina · Labour exploitation · Micropolitics · Split role · Delegated illegality · Alternative livelihoods · Politics of conservation

Critical analyses of nature conservation policies have largely focused on the legitimacy of global power to control local resources as well as local resistance to global interventions. They may also eclipse the micropolitics through which global discourses are 'refracted, reworked and sometimes subverted in particular localities' (Moore 2000: 655). In the context of J&K, there is inadequate attention to understanding the various forms of micropolitics and local practices that condition and shape the global interventions such as wildlife conservation. It is important to understand here how global interventions are manipulated and shaped at local level, and to take notice of various modes and strategies that local powerful actors devise to compensate their losses whilst directing the costs of conservation to the poorer members of the community. The analyses of nature conservation policies, thus, need to go beyond the 'idealised contrast between global hegemony and local resistance' (Singh 2008: 19), and examine different forms of contestations, compromises and costs hidden within local complexities, more so in conflict affected regions.

© Springer International Publishing AG 2018
S. Gupta, *Contesting Conservation*, Advances in Asian Human-Environmental Research, https://doi.org/10.1007/978-3-319-72257-3_5

95

96   5   The Micropolitics of the Ban on *Shahtoosh*: Costs and Reparations

I have addressed the local forms of politics embedded both in the protest movement and split role of the state in the previous chapter. This chapter extends the discussion further to other forms of micropolitics that have emerged with the imposition of the ban on *shahtoosh*. The following questions are addressed in this chapter: (a) In what different ways did the ban on *shahtoosh* affect the various categories of shawl workers and their livelihoods? (b) How have local power relations within the *shahtoosh* worker community determined the control over shawl production and trade after the imposition of the ban? (c) How did the various categories of shawl workers deal with the new scenario? (d) What initiatives have been taken by the state and non-state agencies to rehabilitate or compensate the losses of affected shawl workers? In this chapter, I assess the differential impact of the ban and differential abilities of various categories of shawl workers to secure their respective interests after the imposition of ban.

The chapter is divided into two sections. In Sect. 5.1, I focus on the various forms of micropolitics embedded in the perpetuation of legends regarding the origin of wool, unpopularity of the ban, declining wages and changing socio-cultural relations, the increasing use of machines and the exploitation of poor shawl workers. I analyse the differential impact of the ban on the various categories of *shahtoosh* workers, and suggest that the poor shawl workers are left with no other alternative but engage with *pashmina* shawl production, a shift which has resulted in a substantial decline in their wages. I argue that although the shawl manufacturers and traders are able to compensate their declining incomes by indulging in illegal *shahtoosh* trade and increasing the price of raw wool by creating wool shortages in the market, these tactics have proved detrimental to the livelihoods and incomes of poor skilled workers. Following this, in Sect. 5.2, I examine the initiatives of state and non-state organisations to support the ban affected shawl workers. I argue that the initiatives in this regard have proven to be very limited and failed to take into consideration the primary concerns of the *shahtoosh* workers. The central argument in this chapter is that the global conservation interventions such as the ban on *shahtoosh* trade are not simply met with local resistance but seep into existing relations of domination and subordination at the local level resulting in differential impact on various categories of resource users.

5.1   Impacts of Banning

As explained in Chap. 3, the categories of workers such as the spinners, weavers and embroiderers perform the most skilled and laborious tasks. Yet, economically these are the most exploited and marginalised categories of workers. Manufacturers, traders, *poiywans* and hawkers, who contribute the least in terms of actual production of shawls, are economically the most powerful actors as they control the entire manufacturing and trading processes. The hierarchical power relations are further reflected in the differential impact of the ban on the various categories of *shahtoosh* workers, not just in economic terms (redundancy, decline in wages and exploitation) but also in political (corruption and nexus building for illegal trade) and social contexts (gender relations, loss of prestige and decline in social status). In this section, I highlight the micropolitics of the ban on *shahtoosh* by considering five factors: (a) the perpetuation of myths regarding the origin of wool and the unpopularity of the ban; (b) the declining wages for different categories of workers; (c) the changing social and cultural relations; (d) the redundancies caused due to the increasing use of machines and (e) the exploitation by powerful manufacturers and traders.

5.1.1 The Origin of Wool and the Unpopularity of the Ban

In the previous chapter, I discussed the legal proceedings in the J&K High Court and presented the arguments of the representatives of J&K state regarding the origin of *shahtoosh* wool. Contrary to the scientific evidence, the state officials, politicians and trade associations held the belief that the antelope is not killed to obtain *shahtoosh*. The majority of my respondents amongst shawl workers were unaware of the reasons behind the banning of *shahtoosh* and considered the ban as unjustifiable. I found that only a few respondents, mainly manufacturers and traders, acknowledged the connection between the ban and the massacre of *chirus*. Others argued that the wool is shed by a goat found in Ladakh and is collected from the bushes and rocks by the herdsmen. I mentioned in Chap. 3 that some shawl workers even associated the origin of wool with musk deer, ibex, rabbit and sheep. Below I present some more narratives to indicate the unpopularity of the ban and diverse myths regarding the origin of *shahtoosh* in Kashmir.

Shamina Begum (age 40) of Rathpora, Srinagar is a spinner. It is her family occupation and she learned the skill of spinning from her mother when she was 13. Since then, she has been spinning and contributing to her household expenses. On the question of the source of wool, she explained:

> The origin of *shahtoosh* wool is a bird found in Ladakh who eats red sand. This bird sheds its feathers while flying which get stuck to the bushes. The herdsmen collect it and sell it to the shopkeepers in Srinagar.

Another point was made by Mohammed Rajab Naqqash (age 48), a *chappawal* of Nowshera, Srinagar who narrated:

> The ban on *shahtoosh* is not justifiable as it based on the wrong reason that wool is obtained after killing an animal found in Tibet. Actually, the wool is collected by shearing goats that live on the Nepalese side and eat white mud. Had the reason behind the ban been true, I would have been the first one to support it.

Fayaz Ahmed Sheikh (age 45) of Tawheedabad, Srinagar is a weaver and has been associated with this occupation for the last 30 years. He suggested:

> No animal is being killed for *shahtoosh* wool. Had it been the case, the animals would have become extinct centuries ago. The mere fact that the supply of wool in Kashmir has increased over the last three decades confirms the fact that the animal is safe. I have heard that the animal looks like peacock!

The above narratives point to the conviction of local *shahtoosh* workers who not only subscribe to the widely popular myths but also consider the justification of the ban put forward by the wildlife conservationist groups as based on false premises. On the contrary, they also see a positive relationship between the increased supply of *shahtoosh* and the population trend of the species. This suggests that when nature conservation interventions collide with layered formations of landscape and livelihood, 'a space opens up for people to challenge the truths in the name of which they are governed' (Li 2003: 5120). The fact that the arguments of shawl workers are in stark opposition to the knowledge of the scientific and conservation community also indicates the inadequate efforts made by these groups to generate awareness about the rationale behind the ban before its imposition in Kashmir. As argued in Chap. 3, it might be the case that the myths regarding the origin of *shahtoosh* are being perpetuated by the powerful actors who tend to gain the most from the *shahtoosh* trade. However, it also shows that the 'rationality' of the global conservation community may fail to change the 'local knowledge' or already existing beliefs of the affected communities, and hence can result in their non-conforming behaviour towards conservation policies. This corroborates the viewpoint of Singh (2008: 13) that international treaties can create an appearance of agreement about environmental management, but 'they do not necessarily shift belief systems or replace existing decision-making frameworks'. In the following discussion, I demonstrate that it is the powerful stream of manufacturers and traders who actually determine the conforming and transgressing behaviours of poorer workers by involving them in the now *illegal* production of *shahtoosh* shawls.

5.1.2 Different Categories, Differential Impact

As indicated in the previous chapter, the ban on *shahtoosh* resulted in the unemployment of a large number of shawl workers in Kashmir. Those who were dealing exclusively with *shahtoosh* have been made redundant and those who had been working with both *shahtoosh* and *pashmina* have seen a sharp decline in their wages. The IFAW and WTI (2003) in their survey of *shahtoosh* workers after the ban claimed that 55% of the workers reported a complete shift from *shahtoosh* to *pashmina*, 16% from working exclusively with *shahtoosh* to working with both kinds of wool, 11% were left unemployed, while 10% shifted from working with both *shahtoosh* and *pashmina* to *pashmina* only.[1] Apart from this, no census survey of *shahtoosh* workers has been done by any governmental agencies after the imposition of the ban.

Although the wildlife conservation agencies consider the *pashmina* shawl industry in J&K as the most viable alternative for *shahtoosh* workers, the shawl workers argue that it is economically much less attractive for three reasons. First, *shahtoosh*

---

[1] It is to be noted that this survey was conducted by pro-ban NGOs. As such, their conclusions and claims need to be taken with caution.

5.1 Impacts of Banning

production provided subsistence livelihoods to separators, spinners and weavers because unlike the case of *pashmina*, machines cannot be used for dehairing, spinning and weaving *shahtoosh*. Second, in the case of *shahtoosh*, the wool was provided by the manufacturers to the spinners (owing to the high cost of *shahtoosh*), but in the case of *pashmina*, the spinners are required to buy the raw wool from the manufacturers or *poiywans*, and get paid only for their labour, resulting in the loss of their net incomes. Third and most importantly, *pashmina* shawl production provides much lower wages which are less than half of those earned from *shahtoosh*. Moreover, *pashmina* wool processing is an already saturated sector with very limited opportunities for new entrants. This has brought a large decline in the average incomes of all the categories of *shahtoosh* workers, although to different degrees. Since there is no alternative available, the shawl workers have shifted from working exclusively with *shahtoosh* to *pashmina*, although some of them still process *shahtoosh* supplied by the manufacturers or *poiywans*. In the following discussion, I explain the economic impact of the ban by presenting narratives of various categories of *shahtoosh* workers. I argue that different categories of *shahtoosh* workers have experienced differential vulnerabilities and costs.

Before the ban, the job of dehairing *shahtoosh* was performed by separators. In the last few years, the manufacturers have started increasing the use of machines to dehair *pashmina* and reduce the production cost of shawls to compensate their losses due to the ban. This has rendered the category of separator jobless with no alternative employment available. As Taja Begum (age 55), a separator of Nawa Kadal, Srinagar explained:

> I used to clean 200 grams of *shahtoosh* per day and earned Rs. 250 for it. Although I did not receive this amount of wool everyday, my monthly income with *shahtoosh* was Rs. 1000 per month [before the ban]. With this income, I supported my family by contributing to the household expenses. After the ban, I get no wool to clean and the job of dehairing *pashmina* has also now been taken over by machines.[2]

The spinners now get machine-carded *pashmina* and, at times, raw wool for spinning from the local *poiywans*. These spinners, on average, spin 10 g of raw *pashmina* a day and make 110–120 knots for which they earn Rs. 130 at the rate of one rupee per knot. The *poiywans* deduct Rs. 60 as cost of raw *pashmina* and pays Rs. 70 for their labour. Their earnings with *pashmina* do not exceed Rs. 500–600 per month. The income was three times more with *shahtoosh*. Some of the spinners informed me that, occasionally, they still spin raw *shahtoosh* wool for their employers. Besides separators and spinners, the ban has also hit weavers by reducing their incomes to a third of what they earned previously. The average monthly income of a weaver was Rs. 6000 with *shahtoosh* which has now declined to Rs. 2500 working with *pashmina*. For example, Shoukat Ahmed Sheikh (age 35) of Tawheedabad, Srinagar is a weaver. *Shahtoosh* shawl weaving was his family occupation and he was associated with it for the last 20 years. Explaining his poor economic condition after the ban, he recalled:

---

[2] For a comparison, note that the minimum wages for casual labour in J&K in 2001 were approximately Rs. 75 per day or approximately Rs. 2000 per month.

In 2002, when the state government imposed the ban on the production of *shahtoosh* shawls, I became redundant. I approached two *vastas* [manufacturers] but I was not offered any work. The *vastas* now provide *shahtoosh* only to a few trusted weavers who have been known to them for years. After six months of no work at all, I finally got an assignment from the third manufacturer I approached, to produce four *pashmina* shawls. I shifted to *pashmina* weaving but there has been a considerable fall in my income. I earned Rs. 5000–6000 per month with *shahtoosh* but now my income does not exceed Rs. 2500 per month. With this low income, it is not possible to feed my five kids, a wife and a mother. I feel completely depressed whenever I think of my present situation [...] You see, I made such a big house with the income from *shahtoosh* and now if I have to think of its maintenance, I cannot afford it.

As with other categories of workers, the ban has also negatively affected the wages of warp-threaders (*barangors*). Tasleema Gulzar (age 37), a *barangor* of Syedpora, Srinagar who has been associated with *shahtoosh* for the last 23 years narrated:

My husband is a *rangrez* [dyer] and with the combined incomes, we used to bear the expenses of the family. After the ban, the competition among the *barangors* to get the *pashmina* warps has increased in the locality. Though it has benefited the *vastas* and *poiywans* in terms of bargaining power, this has resulted in our lower wages. Earlier, for passing 200 *shahtoosh* threads through a comb, I earned Rs. 15 but now for passing 200 *pashmina* threads, I get only Rs. 7 even though it takes the same time. This has reduced my monthly income from Rs. 2000 with *shahtoosh* to Rs. 800 with *pashmina*. I still get one or two *shahtoosh* warps in a month but it does not compensate my loss.

Mohammed Altaf (age 40) of Merjapora, Srinagar has been associated with *shahtoosh* shawl clipping for the last 18 years. He explained that although clipping shawls is not his family occupation, because of high demand of clipping *shahtoosh* shawls in the 1980s, he chose this occupation. While he earned Rs. 80–100 for clipping a *shahtoosh* shawl, *pashmina* fetches only Rs. 40–50 per shawl. Therefore, this change has resulted in the decline of his income from Rs. 9000 to Rs. 4500 per month. He informed me that although he gets some *shahtoosh* shawls to clip occasionally, the *vasta* pays him less on the excuse of 'illegality'.

Syed Mohammed Yusuf (age 75) of Khanyar, Srinagar is a senior darner (*raffugar*) who has been working in this occupation for the last 50 years. Although proud of his skill and the shawl industry, he bemoaned the ban on *shahtoosh* and resultant fall of his monthly income from 6000 (with *shahtoosh* and *pashmina*) to 3000 with *pashmina*. He exclaimed that while the prices of all commodities in the market are going up, the wages of poor artisans in Kashmir are going down.

Mohammed Rajab Naqqash (age 48) of Nowshera, Srinagar works as designer as well as a printer. He was a carpet designer until 1986, but since shawl designing was more profitable, he shifted to it 20 years ago. After the ban, he started printing *pashmina* and *ruffle* shawls primarily, although he admitted that he gets *shahtoosh* shawls occasionally. Explaining his inability to venture into another occupation, he commented:

We have been associated with this skill throughout our lives. To learn a new skill, we need another life. After the ban, we received no support from the government [...] Currently, I earn Rs. 20 for printing a *pashmina* shawl but with *shahtoosh*, I used to earn Rs. 50. This has resulted in the decline of my income from approximately Rs. 15,000 to 5000 after the ban.

5.1 Impacts of Banning

The incomes of embroiderers show no uniform pattern, they get wages according to the labour needed for embroidering each shawl. The income also varies according to the expertise, seniority and skill of the embroiderer. Although it takes the same time to do embroidery on a *pashmina* or *shahtoosh* shawl, the wages they get for embroidering *pashmina* are approximately one-third of those for *shahtoosh*. The average incomes of embroiderers with *pashmina* range from 3000 to 4000 per month as compared to 8000 to 10,000 with *shahtoosh*. Although after the ban, they have primarily started embroidering *pashmina* shawls, some of the respondents informed me that they do occasionally get *shahtoosh* shawls from the manufacturers and *poiywans*. However, they argued that it is the manufacturers who benefit from this illegal production and trade because the embroiderers, most of the time, are paid less on the pretext of illegality.

Abdul Azad (age 50) is a washer (*dhobi*) and the only earning member in his family. Since the ban, he has been making his living from washing *pashmina* and *ruffle* shawls. He earns Rs. 50 for washing a *pashmina* shawl but in the case of *shahtoosh*, the same job fetched him Rs. 100–150. He informed me that his income has fallen from 20,000 per month with *shahtoosh* to Rs. 12,000 with *pashmina* and *ruffle* shawls. Similarly, Mohammed Ashraf (age 23), a dyer (*rangrez*) of Nowshera, Srinagar observed a fall of his income from 9000 to 2000 after the ban.

According to the shawl workers, the agents (*poiywans*) have compensated their loss of incomes by adopting various strategies including the illegal distribution of *shahtoosh*. The agents, however, claim that they have witnessed a steep fall in their incomes after the ban. Jan Mohammed (age 45), an agent of Nowhatta, Srinagar explains the ways by which he is trying to reduce the production costs and cope with his declining income.

> I collect the machine-carded wool from the *vasta* and go to the villages to distribute it to the *katun wajens* [spinners]. I prefer going to villages because I can bargain and get the wool spun at lower rates. This way, I am preferred by the *vasta* in comparison to other *poiywans* in the locality. While working with *shahtoosh*, I used to buy threads from the *katun wajens* at 10% profit margin and sell those to the *vasta* with a margin of 15% but with *pashmina*, I am able to get only 10% as commission. As a result, my income has declined from Rs. 25,000 to 7000 approximately.

The decline in the incomes of hawkers shows no uniform pattern. Bashir Ahmed Sofi (age 45) of Waniyar is a small-scale manufacturer as well as a hawker. Every year, he goes to Delhi for 6 months to sell shawls and does weaving along with his subordinate workers in his *karkhana* for the rest of the year. He explained:

> I have been associated with this job for the last 25 years. I work with my uncle who is a senior hawker in the locality. I go with him to Delhi to sell shawls. Besides the shawls we produce in our *karkhana*, we also borrow shawls from the big *vastas* in Srinagar to sell [...] Before the ban, our annual sale was more than ten *lakhs* every year and we made good profits out of *shahtoosh* business outside the state. We are not able to make such profits with *pashmina* and *ruffle* shawls now. Our standard of living has been constantly going down since the ban.

As explained earlier, the manufacturers are the most powerful actors, controlling the whole production process and making most of the profit. Most of them, however, were reluctant to provide any estimates about their incomes. Although the narratives of various shawl workers suggest that *shahtoosh* wool is still provided by the manufacturers and their agents for spinning, weaving and embroidering, most of the manufacturers reported that the production has stopped entirely in the state since the ban. I present below the narrative of Habibullah Sofi (age 56) of Mondibal, Srinagar who is a manufacturer and also works as an embroiderer and trader.

> My family has been involved in *shahtoosh* shawl production and trade for the last four generations. I employ *poiywans* from different localities in Srinagar to supervise the production process [...] Before the ban, I used to manufacture *shahtoosh* shawls and sell the best quality ones to the exporters in Delhi as well as to the families of big businessmen and politicians [...] Some of my customers own nine to ten *shahtoosh* shawls. Such is their fascination for these elegant shawls. After the ban, the demand for *shahtoosh* has fallen and we have stopped its production. The profit margins with *pashmina* are much less and my monthly income has declined to 40,000, which is even less than one-third of what I used to earn with *shahtoosh*.

During my stay in Srinagar, I also interviewed a *bhotia* who is a local raw wool trader as well as the owner of a dehairing mill.[3] According to him, due to the increasing demand of machine-carded wool, he bought a dehairing machine in 1999 and started supplying wool to the manufacturers in Srinagar. The well-to-do *poiywans* also approach him directly to purchase large quantities of wool. Alongside controlling the raw wool trade, he is also involved in shawl manufacturing. Although he argued that the ban has negatively affected his income, shawl workers including weavers and embroiderers in the locality informed me that not only does he immensely benefit from the sale of illegal *shahtoosh* in the market, but he even creates artificial shortages of wool supply and increases the price of *pashmina* to generate profits. They informed me that he has strong links with the politicians and police agencies in Srinagar and is also well-connected to wool suppliers in Tibet, Nepal and Ladakh.

In the above discussion, I have highlighted that the *shahtoosh* worker community is highly heterogeneous and comprises various actors with diverse interests and power. Problematising the notion of 'community', Agrawal and Gibson (1999) suggest a stronger focus on the divergent interests of multiple actors within communities and the interactions or politics through which these interests emerge. I have examined how the ban had differential impact on the various categories of shawl workers and also how different categories of workers have responded to the ban in order to secure their respective interests. While the ban has rendered the category of separators jobless with no alternative available, it has resulted in the sharp decline of wages of poor workers such as spinners, weavers, embroiderers, warp-threaders, darners, clippers and dyers whose unique skills have contributed the most to *shahtoosh* shawl manufacturing. The other categories such as *bhotias,* manufacturers

---

[3] He was reluctant to provide any information on his income. In conversation with one of his subordinate machine-operators, I was informed that six kilograms of *pashmina* is being dehaired in the mill everyday which suggests that he must be earning more than one *lakh* rupees per month. However, it was not possible to get any estimate about his income from the illegal trade of raw *shahtoosh*.

5.1 Impacts of Banning

and *poiywans* primarily engaged in trading wool and shawls have also witnessed a decline in their incomes but have simultaneously adopted various strategies to compensate for their losses. Reflecting on the different capacities of various social actors to voice and stake their claims, Leach et al. (1999) maintain that if powerful groups do not achieve their desired ends through open negotiation, they are likely to do so through other means. I have shown that the manufacturers and *poiywans* are taking up multiple jobs of weaving, supervising, trading, and have started distributing the wool to rural artisans for getting the yarn spun and woven at the lowest possible cost. Moreover, after losing the legal battle, powerful actors such as *bhotias* and manufacturers have opened up new avenues for making profits by the illegal *shahtoosh* trade, creating wool shortages, increasing use of machines, and exploiting poor shawl workers. I discuss in detail these various strategies in the rest of this section.

5.1.3 *Machines and Adulteration*

The use of machines for dehairing and spinning is one of the biggest challenges to the livelihoods of poor shawl workers in Srinagar.[4] The shawl workers believe that the increasing demand of Kashmiri shawls in the 1980s has led to the mechanisation of spinning process to save time and labour costs, and to the deterioration in the quality of wool (by adulteration) to lower the production costs. The dehairing of raw wool initially started taking place with the help of machines in Amritsar, Punjab in the early 1990s. In 1998, the Government of India provided dehairing machines to the Ladakh Hill Development Council to be sold to *bhotias* at subsidised rates. This set the trend for the use of machine-cleaned wool in Kashmir. By 2000, most of the local wool dealers and big manufacturers in Srinagar had started purchasing their own machines for dehairing and spinning wool, and after the ban on *shahtoosh*, manufacturers have increased this use of machines to compensate their losses by lowering production costs. Although the machines have posed a direct threat to the livelihoods of the separators, spinners and weavers, they have benefited the manufacturers and raw wool traders.

A separator in Srinagar informed me that until the late 1990s, the local wool dealers and manufacturers had to send the raw wool to Amritsar for carding and cleaning but now they not only save the transportation costs, they have even started adulterating pure *pashmina* with poor quality wool at the time of its dehairing and processing to make swift profits.[5] The separators and spinners argue that manual dehairing is better than machine-carding because the machines produce short fibre,

---

[4] Similar processes of machines taking over some of the artisanal jobs have occurred world over and have been documented by several scholars. For example, Mohamad (1996) notes that spinning as a process died out by the 1920s among yarn preparers in Malaysia who lost their livelihoods and were displaced because of the availability of machine-spun yarn in the market.

[5] Interview with Haleema Begum (age 58) of Syedpora, Srinagar (12 November, 2006).

which cuts frequently and is very difficult to spin. As a result, poor spinners are able to produce less number of threads and hence earn lower wages. The spinners argue that the short length of the thread also gives the *poiywans* an excuse for deducting their wages. Moreover, they argue that machine-carding also leads to loss of the durability, softness and appearance of yarn.

The machine weaving and embroidery of shawls in Amritsar has also posed a threat to the incomes of weavers and embroiderers as the traders get the shawl woven and designed at much cheaper rates in Amritsar than getting them hand-woven and embroidered in Kashmir. These machine-made shawls from Amritsar are sold as Kashmiri hand-made shawls, which has resulted in the loss of the distinctive Kashmiri shawl identity. The customers are initially unable to see the difference because of the fine finishing quality of the machine-made shawls. However, the shawl workers argue that the overall inferior quality of machine-made shawls emerges after a few months of use. The growing dissatisfaction of the customers with the quality of the product is resulting in a declining demand for Kashmiri shawls in the markets outside of the state. Taking this point further, Azad Ahmed Khan (age 46), a weaver of Nowshera, Srinagar commented:

> Misfortunes do not come alone. The ban on *shahtoosh* was implemented along with the rising use of machines for manufacturing *pashmina* shawls which have made several shawl workers redundant in Kashmir. In 2004, I decided to purchase cheap machine-made shawls and blankets from Amritsar and sell those in Calcutta in the name of Kashmiri hand-made products [...] I found many people outside the state doing this. In the winter, I go to Calcutta and for rest of the year, I do *pashmina* weaving if I am lucky to get some work from a *vasta*.

According to my respondent shawl workers, since it is not possible to spin and weave pure *shahtoosh* on machines, many experiments of mixing *shahtoosh* with *pashmina* were initiated by the wool traders in order to make the *shahtoosh* wool suitable for machine processing. When the demand for *shahtoosh* shawls was at its peak in the 1980s and the early 1990s, some manufacturers started selling shawls made of a mix of *shahtoosh* and *pashmina* wool in the name of pure *shahtoosh* shawls.[6] The use of adulterated wool by manufacturers has created various problems for artisans.[7] For example, according to a clipper, there is always a risk of adulterated patches getting torn while brushing off the shawl to remove its unwanted knots, which gives manufacturers and *poiywans* an excuse to deduct their wages. Similarly,

---

[6] Here, we can compare the case of woollen shawls in Kashmir to silk fabric in Malaysia where rising demand of silk products led to deterioration in their quality. Mohamad (1996: 169) notes that the qualitative deterioration in production was brought about when the use of natural fibre yarn was reduced in silk weaving. At the beginning of the twentieth century, the international trading pattern in silk had taken a turn when supplies could not cope with rising demands for silk all over the world. For local weavers the price of silk went up beyond consideration, and they resorted to substitutes of cheaper, lower quality silk yarn to sustain production. It is noteworthy that unlike silk weavers, in the case of *shahtoosh* and *pashmina*, it is manufacturers (and not weavers) who started adulteration practices to meet the market demands and make quick profits.

[7] The information presented was shared on the condition of anonymity by several shawl workers including spinners, clippers, dyers and embroiderers during interviews in Srinagar (October–November, 2006).

a dyer informed me that it is difficult to get a uniform colour while dying the adulterated shawls. He argued that the manufacturers order him to dye the more adulterated shawls black to retain the uniformity of its colour. Besides these problems, embroiderers informed me that often, they get shawls that have *shahtoosh* patches in the middle with *pashmina* borders. These borders are to be embroidered to hide the adulterated portions of the shawl.

Following the trend of adulteration initiated by raw wool traders and manufacturers, the poor spinners started retaining the threads made from pure wool while returning those made from adulterated wool (purchased from the market) to their *poiywans*, and selling the pure quality threads to some other *poiywans* in order to make extra money. Unlike the manufacturers and raw wool traders, these poor spinners could not succeed in their strategies as the manufacturers and *poiywans* started checking the uniformity of each thread carefully and assessed if the thread was spun from the same wool which was provided to them. It is to be noted that the adulteration tactics of the manufacturers and traders succeeded as most customers cannot detect the difference between pure and adulterated wool, but the tactics adopted by the spinners failed because *poiywans* and manufacturers could easily identify and scrutinise the quality of the yarn. This indicates that even while indulging in adulteration activities, it is the already powerful actors who thrive at the cost of poorer workers.

In the previous chapter, I mentioned that although the manufacturers and traders had succeeded in delaying the ban for selling their products being processed at that time, the interests of the poor with regard to rise in wages and the ban on machines remained unaddressed. The workers' demands were abandoned by the manufacturers and politicians once the ban on *shahtoosh* became inevitable.[8] On the contrary, the shawl workers witnessed loss of jobs and a further decline in their wages. We also noticed that the government provided machines for dehairing at subsidised rates to the wool traders in Ladakh in the name of 'development'. This suggests that wildlife conservation policies as well as development schemes can be 'strategically manipulated by different actors with different ends in mind' (Rossi 2004: 23) and most of the time, actors in weaker bargaining positions fail to make their demands and interests prevail. The use of machines and saturation in the *pashmina* industry, lack of direct access to markets and increasing wool prices have all proven to be detrimental to the poor artisans who have been struggling for their sustenance in the wake of ban. In the following discussion, I present the various strategies of control and exploitation of poor shawl workers, adopted by the *poiywans*, manufacturers and raw wool dealers respectively.

---

[8] Venkatesan (2009: 139) notes that the weaving industry in Tamil Nadu, India is characterised by uneasy balances and short term alliances because of the simultaneous existence of both cooperation (between workers and traders) and competition (with other workers/traders). Likewise, in the case of *shahtoosh*, we witnessed the co-option of workers by the manufacturers and traders during the protest against the ban amidst conflicting interests of manufacturers and artisans.

5.1.4 Decreasing Wages, Increasing Prices: Strategies of Labour Exploitation and Control

In rural areas, spinning, weaving and embroidering are subsidiary occupations as the primary source of income is agriculture and horticulture. The *poiywans* now prefer to provide raw wool to the rural shawl workers than to the local spinners and weavers in Srinagar for its processing. They bargain with rural artisans to get the wool spun and woven at the lowest rates possible. This shift of jobs to rural artisans has further marginalised the ban affected categories of spinners and weavers in Srinagar.[9] Shedding more light on it, Mumtaza Begum (age 55), a spinner of Nowshera, Srinagar added:

> Because we were demanding a rise in wages and started forming workers' associations, the *poiywans* have begun to employ spinners, weavers and embroiderers residing in rural areas. These rural artisans agree to work on lower wages as they have other sources of livelihood, including agriculture and livestock. Rather than helping us in these difficult times after the ban, they [*poiywans*] have devised new ways of increasing their own profits. They even provide raw *shahtoosh* to the rural artisans for processing.

Another interview with Mushtaq Ahmed (age 45), an embroiderer of Khanmulla village, indicates some other exploitative practices adopted by these middlemen. He narrated:

> The *poiywan* deducts 15% of my wages as his commission for providing me work. He provides threads for embroidering the shawl for which he charges me three times more than they are sold in the market. I am aware of this but if I start buying my own thread from the market, he will refuse to provide me work in future [...] These days, getting work is very difficult. Since I have a large family to support, I do not want to lose this source of income.

The shawl workers explained that like the *poiywans*, manufacturers have resorted to several practices of reducing their wages and exploiting them. The most common practice is to find defects in their products and deduct their wages. For example, the manufacturers or their agents pass needles through the woven shawls to assess the quality of weaving. If the needle does not pass through the holes, the quality of weaving is assessed to be good but if it passes through, they either ask the poor weavers to pay for the whole shawl or deduct their wages. Similarly, the manufacturers and *poiywans* have made another excuse for paying low wages to the spinners by checking the length of the thread spun. The spinners argued that the short length is primarily due to the low quality and adulterated wool which cuts frequently because the wool today is mainly machine-carded and not manually dehaired. Even though this wool is provided by the manufacturers and their agents, they put the blame on these poor women and pay them lower wages in return. If they are unable to find any defects, the *poiywans* make profits by selling these spinners raw *pashmina* at higher prices. Since they cannot refuse to buy raw wool from the same

---

[9] Hareven (2002) presents similar experiences of the silk weavers of Kyoto who are losing their jobs since the manufacturers have started sub-contracting *obi* (a part of Japanese traditional kimono dress) weaving to rural weavers who are willing to work for them for lower wages.

5.1 Impacts of Banning

poiywan or manufacturer for fear of losing work, the exploitation of these poor workers continues.

The payments made by the manufacturers and *poiywans* to their employed workers are irregular and can be subject to monthly or annual delays. The shawl workers understand this as a way of establishing control over them so that they do not start working for other employers.[10] They argue that the practice of irregular payments for their labour has increased manifold after the ban on *shahtoosh*. Some shawl workers explained that even when they get *shahtoosh*, they are paid much less than they were before the ban on the excuse of its *illegality*. However, they argue that the manufacturers make profits and increase their incomes out of the same illegal production by selling the products at rates much higher than those before the ban.

Since there is growing competition for jobs in the already saturated *pashmina* industry, this has increased the bargaining power of both manufacturers and *poiywans*. The *poiywans*, who act as agents of manufacturers, have increased their commission for providing work to poor artisans. I further demonstrate exploitation of poor shawl workers in the following two narratives:

> 25 years ago, I embroidered the picture of Indira Gandhi [former Prime Minister] on a shawl. I designed it and stitched it. I sold it to the *vasta* in my locality. I was paid Rs. 90 for it. The piece was selected by the Handicrafts Department in Srinagar for an exhibition in Delhi where the *vasta* was awarded one *lakh* rupees for that master-piece. We neither get good wages nor recognition for our skill. After the ban, the exploitation of the *vastas* and *poiywans* has further increased as they have adopted several new ways to exploit us. They count the number of flowers embroidered and pay only Rs. 40–50 for each but when they sell it to the traders or exporters, this count does not matter. They sell the designs produced by us at much higher prices […] Two months back, a *poiywan* asked me to work on a shawl for which the *vasta* will pay Rs. 6500. He said that out of this amount, Rs. 1500 will be his commission for providing me work. Although I knew that he would earn this money for nothing, since I wanted work, I agreed to it. Such is the fate of craftsmen in this industry whose products are famous worldwide […] When a *ragbar* [embroiderer] dies, his children inherit nothing but his spectacles, *neut*,[11] needle and a debt of some thousands he borrowed from the *vasta* to meet his household expenses.
>
> Abdul Qazi (age 72), embroiderer, Syedpora, Srinagar
>
> My *vasta* has not paid me anything for the last year. Whenever I go to him for my wages, he says that he will pay me next month. The *poiywans* exploit us even more. Earlier, my *vasta* used to pay Rs. 60 for darning a shawl, but now I am paid only Rs. 50 because Rs. 10 is the commission of the *poiywan* for providing me work. Since it has become very difficult to get work, I cannot refuse to do it even at these low wages for the fear of being replaced by others in the locality.
>
> Syed Mohd. Yusuf (age 75), darner, Narwara, Srinagar

---

[10] A similar example of silk weavers of Kyoto is discussed by Hareven (2002) who notes that as employees of capitalist manufacturers, Nishijin weavers have little control over the type of *obi* they produce, terms of employment, working conditions and pay rates. Even when they weave in their own households, they are neither independent artisans nor small entrepreneurs.

[11] An iron or steel ring worn on the upper part of the forefinger by embroiderers in Kashmir to avoid getting hurt by the needle while stitching the shawl.

As we observed, the workers do not refuse to work for low wages because of increased competition in getting work after the ban. These poor shawl workers also argued that the other reason for not leaving their old employers is that they are the ones whom they depend on for borrowing money in times of need. In return for the help offered to these shawl workers, the manufacturers make profits by exploiting their skills and labour in the following years and thereby controlling them.[12] Thus, their poor economic conditions and indebtedness results in amplified exploitation by their employers.

I have mentioned in Chap. 3 that some of the weavers in the 1980s started taking up trading jobs and small-scale manufacturing. After the ban, more weavers ventured into the business of selling *pashmina* shawls outside the state to supplement their declining incomes. However, due to inadequate marketing skills, illiteracy and lack of exposure to outside world, very few of them could succeed in selling shawls directly to customers and fashion outlets outside J&K.[13] The weavers informed me that the manufacturers always discourage them from taking up trading functions, explaining that this involves huge risks.[14] Due to their poor access to the market, some of the weavers argued that if somehow they manage to produce a shawl and go to the manufacturer directly (bypassing the *poiywans*) to sell it, he never pays the full price. Most of the time, he pays them half of the price in cash and for the rest, he provides raw wool. This way, the manufacturer is able to maintain his market position as well as establish control over his subordinate workers. As Bashir Ahmed (age 40), a weaver of Narwara, Srinagar explained:

> I produce some shawls but it is not profitable because I do not have links to the shopping outlets and big businessmen outside the state. Since we are poor, the traders and customers outside do not trust us. Moreover, owing to the militancy in the state, our Kashmiri Muslim identity becomes another barrier to explore trading opportunities outside J&K. So, even if we produce shawls, we are left with no choice but to sell them to the *vasta* […] He is shrewd, never pays us cash but provides raw material in exchange for our products.

The shawl workers also reported that there have been fluctuations and increases in the prices of raw *pashmina* and *shahtoosh* in the last three decades. This could possibly be due to the fact that earlier there were only a few raw wool traders in Srinagar. Owing to the profits associated with wool trading, many manufacturers have taken up trading functions resulting in increased competition and fluctuating

---

[12] In the context of mat weavers in Tamil Nadu, Venkatesan (2009) explains that traders give loans and advances to the weavers in times of need, and control their production. As one of her respondent weavers remarks, 'they trap us like fish in a net. They strew money [for bait]' (ibid.: 154).

[13] This points to the Simmel's (1950) argument that in trade, intelligence always finds expansions and new territories, an achievement which is very difficult to achieve for the original producer with his lesser mobility (cited in Venkatesan 2009: 143).

[14] By discouraging new entrants in manufacturing and trading activities, the powerful actors seek to maintain their privileged position. Jha and Misra (2006: 40), in the case of the silk weavers of Bhagalpur, Bihar, note that the monopoly of big businessmen threatens the very existence of the 'smaller fry within the community in the same way as the big fish devours the small one', a process which can also be noted in the example of *shahtoosh*.

5.1 Impacts of Banning

wool prices.[15] The respondents revealed that most of the time, an artificial shortage of wool is created by Ladakhi wool suppliers and local traders to create demand in order to sell the wool at higher rates. Although increasing wool and shawl prices have benefited the already rich *bhotias* and local wool traders, it has proved detrimental to spinners and weavers because unlike the case of *shahtoosh*, the raw *pashmina* needs to be purchased from manufacturers or *poiywans*.[16] Explaining this, Saleema Rashid (age 60), a spinner of Syedpora, Srinagar narrated:

> In the last ten years, the price of *shahtoosh* and *pashmina* wool has increased rapidly. I used to buy 60 grams of *pashmina* for Rs. 30 but now I get 10 grams for Rs. 60. The traders sell *shahtoosh* illegally for more than 40,000 per kilogram. Though the prices of wool and shawls are increasing and benefiting the rich, our incomes are on constant decline.

The discussion above suggests that on the one hand, the ban has resulted in plummeting incomes for the poor artisans, while on the other, it has motivated the powerful actors such as traders, manufacturers and *poiywans* to maintain their market position by creating artificial wool shortages, adulteration, increased use of machines, exploitation of poor shawl workers and indulging in illegal *shahtoosh* trading. The fact that 'illegality' increases prices, justifies the risk the powerful traders and manufacturers take in breaking the ban by forming alliances and bribing the law enforcement officials. Arguably, a blanket ban rather than helping in conservation, may lead to overexploitation of resources (Ramnath 2002).[17]

MacGaffey and Bazenguissa-Ganga (2000: 5), in the context of Congo, suggest that traders 'transgress and contest the boundaries of the law' through various mechanisms including their 'second economy' activities. We observe that after failing to resist the ban through open protests and negotiations, the powerful traders showed their non-compliance to international conservation regulations by indulging

---

[15] In the 1980s, one kilogram of raw *pashm* was sold for Rs. 400, which rose to Rs. 1100 in the 1990s. In 2006, raw *pashm* was sold for Rs. 1600 and machine-carded for Rs. 6000. This rise of price was even sharper in the case of *shahtoosh* which increased from Rs. 7000 per kilogram in the 1980s to Rs. 16,000–18,000 in the early 1990s and Rs. 30,000 in 1998. According to the local spinners and weavers, in 2006, one kilogram of *shahtoosh* was being sold for 38,000 per kilogram in Srinagar. This sudden rise in the price of *shahtoosh* from Rs. 16,000 to Rs. 38,000 was because of the shortage of wool in the market due to the ban on the *shahtoosh* trade. Similarly, the rates for shawls also increased. In 1996, one plain *pashmina* shawl was sold for Rs. 3000 but by 2006, the price rose to Rs. 5000–6000. In the same way, the price of one plain *shahtoosh* shawl rose from Rs. 15,000 in 1996 to as high as Rs. 60,000, in 2006, owing to the 'illegality' involved. This shows that in a period of 10 years, the price of a *pashmina* shawl has doubled but the price of a *shahtoosh* shawl has increased four times.

[16] This is comparable to the case of silk manufacturing in Bhagalpur, Bihar, as noted by Jha and Misra (2006). The ownership of the finished product remains with the supplier of silk yarn i.e. traders and manufacturers. As such, the poor weavers are 'forced to live a life of destitute masterminded by exporters and dealers through the process of manoeuvrings' such as controlling the supply and price of silk yarn (ibid.: 67).

[17] In the context of northeast India, Ramnath (2002) notes that blanket ban on timber felling and the laws to promote conservation have accelerated deforestation. For the poorer people dependent on these forest resources for sustenance, the ban and its continuing operation has generated untold miseries but it is to the advantage of rich and powerful landowners.

in the illegal *shahtoosh* trade. In Chap. 4, I highlighted that the traders and manufacturers protested against the ban by proclaiming that it is inimical to the interests of J&K and thus morally unjustifiable. However, it is important to note that the 'moralistic construction' (Singh 2008: 13) of manufacturers and traders that the ban is against the *shahtoosh* worker community may also obscure their exploitative practices and huge profits made out of illegal trade which excludes the majority of the affected population but benefit these powerful groups.[18]

The privileged groups, often, are provided institutional security in harnessing state resources and capitalising on environmental opportunities for private gain, resulting in divergent vulnerability profiles for various sections of the affected populations (Collins 2008). Although the J&K state did not provide any institutional security to wool traders and manufacturers, it did help the manufacturers and traders to secure their interests through a weak implementation of the ban. Indeed, the illegality of trade has resulted in increased vulnerability of the poorer *shahtoosh* workers who had remained exploited and marginalised even before the trade became illegal. Under such circumstances, the poor who can least afford to bear the costs of environmental conservation may also ignore or 'deliberately contravene environmental policy measures to defend their livelihoods' (Wiggins et al. 2004: 1940).

Scholars such as Chatterjee (2004) argue that in most of the world, poorer people transgress the strict lines of legality in struggling to live and work. Likewise, Leon (1994) suggests that due to lack of any alternatives, the natural resource dependent populations adopt 'avoidance strategies' by indulging in illegal activities as the only viable possibility to sustain themselves. Bloomer (2009: 63) refers to such activities as 'coping strategies' as opposed to being seen as 'criminal opportunism'. Peluso (1993) suggests that the *Maasai* in Kenya started illegal wildlife hunting for sustenance when their principal livelihood of livestock grazing was restricted by reserve authorities (see also Brown and Marks 2007). While lack of alternative livelihoods and poverty could be the reason for the illegal poaching of *chirus* by Tibetan herdsmen, the arguments presented by these scholars provide only partial explanations to the involvement of poor workers in the illegal *shahtoosh* production. For the poor *shahtoosh* workers, the question of sustenance is undeniably significant. Yet, their poor economic conditions and lack of alternative employment opportunities are not the only determining factors in their continuing involvement in *shahtoosh* production. Because the workers do not own either the yarn or the finished products, their transgressing behaviour and involvement in illegal production is rather controlled and determined by the manufacturers and *poiywans* who employ these workers to make huge profits. I refer to this process as *delegated illegality*, wherein the power-

---

[18] Singh (2008) argues that illegal and unregulated trade in wildlife has been characterised by conservation groups as a great risk for the remaining wildlife populations in Laos. While conservation organisations construct an image of regulation through CITES as a global necessity, their assumed morality provokes counter accusations by the affected groups about the immorality of impositions by western conservationists. However, both the competing representations focus the gaze outwards without considering the uncertainties and divisions that occur within concerning the increasing vulnerability of the poor populations and benefits acquired by the elite through the continued high value trade in wildlife.

5.1 Impacts of Banning

ful actors delegate and assign illegal tasks to their entrusted subordinate workers. For the poor *shahtoosh* workers, the question is not to choose between legality and illegality i.e. processing *pashmina* or *shahtoosh*, rather it is the matter of compliance to the orders given by their employers for whom they have been working for generations.

5.1.5 Declining Social Prestige and Cultural Heritage

The *shahtoosh* workers argued that the ban has not only resulted in the loss of their jobs and affected their incomes adversely, it has also resulted in the loss of traditional skills and cultural heritage of spanning 600 years. They noted that besides being a large source of revenue for the state and livelihoods for its population, the *shahtoosh* shawl industry has given recognition to Kashmiri handicrafts worldwide. The shawl workers argued that there is no substitute for this ancient art and, therefore, the industry should be preserved.

Since the majority of the shawl workers had *shahtoosh* as their family occupation, they attached great cultural value to it. Moreover, the fact that *shahtoosh* weaving and production of shawls was a lucrative profession, these workers argued that people in Kashmir have, at times, preferred the *shahtoosh* production and trade to other jobs. They believe that there are greater chances of class mobility working with *shahtoosh* than any other craft. Hence, those who were engaged with *shahtoosh* considered themselves privileged and enjoyed high social status amongst other artisans in Kashmir.

According to the *shahtoosh* workers, the decline in their incomes after the ban has resulted in their low standard of living in several ways. They opt for buying cheaper food products now, struggling to save their incomes for their daily household expenses. They complained about not being able to provide better education for their children and a lack of proper health facilities due to their poor economic conditions. In order to cope with the loss of jobs or sudden decline in their wages, these shawl workers informed me that they had to do distress selling of their land, property and jewellery items.[19]

The decline in their wages has also resulted in the loss of prestige they enjoyed earlier, working with *shahtoosh*. They have experienced a strong sense of deprivation since the ban as they are ranked lower with regard to social status. They observed that the young weavers cannot afford to get married since they are not sure how they will feed their families. Moreover, they even argued that no one wants his daughter to get married to a *pashmina* weaver nowadays. Nisar Ahmed Baba (age 33), a weaver of Rathpora, Srinagar remarked:

---

[19]Although the poor weavers had to sell their properties to meet the daily expenses after the ban, the big manufacturers have started investing their savings in purchasing land and other non-movable assets to compensate for the present and anticipated decline in their incomes. I also observed that manufacturers owned some of the best houses in the town.

> Before the ban, I was respected in my locality. People used to greet me as *salaam sahib* owing to my prosperity but after the ban, we are struggling even to bear the daily household expenses. The other name for our life now is compromise as we practically experience it at every step [...] These days, even a wage-labourer earns more than we do.

Both the male and female workers argued that after the ban, they have avoided going to any community gatherings, marriage ceremonies or making social visits since they need to work double the number of hours with *pashmina* (in comparison to *shahtoosh*) in order to feed their families. The female spinners explained that they enjoyed more space and independence in the family before the ban. Most of them used to keep a record of their wages with the *poiywans* and withdraw money at the time of their daughter's weddings to buy jewellery and other articles. Since the ban, they have not been able to make such long-term savings as it is difficult to bear the expenses of the household with the sole incomes of their husbands.

Before the ban, the young girls in the family used to spin and save their incomes for their marriage expenses and also contributed to the household expenditures. Since *pashmina* production is not as profitable, many of these young spinners have left their jobs, aspiring to join other occupations. Some young spinners, however, explained that since their family members consider working outside their homes as a cultural loss, it restricts their mobility and choice of occupation. As non-earning members, their position within the household becomes weakened. Female separators who have lost their jobs since the ban are now financially dependent on their parents or spouses for their personal expenses. The use of machines has also had an adverse impact. They argued that the loss of their jobs has negatively affected the relationship with their spouses. Their personal expenses have been restricted and space for decision-making has been limited. Habla Begum (age 44), a female embroiderer of village Ganderbal explains, in the following words, the problems she has been facing after the ban due to the increased exploitation by the manufacturer:

> After the ban, the *vasta* have become even more exploitative. Nine months ago, I finished embroidering a *pashmina* shawl after six months of hard work. The *vasta* refused to pay me anything for my labour saying that my work was not satisfactory. Since I am a woman, I could not fight for my wages. My in-laws abused me and demanded the money for the work I did for six months. They refuse to accept that I had not been paid even a single penny, and believed that I was making excuses to keep the money for myself. I was humiliated to the extent that they threatened to expel me from the house unless I gave them the due amount. As I was unable to pay them the money, I have had to come to my father's house and I have been staying here for the last five months with the hope that one day the *vasta* will pay for my labour and I will go back to my husband [...] I am completely depressed now.

The above discussion suggests that the differential impact of ban is not only based on economic inequalities but also gender relations. It is important to note that female workers are not only exploited by their employers but, in addition to this, also by their own family members. Harriss-White (2003) suggests that the Indian workforce in the unorganised sector is fragmented across various lines of social differentiation including gender. Gender differences are systematically exploited by the merchant capital to segment the labour force and to resolve issues of labour control as female labour is often considered more docile and easy to discipline

5.2 Rehabilitation and Alternative Livelihoods: Accountability of Whom?

(ibid.). In the *shahtoosh* case, the manufacturers and *poiywans* have followed similar strategies by employing the female spinners and embroiderers in rural areas, and paying lower wages to women artisans for reducing the production cost, as well as minimising the threat of workers mobilisation. Although women constitute a major chunk of the workforce in the *shahtoosh* industry, they are least paid for their skills. Undoubtedly, after the ban, women *shahtoosh* workers, especially separators and spinners have been the worst affected. The situation of women *shahtoosh* workers is comparable to other women weavers and spinners in India and abroad (see for example, Venkatesan 2009; Ramaswamy 2006; Mohamad 1996).[20] It is to be noted that while male *shahtoosh* workers have witnessed a decline in social status and hierarchy, female workers, in addition, have experienced a decline in their position within the family in terms of decision-making power, respect and recognition.[21]

5.2 Rehabilitation and Alternative Livelihoods: Accountability of Whom?

Although no substantial efforts have been made by the state or non-state agencies to rehabilitate and compensate the poor *shahtoosh* workers, some discrete initiatives have been taken by the WTI and J&K state with little success. Below, I critically examine these initiatives and argue that poor people affected by nature conservation policies, especially in developing countries, are left on their own to cope with new circumstances wherein their modes of sustenance and occupations are suddenly shifted from the domain of legality to illegality.

In 2005, the WTI started a project titled, 'A Livelihoods Initiative for Traditional *Shahtoosh* Workers in Kashmir valley' to support the ban-affected spinners and weavers in Srinagar. The project aimed to reinstate Kashmir hand-made *pashmina*

---

[20] In the context of Malay silk weavers, Mohamad (1996: 189) highlights the exploitation of women weavers by Chinese middlemen who pay them 'virtually nothing' for their finished *sarongs* (kilt or skirts). The majority of the labour force in the industry, whether in the past or present, consists of women. Although they are real producers or actual weavers, they do not control the production process and the finished product. As home based workers they have no official identity and they are subjected to extremely low and unregulated wages (ibid.: 195). Unlike the case of the *shahtoosh* weavers, Venkatesan (2009) explains that the *Labbai* Muslim mat weavers in Tamil Nadu, own the final product. However, since they are dependent on traders to sell their products, they are exploited by these traders who find fault on the pretext of colour specifications, fineness of texture etc. Very often, in such cases, the traders either refuse to buy the product or pay them less.

[21] Hill (2010) proposes that membership based organisations (such as the Self-Employed Women's Association in India) can promote the social foundations of recognition and respect that are critical to identity and agency as well as provide workers with real opportunities to develop alternative non-exploitative economic institutions that may deliver improved wages and social security. In the case of *shahtoosh* weaving in particular or the shawl industry in J&K in general, we see an absence of such women's organisations, possibly because of a combination of factors such as strong nexus between traders, manufacturers and bureaucrats, dependence of artisans on their employers for loans and advances, and the ongoing militancy and political instability in the state.

as a viable alternative to *shahtoosh* by providing the Geographical Indication (GI)[22] craft mark to the *pashmina* shawls of Kashmir so as to create their credibility and demand in international markets. The representatives of the WTI explained that although their campaigns to ban *shahtoosh* were primarily funded by the IFAW, they were not able to receive any funds from other international agencies or donors to support the livelihoods project for *shahtoosh* workers. It was only after the WTI received a small grant from the British High Commission, New Delhi for a period of 3 years that the project was initiated. A WTI official in Srinagar, on the condition of anonymity, also informed me that even within the organisation there was little enthusiasm in sustaining the project as many of its members felt that it was not directly related to wildlife conservation and did not fit into the main objective of a conservation organisation.

According to the WTI, because the project started 3 years after the ban on *shahtoosh*, they have faced several problems in convincing the shawl workers to participate in their activities. In 2006, the WTI formed a trust called the 'Charitable Trust for Promotion of Handmade *Pashmina* in Kashmir' to eliminate the role of the middlemen (*poiywans*) by providing raw *pashmina* at lower rates directly to spinners and weavers, and buying back yarn and fabric at a slightly higher price than the market rate. The spinners were paid Rs. 1.20 per knot as compared to Rs. 1 paid by their *vasta* or *poiywan*. Yet, very few prospective 'beneficiaries' came forward to participate in the WTI project. The reason was not just the mistrust for an organisation that has campaigned for the implementation of the ban but also the uncertainty involved in the short life period of such projects. As one couple, Hanifa Begum (age 39) and Mushtaq Ahmed (age 43) of Syedpora, Srinagar explained:

> The WTI project workers offered us better rates for spinning and weaving raw *pashmina*. But they said that they would provide the wool for six months and did not give any long-term commitment. These projects come and go but if we do not remain loyal to our *vasta* and *poiywan*, we will not get any work in future.

The Board of Trustees of the aforesaid trust mainly comprised manufacturers which the WTI has been able to enroll by giving the incentive of publicity to handmade *pashmina* shawls with the GI tag in national and international trade fairs, exhibitions and sales outlets. I also attended a meeting held by WTI officials with the Board of Trustees. The meeting covered a variety of issues. An exhibition was to be organised by the WTI in collaboration with the trust in the famous handicraft market Dilli Haat in New Delhi in 2006. In the meeting, two manufacturers were selected to display their *pashmina* products along with spinning and weaving processes to prospective customers. Another important issue discussed in the meeting was the problem being faced by the WTI to involve spinners and weavers in the 'alternative livelihoods' project. It emerged in the discussions that the workers can only be motivated by increasing their wages, and lowering the cost of raw *pash-*

---

[22] A geographical indication (GI) is a name or sign used on certain products which corresponds to a specific geographical location or origin (e.g. a town, region, or country). The use of a GI may act as a certification that the product possesses certain qualities or reputation, due to its geographical origin.

mina. Some manufacturers contended that the increase in wages of the weavers and spinners will be against their interests since it would increase the production costs while others referred to it as a much needed sacrifice to be made to earn the profits which can be made later with the rise in demand for pure hand-made *pashmina* products if the GI tag is able to open new market opportunities abroad.

A WTI official, on the condition of anonymity, informed me that during the initial interactions of the WTI with the manufacturers, some manufacturers even proposed that the best way of utilising project funds was to hand over the money to the Manufacturers' Association, the members of which would then distribute the money to the artisans working under them so as to compensate their loss of incomes due to the ban.[23] However, when the WTI declined the suggestion on the basis that the grant is for providing alternative livelihoods to the affected spinners and weavers, these manufacturers ridiculed the organisation, stating that the WTI project will fail just as hundreds of governmental schemes have in the past. The WTI official also recalled that in 2005, the WTI team organised a workshop titled, 'Threats and Challenges to Pashmina Craft and Stakeholders Role' at Srinagar. The discussion was mostly dominated by manufacturers, traders, exporters and *poiywans* but artisans remained silent. When asked to speak their views, these artisans were hesitant to share them. The official informed me that the WTI team came to know later that manufacturers and traders had asked the artisans to keep quiet and let the manufacturers deal with the WTI officials because a pro-ban organisation could not be trusted, and they need to be careful in their engagement with the organisation. This suggests that the manufacturers and traders, using their privileged position, tried to hijack the proposed project of the WTI, again marginalising the interests of the skilled workers.

While the WTI's initiative can be considered as tokenism, the efforts made by various governmental agencies and departments (Handlooms Department, Handicrafts Department, Weavers Service Centre, School of Designs and Craft Development Institute) failed to address the primary concerns of the ban-affected workers by means of providing any monetary compensation or creating alternative livelihood opportunities. These agencies are mainly involved in technological upgradation, loans and credit facilities and providing training to spinners, weavers, dyers, designers and embroiderers to improve their skills. The shawl workers argued that they did not require any training in the art which they have been doing for generations.[24] Instead, their main concerns relate to the loss of jobs due to the ban, increasing redundancy due to the use of machines by manufacturers, problems of adulteration, increasing exploitation, low wages and lack of access to markets.

---

[23] Interview in Srinagar, 20 November, 2006.

[24] The futility of the state programmes for artisans can also be seen in examples discussed by other scholars. For example, Venkatesan (2009) notes that mat weavers in Tamil Nadu do not see any relevance of training programmes or workshops started by the government nor do they agree that the provision of new improved looms and natural dyes is able to help them in real sense. The weavers participate in such training programmes and workshops just to avail the stipend provided for training by government departments and NGOs.

While no funds were made available by the state and central governments specifically for ban-affected *shahtoosh* workers, some state agencies such as the Handlooms Department and the J&K Small-Scale Industries Corporation have been providing low interest loans and credit to shawl workers in general. Yet, very few weavers or spinners are able to access these funds due to lack of information, highly cumbersome paper-work and other formalities involved in the process. According to poor shawl workers, it is mainly the manufacturers and some influential senior weavers who are able to avail these opportunities as they are well-connected and well aware of the tactics to attract these funds.[25] Reflecting on his experience with the Handlooms Department, Gulam Hasan (age 44), a weaver of Rathpora, Srinagar explained:

I have been registered with the Handlooms Department since 1992. In 2004, I came to know about a scheme of loans for up to one *lakh* rupees for shawl workers. I applied for it. The officer asked me the names of the instruments used in weaving and tested my weaving skills. He then asked for a bribe of 10,000 rupees and an undertaking by a government officer in support of my application for a loan. I did not know any government official and dropped the idea [...] but it is very easy for *vastas* to get loans from government departments as well as banks. Last year, my *vasta* asked me and some other subordinate weavers and spinners to come to his house and work together to show that he runs a *karkhana*. We all gathered on the fixed day, two officers came and took pictures. A few months later, I came to know that the *vasta* has got a loan of ten *lakhs* for the *karkhana* that exists only in pictures.[26]

Imtyaz Ahmed Sofi (age 42), a weaver of Eidgah, Srinagar explained his disappointment with the state initiative to promote Kashmiri shawls in craft fairs outside J&K:

The Department of Handicrafts started a programme in 2000 under which 17 craftsmen from Srinagar are sent to Delhi every year for two weeks to sell their products in craft fairs and exhibitions. The selection is supposed to be made on the basis of a lottery system. I applied twice but did not succeed. The people who get selected for such programmes are the sons or relatives of the manufacturers and politicians. Even if some weavers get selected, they do not have enough stock of shawls to exhibit and sell [...] they sell their seats to *vastas*.[27]

---

[25] Referring to the difficulties the poor face in accessing government aids, Jha and Misra (2006) present the case of the silk weavers of Bhagalpur, Bihar, who benefited little from the provision of bank loans started by the state and central governments after the communal riots in 1989 to compensate their loss of handlooms and livelihoods. Corruption on the part of government officials and middlemen were the main reasons for the failure of these rehabilitation and compensation schemes. As a result, the majority of weavers ended up working as labourers.

[26] A similar observation is made by Venkatesan (2009: 139) that the mat weavers of Tamil Nadu may profess a commitment to the weaving industry, whether they weave or not, so that they can benefit from craft development schemes. As a result, non-weavers often access governmental funds normally meant to support weavers. If those who actually weave are able to receive such funds, they rarely invest the money into weaving. Rather, in most cases, these funds are used to meet other requirements of the household.

[27] For handloom weavers in Varanasi, Showeb (1994) explains that even when a weaver knows the art of weaving quite well, he is often not able to rise economically because he fails to get financial backing for launching his own enterprise independent of the control of traders. Moreover, because of his low economic status he has poor connections in the raw material market as well as in the

5.3 Conclusion 117

Mohd. Maqbool Mir (age 75) of Hamchi, Srinagar is an embroiderer who explained his lack of interest in a training programme by the School of Designs in the following words:

> I have heard that the School is providing training to the shawl embroiderers these days. These programmes are futile as we know better designs than the young experts in the schools. The government needs to plan programmes which can help us overcome the real problems we face — low wages and exploitation.

The discussion suggests that the state did not show any serious concerns towards rehabilitating *shahtoosh* workers either by initiating targeted programmes or providing compensation to the affected populations. However, it has served the powerful actors such as manufacturers and traders. They are able to bribe the state officials and adopt tactics to demonstrate their direct association with the production process in order to avail the opportunities or provisions in the form of loans and participation in craft fairs and exhibitions meant for artisans. I have argued that there is lack of confidence and interest amongst the shawl workers in the government schemes due to the inadequacy of programmes in addressing their primary concerns, the weak possibilities they see in receiving government aid and their inability to bribe the authorities concerned. This shows that neither the state nor any non-state agencies have assumed the accountability to recompense the losses incurred to the poor artisans in the wake of wildlife conservation.

5.3 Conclusion

From this chapter, I draw three main conclusions. First, the *shahtoosh* worker community is highly heterogeneous wherein shawl manufacturers, traders and skilled workers have divergent interests and different powers. As a result, the ban has had differential impact on the various categories of shawl workers who have showed differential abilities to challenge, contravene and manipulate the new regulations of resource use. We observed that the ban has resulted in the loss of jobs for the separators and those working with *shahtoosh* exclusively, and a sharp decline in the incomes of the poor artisans such as spinners, weavers, embroiderers etc. who perform the most arduous jobs in the production process. These poor shawl workers fail to set up new businesses or enter into new trades due to a lack of financial capital and access to market opportunities. The rich and powerful categories of manufacturers, *poiywans* and wool traders have also experienced loss in their incomes but they have been able to devise new strategies and tactics to maintain their monopoly and control over the production process. The manufacturers have increased the use of machines and have adopted several adulteration practices to reduce the production costs. They have also been exploiting the skilled workers by deducting their wages

---

selling market of silk fabrics. Likewise, Jha and Misra (2006) observe that the lack of capital and monopoly of big businessmen are the biggest constraints faced by the silk weavers of Bhagalpur, Bihar and primary reasons for their abject economic conditions.

and paying them irregularly. The new strategy adopted by the *poiywans* concerning the distribution of wool to rural artisans has further marginalised the spinners, weavers and embroiderers of Srinagar who are already hit hard by the ban. Similarly, the wool traders have been creating artificial wool shortages and increasing the price of wool. The impact of the ban can also be observed through the changing gender relations and loss of social status or prestige enjoyed by the *shahtoosh* shawl workers before the ban.

The second conclusion is that the conservation community has failed to change both the existing belief systems of the local shawl workers and their behaviours to conform to the international regulations. There is widespread ignorance about the origin of *shahtoosh* and the unpopularity of the ban in the Valley. Not only do the shawl workers subscribe to the age old myths, but some of them even see a positive relationship between the increased supply of *shahtoosh* and the population of *chirus*. We have observed that the behaviour of the shawl workers is actually governed by the rationality of their employers who supply *pashmina* and *shahtoosh* wool to them. Since the ban creates demand and illegality increases prices, it motivates the powerful actors to violate the ban and accrue profits. While the manufacturers, *poiywans* and wool traders showed their non-compliance by indulging in the illegal *shahtoosh* trade for profit making, the involvement of poor artisans in illegal production is primarily determined by the will and interests of their employers, a process which I referred to as *delegated illegality*. Further, by involving the subordinate workers in illegal production and creating 'moralistic construction' by protesting against the ban, I have argued that these powerful actors are able to cover both their exploitative practices and strategies adopted to compensate their loss of incomes.

The third conclusion is that in the absence of any alternative employment opportunities, the poor shawl workers are more or less locked into the system with little chance of escape. The alternative livelihoods project initiated by the WTI has created little incentives for the shawl workers to cooperate since they see *pashmina* as a less attractive option. The lack of long-term assurance in the WTI project and the dependency of artisans on their manufacturers for loans and advances in times of need discourage these artisans from participating in the project. The increased competition and saturation of the *pashmina* industry has, in turn, increased the bargaining capacities of the powerful categories resulting in further exploitation of shawl workers. Moreover, there have been no initiatives taken by the state to recompense the loss of livelihoods of the *shahtoosh* workers. The few initiatives taken by the Handlooms Department to support shawl workers in general have mainly served the already powerful. The fact that the *shahtoosh* workers are concentrated in one geographical region, Srinagar, means that a well designed and targeted strategy for providing alternative livelihoods by the government or international agencies at the time of the ban could have substantially reduced the miseries of these workers.

The discussion presented in this chapter motivates us to look beyond the dichotomous positions of global control *versus* local resistance in the context of nature conservation, and analyse how international conservation interventions seep into the existing relations of domination and subordination at ground level. It also signals the need to analyse how environment and politics are intertwined and differentially articulated in particular localities. It is demonstrated that the nature conservation regulations and interventions, instead of determining the local realities become refracted and reshaped through various forms of microlevel politics and practices. It can be, thus, argued that power rather than being concentrated at a particular site, is dispersed and fluid, and that power relations among various actors play decisive role in influencing the outcome of nature conservation policies.

Overall, the case study of *shahtoosh* has provided insights into the history, growth and decline of *shahtoosh* shawl production in J&K. It has also examined the power and agenda of international wildlife conservation, the paradoxical relationship between the centre and the state, the split role of the J&K state and the limited space for protest by the citizens. It has highlighted the exploitative relationships within the shawl industry and the nexus between powerful actors such as government officials, politicians and big manufacturers or traders that works against the interests of artisans. Indeed the ban on *shahtoosh* has coincided with the historical attempts of powerful actors to dominate and exploit poor shawl workers along with extending the coercive and violent powers of the state to control its dissenting populations. Arguably, the global concern of *chiru* conservation is proven to be incompatible with the livelihoods of *shahtoosh* workers, despite the fact that these workers are not involved in the killing of the antelope. Although there are valid ethical arguments in favour of wildlife conservation, it is the poor and powerless who bear the burden of nature conservation costs.

While there are overtly conflicting interests of diverse actors involved in the controversy over the ban on the *shahtoosh* trade, not all nature conservation interventions are characterised by such evident conflicts. Therefore, it is worthwhile to assess how the nature conservation agenda unfolds itself in the situations where there is an *apparent* convergence of the interests of different stakeholders. I address this question through the case-study of forest management programme involving local communities in J&K in the following chapters.

References

Agrawal A, Gibson C (1999) Enchantment and disenchantment: the role of community in natural resource conservation. World Dev 27(4):629–649

Bloomer J (2009) Using a political ecology framework to examine extra-legal livelihood strategies: a Lesotho-based case study of cultivation of and trade in cannabis. J Polit Ecol 16:49–69

Brown T, Marks S (2007) Livelihoods, hunting and the gamemeat trade in northern Zambia. In: Davies G, Brown D (eds) Bushmeat and livelihoods: wildlife management and poverty reduction. Blackwell, Oxford, pp 92–106

Chatterjee P (2004) The politics of the governed: reflections on popular politics in most of the world. Columbia University Press, New York

Collins TW (2008) The political ecology of hazard vulnerability: marginalisation, facilitation and the production of differential risks to urban wildfires in Arizona's White Mountains. J Polit Ecol 15:21–43

Hareven T (2002) The silk weavers of Kyoto. University of California Press, London

Harriss-White B (2003) India working essays on society and economy. Cambridge University Press, Cambridge

Hill E (2010) Worker identity, agency and economic development: women's empowerment in the Indian informal economy. Routledge, London

IFAW and WTI. 2003. Beyond the ban: a census of shahtoosh in the Kashmir valley. New Delhi: International Fund for Animal Welfare and Wildlife Trust of India

Jha UM, Misra DC (2006) Economics of silk weavers. Sunrise Publications, New Delhi

Leach M, Mearns R, Scoones I (1999) Environmental entitlements: dynamics and institutions in community based natural resource management. World Dev 27(2):225–247

Leon M (1994) Avoidance strategies and governmental rigidity: the case of the small-scale shrimp fishery in two Mexican communities. J Polit Ecol 1:67–81

Li TM (2003) Situating resource struggles: concepts for empirical analysis. Econ Polit Wkly 29:5120–5128

MacGaffey J, Bazenguissa R (2000) Congo-Paris: transnational traders on the margins of the law. International African Institute in Association with James Currey and Indiana University Press, London

Mohamad M (1996) The Malay handloom weavers. Institute of Southeast Asian Studies, Singapore

Moore DS (2000) The crucible of cultural politics: reworking "development" in Zimbabwe's eastern highlands. Am Ethnol 26(3):654–689

Peluso NL (1993) Coercing conservation: the politics of state resource control. In: Lipshutz R, Conca K (eds) The state and social power in global environmental politics. Columbia University Press, New York, pp 199–218

Ramaswamy V (2006) Textiles and weavers in South India. Oxford University Press, New Delhi

Ramnath M (2002) The impact of ban on timber felling. Econ Polit Wkly 30:4774–4776

Rossi B (2004) Revisiting Foucauldian approaches: power dynamics in development projects. J Dev Stud 40(6):1–29

Showeb M (1994) Silk handloom industry of Varanasi: a study of socio-economic problems of weavers. Ganga Kaveri Publishing House, Varanasi

Simmel G (1950) The sociology of George Simmel (trans and ed: Wolff K). The Free Press, New York

Singh S (2008) Contesting moralities: the politics of wildlife trade in Laos. J Polit Ecol 15:1–20

Venkatesan S (2009) Craft matters: artisans, development and the Indian nation. Orient Black Swan, New Delhi

Wiggins S, Marfo K, Anchirinah V (2004) Protecting the forest or people? Environmental policies and livelihoods in the forest margins of Southern Ghana. World Dev 32(11):1939–1955

Chapter 6
Forests, State and People: A Historical Account of Forest Management and Control in J&K

> *The poor would eat in abundance from the forests,*
> *Animal, birds, and flowers alike.*
> *Then came the order that for hunting*
> *One needs a licence.*
> *How much will they loot,*
> *What will they answer to God?*
> *Our ruler is so greedy that*
> *Even Mulberry leaves have built him a treasure,*
> *Even then he is not satisfied.*
>
> — Ali Shah Khoiyami (1904) (From *Akhir Zamaan* (The End of the World), an unpublished manuscript in Kashmiri. The English translation of the poem can be found in Zutshi (2004: 137))

Abstract In 1891, the Forest Department (FD) was established in the princely state of J&K which epitomised centralisation of power with regard to access and control of forest resources. Just over 100 years later, in 1992, a regulation for 'Joint Forest Management' was promulgated with the aim to conserve and use forest resources based on the principles of decentralised management and people's participation. In this chapter, I present the story of forest management in J&K beginning from the practice of 'communal' management in the pre-colonial period to 'custodial' in the colonial era and, ultimately, to 'social and participatory' management in the late 1980s.

Keywords Forest management · Forest history · Jammu and Kashmir · Joint forest management · National afforestation programme · Scientific forestry · Colonial period · Post-colonial period

While being wary of historical determinism, it is pertinent to understand the various political and scientific influences in the past that have shaped the present discourse and practice on forest conservation and management. At each and every historical stage, forest management has been a 'political process' wherein resource access, control and

---

© Springer International Publishing AG 2018

S. Gupta, *Contesting Conservation*, Advances in Asian Human-Environmental Research, https://doi.org/10.1007/978-3-319-72257-3_6

121

use have been 'bitterly contested' (Bryant 1997: 194). Moreover, forests are not only 'physical domains', but also 'contested social spaces' in which different groups of people attach different meanings (Beinart and Hughes 2007: 119). As Sivaramakrishnan (2003: 277) suggests, there is a need to constantly question what is presented as 'natural' or 'inherently systematic' to see its historical production. A historical understanding of the issues of access, control and management of forests, thus, can provide useful insights for analysing the agenda, power and interests of various stakeholders involved in the contemporary politics around forest conservation and use in J&K.

In 1891, the Forest Department (FD) was established in the princely state of J&K which epitomised centralisation of power with regard to access and control of forest resources. Just over 100 years later, in 1992, a regulation for 'Joint Forest Management' was promulgated with the aim to conserve and use forest resources based on the principles of decentralised management and people's participation. In this chapter, I present the story of forest management in J&K beginning from the practice of 'communal' management in the pre-colonial period to 'custodial' in the colonial era and, ultimately, to 'social and participatory' management in the late 1980s.

I begin with a brief discussion on forest management in the early colonial period and demonstrate that the state interest in forest control was limited only to the collection of forest dues from local communities and forest management had largely remained communal in nature. Following this, in Sect. 6.2, I analyse how, by the end of the nineteenth century, the forest citizens had to cope with a new political and economic context when British intrusion in forest control and management went far beyond those of previous rulers. The principles of 'scientific forestry' determined forest resource utilisation with the aim of increasing the sustainable productivity of forests and various measures were taken to restrict the access of local communities. In Sect. 6.3, I explain how the colonial system of forest management perpetuated in the post-colonial period, and also highlight the shift from 'custodial' to 'social' management of the forests in the state. The discussion presented in this chapter rests on the argument that 'scientific forestry' served to legitimise the role of forest bureaucracy in the control and use of forests. It is argued that although the rationality of scientific management succeeded in curtailing the rights and access of local communities dependent on forest resources for sustenance, it proved subservient to the imperial, commercial and varying national interests in the colonial and post-colonial periods respectively.

6.1 Forest Management in Early Colonial Period

While rich and extensive accounts of forestry in pre-colonial and colonial periods are available for different provinces of British India (and some princely states), little is known about the history of forestry in J&K before the British influence, primarily

6.1 Forest Management in Early Colonial Period

due to a lack of any systematic records or studies. As mentioned in chapter two, the princely state of J&K came into existence in 1846 when Maharaja Gulab Singh, the Dogra ruler of Jammu accepted British suzerainty. The forest management in the state during the second half of the nineteenth century can be partially understood on the basis of reports of British forest officials in the service of J&K princely state. These are mainly annual assessment reports without any reference to the previous systems of forest management.[1] Some inferential description on the previous system of forest control and administration in the state, however, is provided in the forest reports and working plans made by the state forest officials after independence. The understanding of the history of forestry in the state during the early colonial period, therefore, remains sketchy.

Historically speaking, the forests of the state were controlled by the princely rulers and *jagirdars*, the chiefs to whom lands were granted by the rulers in recognition for their services and loyalty to the state.[2] Prior to the establishment of a fully fledged Forest Department (FD) in J&K in 1891, the administration of forests was under the control of civil authorities of Revenue Department.[3] The *Wazir-i-Wazarat* was in charge of the district while *Tehsildars* under him managed the affairs of the *tehsils* or sub-districts. *Girdawars* worked under *Tehsildars* and, in turn, were assisted by *rakhas* at village level. The *girdawars* used to collect the forest dues or *rasums* on the various forest products consumed by the villagers.[4] Due to scarce records available on forest management before the formation of J&K princely state, it is not clear whether such an administration existed before the nineteenth century. Some forest reports of the post-independence period suggest that the practice of collecting forest revenue gained momentum under the reign of Maharaja Gulab Singh (1846–1857), the first ruler of the state of J&K. The existence of formal administration of forests by the state, therefore, can be inferred to have started gaining roots during this time. Since the forests were available in abundance (owing to small population size and a limited set of uses), the administration of forests only involved the collection of revenue and the protection of forests was not a significant concern to the princely state.[5]

A recent report on participatory forest management in J&K state by the FD suggests that an informal community system for forest management known as *bradari* existed in all the big villages and hamlets in the mid nineteenth century (Patnaik and

---

[1] It can be noted that the denial of existence of any previous systems of forest management could be a deliberate strategy on the part of British colonial officers to establish the supremacy of 'scientific management'.

[2] *Jagirdars* had the power to govern the forest areas that fell within their territories and earned part of their incomes from taxes imposed on the local communities for forest use.

[3] Working Plan for Jammu Forest Division 1998–99 to 2007–08 (p.36), prepared by the Conservator of Forests, Manoj Pant (2008).

[4] While there were no clearly defined arrangements for the collection of *rasums*, it is reported that they were collected from individuals in some cases and collectively from the village communities in others (ibid.).

[5] Ibid.

124 6 Forests, State and People: A Historical Account of Forest Management and Control…

Singh n.d.).[6] The *bradari* consisted of all the village residents with an informal body of senior members to take decisions on local disputes over forest use. The village residents were obliged to meet requirements of fodder, fuelwood and minor forest produce from the village forests without interfering with those of other villages. Any disputes regarding access to forest resources were, most of the time, resolved by the community locally. It is reported that under the *bradari* system, the community members showed concerns towards protecting and managing forests. There was a common practice of growing trees around places of worship known as *banis* (sacred groves) as well as near *boalies* (water springs).[7] The village communities attached religious value to the forests and were aware of their significance in meeting the everyday requirements of fuel, fodder and small timber. There are also no records of forest destruction and its adverse consequences for village communities in the pre-colonial period, nor is there evidence for serious conflicts over control for forest resources. It is claimed that even with limited state intervention, local communities were able to protect forests effectively (Patnaik and Singh n.d.: 53).[8]

The state management of forests did exist in the pre-colonial period, but as in other parts of India, 'it was restricted in scale and application' (Gadgil and Guha 1992: 52). Bryant (1997: 194), notes that in comparison to the previous systems of forest regulation, the British forest management was new in terms of 'the method and scale of the attempt' as also in the extent of 'coercive and administrative powers' exercised by the colonial state. Likewise, in J&K, since the late nineteenth century, forest users witnessed a drastic change in their access to resources as the British intrusion in terms of control, management and exploitation of forests went far beyond those of previous rulers. In the following discussion, I critically examine the new system of forest management during the British rule in J&K.

6.2 Local Access Versus Commercial Interests: The Politics of Scientific Forestry in the Late Colonial Period

As stated in the first section, the forests were managed by the Revenue Department of the J&K princely state (under British suzerainty) until 1891. Since the prime concern of the colonial administration during that time was the expansion of agriculture, the department transferred vast areas of previously uncultivated land to the local communities for agricultural purposes. There was limited demand of timber in

---

[6] This report, titled 'Study on Participatory Forest Management in J&K State' is authored by P. Patnaik, Principal Chief Conservator of Forests and S. Singh, Chief Conservator of Forests.

[7] Working Plan for Doda Forest Division, 1978–79 to 1987–88 (GoJ&K 1988).

[8] I may mention that such claims regarding communal management are based on the interpretations of present day state forest officials rather than any authentic historical evidence, hence need to be considered with caution.

6.2 Local Access Versus Commercial Interests: The Politics of Scientific Forestry in… 125

the state and, therefore, the contractors were invited from outside J&K for commercial extraction of timber. These contractors were granted *pattas* or written permits which allowed them to select forest patches for felling according to convenience (Patnaik and Singh n.d.: 49). However, the management of forests by the Revenue Department was considered 'haphazard and destructive' by the forest authorities.[9] As H.F. Cleghorn (1861: 5), a colonial forest official (also known as the 'father of scientific forestry' in India) stated:

> It is only of late years, that attention has been drawn to the importance of conserving tropical forests […] The matter of complaint was that throughout the Indian empire, large and valuable forest tracts were exposed to the careless rapacity of the native population, and especially unscrupulous contractors and traders, who cut and cleared them without reference to ultimate results, and who did so, moreover, without being in any way under the control or regulation of authority.

The government of British India realised that more systematic surveys of forest resources and management of timber harvest would be essential if the increasing requirements of timber for railway constructions in India and ship-building in Britain were to be met. At that time, most North Indian rail lines were built with deodar (Cedrus deodara), growing in the high mountains of Himachal and J&K (Tucker 1982: 116). Therefore, at the behest of the British Resident in J&K, the first step towards forming a separate forest administration was taken in 1857 when Mahal Nawara, a forest section was created within the Revenue Department of the princely state (Kawosa 2001: 36). The idea behind creating this section was to generate more revenue from forest resources. Under the system of forest working in vogue then, the contractors were allowed to work in any way that suited them but with the condition that half of the produce extracted is handed over to the Revenue Department as state's share (Negi 1994: 48). This system continued until 1883 when Ain-i-Jangalat, a preliminary regulation for controlling forest working was introduced and the mode of collection of the state's share was changed from kind to cash (ibid.).

In British India, during this period, there were growing concerns within bureaucratic circles over the management of forests by an autonomous organisation, independent of Revenue Department. The British colonialists felt that the state ought to pursue active interventionist policies based on the principles of scientific forestry. It was held that there was a need for systematic management of forests by classifying the forest areas according to their use.[10] Dietrich Brandis, the first

---

[9]Annual Report of the Forest Department, J&K State 1893–94, prepared by the Conservator of Forests, J.C. McDonell (1894: 2).

[10]The British government in India passed the first Forest Act in 1865. The Act empowered the government by declaring any land covered with trees, brushwood, or jungle as government forest, provided that such notification should not abridge or affect any existing rights of individuals or communities (Khator 1989: 13). Subsequently, a more comprehensive Act was passed in 1878 which classified forests into three categories: a) reserved forests, b) protected forests, and c) village forests. The distinction between these forests was based on the people's rights over forest produce. The reserved forests, as the name implies, were to be totally controlled by the government, and the people were denied any access to them. The protected forests could be accessed by the people but

Inspector General of Forests in India, explained the strategy of systematic management as 'a separation of forest domain of the state from the large mass hitherto more or less open to the public' (cited in Pathak 2002: 90). The implication was that there existed a conflict between the people and the forests and that the conflict had to be managed (Khator 1989). For this, the new silvicultural systems were inaugurated, and science based laws and regulations for the utilisation of forests launched. The main reason suggested for such a shift was that forests had a long gestation period and therefore could not be left to private agencies. It was also believed that a systematic management based on scientific approaches would yield rich economic dividends and satisfy the timber demands. As Khator (ibid.: 13) argues, for the first time, the British declared forests as 'valuable good' – goods which needed protection and proper distribution. Hence, it seems that the deployment of science for the pursuit of systematic management of forests coincided with the commercial and imperial interests.

The government support for scientific forestry in British India began in the mid nineteenth century in a modest way with the appointment of mainly medical surgeons trained in natural history to the posts of Conservator of Forests in various provinces of India. Experts were being deployed as 'scientific soldiers' and new agencies to manage forests were established, thus shifting control and access over forests from 'communal' to 'custodial' (Rajan 2006). Initially, the forest officials were charged with three main tasks: to satisfy the complaints and demands of the lessees of the forests, to assure the government the provision of its full timber demands for the dockyards, gun-carriage factories and public works, and to restrict practices such as shifting cultivation (Stebbing 1926).

The mid nineteenth century India was marked with inter-departmental disputes and tensions since the scientific management of forests created several forms of conflicts within bureaucracy over the question of control of forest resources (Bryant 1997: 209). Within the state of J&K, the forest authorities were pressing for the creation of forest reserves and enactment of legislation for declaring forests under their exclusive control (Pathak 2002). Subsequently, along the lines of other provinces of British India and some princely states, a FD was established in J&K in 1891 with the stated goal of systematic forest management based on the principles of 'scientific forestry'. J.C. McDonnell of Indian Forest Service (IFS) was appointed as the first Conservator of Forests in the princely state of J&K. The Annual Report of the J&K Forest Department (1893) suggests the persistence of conflict between the Revenue and Forest Departments in the initial years of its formation. For example, it stated that the management of minor forest products and medicinal plants, such as kuth (*Sassurea lappa*), kaur (*Picorrhiza kurroa*) and rasaunt (*Berberis lyceum*) which continued under the Revenue Department ought to be credited to the FD. It also stressed that the authority to grant permission for grazing cattle, which remained with revenue authorities, need to be brought under the FD. The report suggested not only that the expenditure for the present grazing establishment could be

---

restrictions applied as to what could be taken away from the forests. The village forests served as community forests providing daily subsistence to the people.

6.2 Local Access Versus Commercial Interests: The Politics of Scientific Forestry in... 127

saved, but that the FD would be better able to manage these affairs. Although the entire management of grazing and control of minor forest produce continued to be administered by the Revenue Department, the FD was allowed to close one-fifth of the total forest area in the state in the following years.[11]

During the initial years of its formation, the main objectives of the department were demarcating forest areas and preparing working plans.[12] The forests were divided into beats, compartments, ranges, divisions and circles, and were to be managed by forest guard, forester, range officer, divisional forest officer and conservator of forests at respective levels.[13] Under the working plans, the number of trees to be felled and the period allowed for their removal was specified. It was ensured that only the mature trees (with the prescribed diameter trunk for each timber species) were felled which could be replaced by the growing stock of young trees.[14] The FD engaged its own contractors for timber extraction and the rates of the work were fixed by auction or tender. Some training programmes were started to educate the subordinate forest staff in scientific forestry and silvicultural practices. Furthermore, the working plans promoted the establishment of nurseries and new plantations with the aim of increasing the supply of timber and revenue generation. It was held that any expansion of governmental forests could be paid for by the revenue collected within a year of their management (Barton 2002: 67). The forest officials maintained that the money thus raised would become available to build access roads and canals for floating timber, simultaneously serving the purpose of building up the communications in India for the maintenance of colonial rule as well as for the stimulation of industry (ibid.). Subsequently, the construction of roads that opened to the forests were undertaken in J&K along with the building of railways, with the aim of making transportation of timber and other forest products easier.

The process of demarcation and the preparation of working plans were also accompanied by the task of enclosing demarcated areas in order to restrict the access available to the villagers and livestock previously (see Bryant 1997).[15] The FD, however, had to face a great deal of opposition, first, from the *jagirdars* who

---

[11] Progress Report of Forest Administration in the J&K State 1910–11, prepared by the Conservator of Forests, R.C. Milward (1911).

[12] The Annual Report of the J&K Forest Department (1894: 2) mentioned the significance and need of scientific management stating that in Jammu, there are enormous areas of forests, the best and extensive ones lying in the Nowshera and Kotli tehsils. As soon as the trained officers are available, they should be demarcated and rough plans made for working.

[13] Each compartment comprised four to five beats. Such five compartments formed a range, five ranges made a division and five divisions comprised a circle.

[14] For the purpose of timber felling, three main types of methods were adopted depending upon the condition and status of forests. For the better stocked forests, 'selection felling' was practised, for poor quality forests, 'improvement felling' was prescribed and for the forests not yet under working plans, 'unregulated felling' system was adopted (Progress Report on Forest Administration in the J&K State 1910–11).

[15] Bryant (1997: 215) states that this kind of restriction had several adverse effects on the peasantry in the entire British India since forest officials made no provision for the long-term production of 'minor' forest products.

defended their previous rights of felling the forests within their territories, and second, from the villagers who had been accustomed to using forest areas for fulfilling their daily requirements of firewood, fodder and timber.[16] Although denying the ownership rights to the *jagirdars*, the resistance by villagers was dealt with by assuring them that their needs would be met by keeping some grazing grounds open for their cattle and providing them with free grants of timber.[17] In the case of *jagirdars*, it was held that timber and firewood could be provided to them as a temporary concession at the free will of the State government.[18] While the various forest reports indicate grants of free timber to the villagers in the initial years, the practice soon changed into 'privileges' or 'concessions'.[19] In addition, the FD maintained that the privileges or concessions in the case of forest produce or access to pastures could be suspended for the offender or even for the whole village in situations when trees in the demarcated forests are felled without permission from the forest officials, and when fire is caused wilfully or due to negligence of villagers.[20]

In 1901, another significant shift can be noted when the practice of free grants was discouraged and timber started being sold to the villagers.[21] The Annual Report of the Forest Department, J&K (1901–02: 14) stated that 'initially, there was outcry but people soon became accustomed to the rule and began purchasing trees'. It further stated that the selling of trees was becoming a good source of revenue to the Department and was bound to increase in the future. Moreover, the forest authorities held that the practice of granting free timber had resulted in overexploitation of the forests in the previous years, and therefore, needed to be checked.[22] Explaining the need and significance of shifts from 'privileges' and 'concessions' to the 'sale' of timber and forest products to the villagers, J.C. McDonell, the Conservator of Forests, J&K, in the Annual Report of the Forest Department (1901–02) stated:

---

[16] This has been noted in the Annual Forest Report of the J&K State 1893–94.

[17] Annual Report of the Forest Department, J&K State 1893–94.

[18] Annual Report of the Forest Department, J&K State 1902–1903, prepared by the Conservator of Forests, J.C. McDonell (1903).

[19] See Progress Report of the Forest Administration, J&K 1909–10. It is important to note that 'concessions' and 'privileges' are distinct from 'free grants' as under the former system, timber is granted to the villagers after the payment of small fraction of its value.

[20] The Jammu and Kashmir Forest Act 1930, Chapter II, Clause 7.

[21] According to the Annual Report of the Forest Department, J&K State 1901–02, prepared by J.C. McDonell (1902), for the first time, 111 blue pine and 40 silver fir were sold to the villagers in Kamraj division of J&K while in Kashmir, 624 pine and fir trees were sold to the village people for house construction (ibid.: 14).

[22] The Annual Report of the Forest Department, J&K State 1901–02 noted that many free grants were made by the *Tehsildars* in the past even to people who had no rights on forest produce. Such a 'misuse' of free grants has also been reported in the Annual Forest Report, J&K State 1893–94. Due to floods in Srinagar in 1893, forests were open to the villagers for 2 months for supplying free timber to repair the damages to their houses (ibid.: 16). According to the report, the experiment was not successful from the forest point of view as it not only resulted in the people taking advantage to cut timber as much as they could but also drew villagers from neighbouring villages who were not affected by floods.

6.2 Local Access Versus Commercial Interests: The Politics of Scientific Forestry in... 129

> There is a need to be careful as the provision of giving forest produce to the villagers at a nominal rate can result in manifest loss to the state when, after all, they have no right to such a concession (p. 26). The practice of selling timber to the local communities is no hardship on the people as they are ten times more well-off than they were in the past owing to their general rise in wages paid by the Forest Department for construction activities as well as due to higher value of their farm produce [...] It is fair for the state to reap some benefits from the general prosperity of which it is the cause. Villagers have also begun to be economical in timber. They have started using bricks for walls instead of all timber, which is far more suited to a cold winter than the old wasteful log huts (p.14).

Analysing the colonial forest policy, Khator (1989: 14) observes that it primarily emphasised the commercial value of forests, as a result of which, the 'economic elite entered the scene' and used forests as means of participation in the growing economy. At the same time, the local people who were forest owners by tradition were transformed into labourers. Forests thus provided a rich ground for the extension of capitalism and imperialism (ibid.). Guha (1989) also documents this change, from forests being used to meet the subsistence needs of locals to meeting the wider commercial demands of timber and resin to the expanding British empire, as also local people's right translated into privileges and, in many cases, denied. As such, the drive to maintain high levels of timber production in the colonial period led to a continuous rise in tree felling in J&K.[23] Further, the commoditisation of firewood and the establishment of various firewood depots in Kashmir by the end of the nineteenth century promoted the commercial interests. Certain areas were set apart and leased to the contractors for felling trees for firewood in return for royalty levied by the FD.[24] This new arrangement resulted in serious problems in the supply of firewood to the people in the state. Often, the contractors created artificial supply shortages in the market in order to get the highest rates possible. An initiative to check such shortages of firewood was, however, taken in 1918 by the FD when it was decided that all firewood coupes will work under the supervision of the Department and fixed rates would be determined.[25]

Another important concern of the forest authorities was getting suitable land free of all concessions for timber or claims for grazing to undertake plantation activities prescribed in the new working plans.[26] In response, forest regulations such as The Closure Rule (1909) and Rules for Closure of Deodar Forest to Nomad Goat (1912) were formulated and major parts of forest areas were closed to grazing. Nevertheless, the FD maintained that the continued grazing by the goats of bakarwals and gujjars (pastoralists) in the forest areas kept open (for grazing) was posing threat to the

---

[23] The revenue from the FD contributed nearly one-third of the total state revenue in the early part of the twentieth century. In 1891, the annual net surplus was about 3 *lakh* rupees and by 1913, it rose to 21 *lakhs*. From then, the progress was rapid and the surplus reached 56 *lakhs* in 1928. For the next 8 years, there was a slump and the surplus averaged 30 *lakhs*, but it reached nearly 50 *lakhs* by 1940. This figure rose to 63 *lakhs* in 1942 followed by 86.5 *lakhs* and 76.3 *lakhs* in 1945 and 1946 respectively (Champion and Osmaston 1962: 420).

[24] The Annual Report of the Forest Department, J&K State, 1901–02.

[25] Progress Report of Forest Administration, J&K State, 1918–19, pp. 23–24.

[26] Progress Report of Forest Administration, J&K State, 1909–10.

130   6 Forests, State and People: A Historical Account of Forest Management and Control...

natural regeneration of forests.[27] It was argued that the heavy grazing by these goats and the cutting of all herbaceous growth for fodder by the local villagers was making it impossible for any seedlings to grow. The FD also attributed the problem of 'overgrazing' to the fact that regulation of grazing still remained under the Revenue Department and, hence, it was impossible for the FD to exercise any form of exclusive control.[28]

The restrictions on the villagers with regard to the access to forests, grazing and felling trees under the new system led to various forms of resistance. An increase in the number of forest fires caused by the graziers was reported in the initial years of enclosing forest areas.[29] The villagers also expressed their opposition to the new regulations by refusing to provide any assistance to the forest staff in putting out forest fires.[30] It is reported that the DFOs had filed legal cases against villagers for such refusals in assisting the FD. Some reports suggest cases of illicit felling of deodar, spruce and fir by local villagers for want of firewood and timber.[31] Various divisional forest reports indicate such cases stressing the need for more staff especially forest guards and foresters to control illegal felling.[32]

Similar to the case of J&K, conflicts between the FD and villagers are also seen in other parts of British India. For example, in the context of Kumaon hills, the Indian Forest Law (1878) brought to light a deep-seated conflict between the subsistence patterns of traditional villagers and the colonial system of timber management. When the law was introduced into the districts, villagers evaded en masse the fee payment system that regulated their use of government forests (Tucker 1982). Saberwal (1999: 89), in the context of Himachal Pradesh and Punjab, also reports the villagers' resistance to the FD through non-compliance of the regulations including choosing the winter grazing grounds on the basis of forage requirements rather than the ones to which they have proprietary rights, the mis-reporting of animal numbers, the evasion of check points, the bribing of forest guards, etc. Chhatre (2003), referring to the Indian Western Himalayas, also narrates similar resistances from the peasants, horizontal tensions across different departments, and vertical tensions between local knowledge professed by provincial bureaucracy and

---

[27] Progress Report of the Forest Department, J&K State 1918–19.

[28] Progress Report on Forest Administration, J&K State 1924–25. Saberwal (1999) in the context of Himachal Pradesh and Punjab, also reports the power struggles between Forest and Revenue Departments over the question of departmental primacy of control over forest lands and its resources. However, here, the instances of conflict are in the form of herder appeals to the Revenue Department in pre-independence India and later to state politicians to intervene on their behalf against the restrictive policies of the Forest Department to curtail their access to forest resources.

[29] Annual Report of the Forest Department, J&K State 1893–94.

[30] See Progress Report of Forest Administration, J&K State 1909–10, Progress Report of Forest Department, J&K State 1910–11.

[31] For example, under the plea of cutting firewood, the Duabgah *hanjis* (boatsmen) were also reported to fell deodars illegally for making boat planks (Annual Forest Report, J&K State 1893–94: 14).

[32] See Annual Forest Report, J&K State 1893–94; Progress Report of Forest Department, J&K State 1910–11.

6.2 Local Access Versus Commercial Interests: The Politics of Scientific Forestry in... 131

central direction emanating from scientific management. Likewise, Guha (1989: 49), speaking for Uttarakhand, documents various peasant oppositions such as the lopping of trees, overgrazing of livestock and the burning of the forest floor for a fresh crop of grass in response to their curtailed access to forest resources. Tucker (1982: 117) rightly suggests, whenever an area was declared as reserved, the transition to the new management system was a 'time of very delicate tensions between villagers and foresters'.

The initial enthusiasm for the systematic management of forests declined in the later years, owing to the outbreak of the First World War (Patnaik and Singh n.d.: 59). In the late 1920s, the forests of the state came under a new management system known as 'uniform felling', which means growing plantations of same age and preferably of single species so that the entire patch could be harvested at the same time. Under this new system, the most accessible forests bearing commercially important species were brought under concentrated felling without any efforts to augment natural regeneration (IFRI 1961: 6, see also Kawosa 2001). Although the previous working plans were based on the principles of scientific forestry, the demands of the War led to heavy removals ignoring even the most basic principles of scientific management. Between the two world wars, the demand of timber increased significantly and forests, including those of blue pine, spruce and fir, were extensively exploited. Tucker (1982) also notes this in other parts of India when massive wartime demands for timber products brought permanent damage to both government and private forests. In J&K, this process continued in the following years as during the Second World War (1939–1945), extensive felling was carried out in total disregard of the working plan prescriptions (Patnaik and Singh n.d.: 59). The gun factory established in Baramulla during the inter-war period for the manufacture of rifles was supplied with abundant timber from the walnut forests, which led to massive depletion of forest cover in the district.[33] After the war, however, the working plans were revised and emphasis was laid again on 'scientific management' by way of stock mapping and collection of statistical data on growth and yield of the forests.

Although the main objectives of the FD were timber extraction and revenue generation, some concern for forest conservation was also shown by the senior authorities. For example, the Progress Report of the Forest Department, J&K State (1909–10) notes that although the timber extraction and free grants are assuming great significance, it is worthy to understand that adequate steps to ensure reproduction be taken in order to maintain the timber supplies. It further stated that although the senior officials recognised the importance of forests and the necessity of young growth, this concern needed to be brought to the attention of junior revenue officials who saw forests as inexhaustible (ibid.: 17).

Sivaramakrishnan observes, by the third quarter of the nineteenth century, scientific forestry was also deeply challenged by responsibility for environmental health, the preservation of nature, and thus with conservation (1999: 279). The very possibility of an emergence of the ideology of forest conservation in the 1850s was, however, in the context of pressing timber shortages (Pathak 2002). For example, a

---

[33] Progress Report on Forest Administration, J&K State, 1920–21.

close connection between the shortage of timber and initiatives for conservation was evident in Punjab in the wake of British conquest (Rangarajan 1996: 26). Punjab was conquered in 1849, and after the denudation of the plains, logs were brought from Kashmir under license from the Dogra ruler. It was feared that there would be an acute shortage of wood in the Rawalpindi hills due to massive requirements for the building of the Muree cantonment. The Forest authorities in Punjab, therefore, began to regulate access to selected forests, especially those of deodar (ibid.: 27). Emphasising the significance of conservation, Stebbing (1926: 12), suggests that 'surplus revenue from the FD was the necessary consequence of conservancy management'. This idea of conservation was addressed to and received by various layers of the administration differently (Pathak 2002). Broader issues such as the climatic effects of deforestation and the importance of forest conservation for the interests of 'humankind' were more influential at the higher levels of administration. At the microlevel, more specific and functional issues dominated. These mainly included revenue generation, timber supplies and appeasing the landholders (ibid.: 87).

Sivaramakrishnan also states that although scientific forestry is primarily concerned with revenue and regeneration, it is 'obliged to deal *in extenso* with questions of rights and property' (1999: 279). Apart from conservation concerns, the need to involve local communities in protecting degraded forests also figured in some forest policies of the state during the British period. The Kashmir Forest Notice (1912) and Jammu Forest Notice (1912) were the two main regulations which outlined the role of village communities in forest protection. Under these regulations, concessions were granted to the village community in lieu of protecting the surrounding forests by helping the forest and police officials in preventing forest offences. Owing to the growing resistance by the villagers to the establishment of closures (fencing of pasture lands and degraded forest areas), the FD in 1930 proposed to form 'village forests' and involve local communities in conservation measures.[34] These village forests were to be formed in each forest range assigning the responsibility of protecting and afforesting wastelands in the village. The *panchayats* or village committees were to be empowered to enforce the rules for the protection of the area and regulate both grazing and concessions on forest products as sanctioned by the FD. This policy, however, could not be implemented effectively since this was the period when the independence movement was going on in the state (as in other provinces of India) and the people were reluctant to cooperate with the government. Yet, this policy can be considered a major initiative in the history of FD, laying the foundation for joint management of forests by state and local communities in the years to come.

To summarise, three main points can be drawn from the discussion presented in this section. First, apart from forming the basis for demarcating forest areas and timber extraction, 'science' found its application in legitimising the role of professional community of foresters for the use and control of forest resources. We have observed that in the process of demarcating forests, claims of ownership by *jagirdars* declined and the rights of local villagers with regard to access to forest

---

[34] The Jammu and Kashmir Forest Act, 1930, Chapter III-A.

resources were curtailed. The 'communal management' of forests changed into a system of 'free grants' of timber to the local forest users at the will of the state. On the pretext of overexploitation of forest resources by the local populations, this system of 'free grants' was further altered to that of 'concessions and privileges' and later into the practice of 'selling' timber and firewood to the local communities.

Second, the rationality of 'science', which stood supreme in restricting the rights of local communities to forest resources ultimately surrendered to imperial and commercial interests. We noted that due to the rise in demands for timber during the world wars, the state forests were exploited in complete violation of the rules and principles prescribed in the working plans. Bryant rightly suggests that scientific forestry may have been compatible with the development of a rational state but in the pursuit 'to make financial ends meet, short-term expediency often proved the more powerful policy influence' (1997: 204). Therefore, we can argue that while conservation imperative and scientific rationality are powerful enough to limit the local people's access to their forest resources, it often proves subservient to the interests of powerful actors or administrators.

Third, to consider the colonial forest policy as based on mere exploitation of forest resources would be a partial depiction of the complex interplay between ideologies and practices of the state. As Sivaramakrishnan argues, scientific forestry was a 'complex, multi-layered discourse formation that was historically and contingently produced' (1999: 280). Besides revenue generation, we also witnessed some concerns of the colonial state for conservation and local community involvement in forest protection which, to a certain extent, helped in accommodating the dissent shown by the affected communities in response to the restrictions imposed and possibly also served as justifications for the interventions by the forest bureaucracy in affairs which were earlier the domain of rulers and village communities. I shall now turn to discuss the perpetuation of the colonial forest policy in the post-colonial period, and analyse the important shifts in ideologies and practices concerning forest management until the late 1980s.

6.3 National Interests Versus Local Needs: The Politics of Forest Management in the Post-Colonial Period

After independence in 1947, the FD of J&K was reorganised under the new conditions caused by the partition of the country and subsequent integration of the state with the Indian Union. The problems which resulted from the partition of India caused serious challenges to the management of forests in the J&K state. Some parts of the erstwhile princely state were now under Pakistani rule, which disrupted forestry operations in the state. The timber sale depots at Jhelum and Wazirabad fell in Pakistani territory and the rivers Jhelum and Chenab lost their usefulness for

134   6   Forests, State and People: A Historical Account of Forest Management and Control...

floating and transporting timber in J&K (IFRI 1961: 6). The disturbed conditions that prevailed at the time of partition caused a temporary cessation of forestry works in the state. In the subsequent years, however, land transport was developed and a new timber depot at Pathankot (in Punjab) was established (ibid.). The timber trade, which had suffered immensely after partition revived and forest working, was resumed with greater vigour by the early 1950s to realise the agenda of nation-building. The rapidly increasing demand from all over the country for industrial wood facilitated the timber extraction activities of the FD in the state. The main demand for Kashmir timber in India was for producing sleepers for the railways. While major demand was for deodar timber (due to its resistance to termites), willows, walnuts and other conifers were also extracted for construction purposes and the manufacturing of rifles, furniture and sports goods.

Although the initial years of independence laid emphasis on raising plantations and increasing the productivity of forests, the concern was more for meeting industrial requirements than conservation *per se*. For example, the main objective of the National Forest Policy (1952) formulated by the Ministry of Food and Agriculture, which laid the basis for most forest regulations in independent India, was the realisation of maximum revenue. The policy clearly pointed out that local interests and priorities should be subservient to the broader national interests. It stated:

> Village communities in the neighbourhood of a forest will naturally make a greater use of its products for the satisfaction of their domestic and agricultural needs. Such use, however, should in no event be permitted at the cost of national interest.

The forest policies and regulations of the Indian government were not extended to J&K in the initial years of its accession (owing to its special constitutional status). Yet, a convergence of interests emerged between those of central and state governments concerning timber extraction and revenue generation. As such, the system of 'uniform felling' which started in the 1920s under colonial rule continued until the late 1970s.[35] The forests with deodar, kail, and fir were felled unabatedly without any conscious efforts to regenerate them. Interestingly, the basic premise of such working in the Forest Department was that most of the mature trees would be removed from the forest areas to allow the younger generation of crops to grow in the following years (Kawosa 2001: 37). The regeneration and maturing of the left over crops after the harvesting operations, however, did not keep pace with the felling of the trees. Moreover, the younger crops were under constant pressure from local communities who were dependent on the forests to meet requirements of fuelwood, housing timber and other forest resources for sustenance (ibid.).

The FD continued to lease the forest areas to the private contractors for timber extraction in lieu of the royalty levied. Owing to their strong links in the bureau-

---

[35] With the increasing demand for timber, firewood and other minor forest products, the forests were exploited at a rapid rate. The outturn of timber in 1947 was 28.00 (000 m^3) which reached up to 451.43 in 1960 and 609.21 in 1980. The outturn for the other major forest products such as firewood was 70.10 (000 m^3) in 1950 which reached to 162.48 in 1960. There was a decline in the extraction of firewood in the next few years due to the availability of alternative sources of fuel and energy and the figure dropped to 82.30 in 1975 (GoJ&K 2005: 24).

6.3 National Interests Versus Local Needs: The Politics of Forest Management... 135

cratic circles, the forest lessees often managed to hold the lease for longer periods than originally granted to them and caused relentless destruction to the state forests (Kawosa 2001: 38).[36] Furthermore, due to heavy snowfall in 1979–80, the forests of the state were damaged considerably. The FD, in response, initiated a 'hygienic operation' and the tasks of additional markings and cleaning the fallen materials in the forest compartments were granted to the lessees on nominal rates (ibid.). Since contractors were already employed for timber extraction purposes, this operation, in actual practice, worked in the interests of timber contractors who started felling even the green trees in these forests under the guise of hygienic clearance (ibid.). The unabated timber felling by contractors (in connivance with forest officials) for over 30 years after independence, generated high revenues but also resulted in the fast depletion of the forest cover.[37]

By the late 1970s, the central government, following the policy of liquidating the autonomy of the J&K state, also started intervening directly in the matters of forest working, which so far, had been the monopoly of the state government. At the national level around this time, the need to dovetail forest conservation with timber extraction was felt by the policy makers and planners. For example, the National Commission on Agriculture in 1976 proposed that for meeting conservation goals as well as maximising timber production for developing industries, there was a need for extending the forestry activity outside the existing forest area through the creation of large-scale man-made forests. Notably, as in the colonial period, the provision of 'free grants' and 'concessions' to local forest users was seen as incompatible with the goals of forest management, and rural populations continued to be regarded as destructors of forests. As the Commission's report, published by the Ministry of Agriculture and Irrigation (1977) stated:

> Free supply of forest produce to the rural population has dilapidated forests. It is necessary to reverse the process. The rural people have not contributed towards the maintenance of regeneration of forests. Having overexploited the resources, they cannot in all fairness expect that somebody else would take the trouble of providing them forest produce free of charge.

The Commission also proposed nationalisation of timber extraction and the setting up of a network of State Forest Corporations (SFCs) in the country. In response, the Government of J&K in 1978 established the State Forest Corporation which was entrusted with the responsibility of extraction and development of forest resources. Rather than leasing the forests to private contractors or agencies, the FD now leased forest compartments to the SFC after reaching an agreement on the payment of a fixed amount of royalty. This system, to a large extent helped to eliminate private contractors who had not only monopolised the forest trade but also indulged in

---

[36] M.A. Kawosa is the ex-Director, Environment and Ecology, J&K Government who played a key role in formulating the Forest Policy of J&K, 1990.

[37] After independence, the revenue from the state forests was on constant rise. There was a sharp decline in the net surplus from 76.3 *lakhs* in 1946–47 to 8.70 *lakhs* in 1947–48 due to turbulent conditions at the time of independence. It gradually picked up and reached 65.25 *lakhs* by 1955–56. The net revenue touched as high as 1536.43 *lakhs* by 1979–80 (GoJ&K 2001: 172).

'unscientific and uncontrolled operations in the forests' (Kawosa 2001: 39). However, the SFC continued to make as much harm to the forests of the state as the private contractors did in the past (ibid.). Forests continued to be exploited at an average annual yield of 10–15 million cubic feet of timber without corresponding investments in their regeneration (ibid.). It is estimated that forests of the state shrunk from about 21,000 square kilometres in 1930 to 13,000 square kilometres by 1990 (ibid.: 42).

It is important to note that by the early 1980s, the narratives of 'deforestation' and 'soil erosion' started influencing international development discourse. Various development and donor agencies drew upon international scientific debates of the 1970s and early 1980s particularly with regard to forest influences on climate, hydrological regimes and soil erosion. Furthermore, the rise of 'alternative development' thinking (e.g. Chambers 1983, 1987; Schumacher 1973) and contributions of studies on the problems of environmental degradation from political economy perspective (see for example, Blaikie 1985; Shiva 1991) led to a significant change in the perception of the role of local communities in problems related to natural resources. The local forest users, who were previously considered as part of the problem of environmental degradation and deforestation now came to be seen as part of the solution by means of their active involvement in forest regeneration and improvement of wastelands. As such, international agencies such as the World Bank, the Swedish International Development Agency (SIDA), the Japan Bank for International Cooperation (JBIC) started promoting afforestation programmes in India through community involvement, with the ultimate aim of increasing tropical forest cover.[38] By the late 1980s, the notion of 'sustainable development' also became an integral part of the mainstream conservation thinking and seeped into various natural resource development policies and programmes across the developing world. This brought to the fore the idea of fulfilling subsistence requirements of forest dependent populations along with meeting the goals of environmental sustainability in various conservation interventions in the following years.

The period of the 1980s also saw a corresponding shift in 'national interest' from revenue generation (that had prevailed since the colonial period) to forest protection and conservation for maintaining environmental sustainability. Moreover, grassroots environmental action fed into this change in thinking about management and use of forest resources. In 1981, a well known environmentalist, Sunder Lal Bahuguna (of Chipko fame), visited the state and started a movement known as 'Kashmir-Kohima Padhyatra' for protecting Himalayan forests and generating awareness against commercial felling. His idea of forest protection infused public opinion with criticism for unabated exploitation of forests for revenue generation in

---

[38] Since the early 1980s, the mainstream international development thinking has largely influenced the national policies in India, including the making of JFM, with its narratives of sustainable development, participation and decentralisation. The international donor agencies mentioned above have also funded JFM projects in many states of India. In the case of J&K, however, these were sponsored by the Indian Ministry of Environment and Forests. Hence, the role of international agencies in J&K can be seen as one of the influencing factors only unlike their role in other states of India.

6.3 National Interests Versus Local Needs: The Politics of Forest Management...

the state. The then Chief Minister, Sheikh Abdullah also gave a boost to the movement by declaring in one of his cabinet meetings: 'we will not earn money by selling our trees' (Patnaik and Singh n.d.: 16).

In the previous section, I mentioned that some attempts to involve local communities in forest protection and regeneration had been made by the forest authorities in J&K since the 1930s, although with little success. A serious concern for local participation in conserving forests, however, figured more prominently in 1982–83 under the 'Social Forestry' programme proposed by the National Commission on Agriculture and sponsored by the World Bank in J&K. The main objective of the programme was to raise plantations along roadsides, regenerate degraded lands and village common areas, and meet the requirements of fuelwood, fodder and small timber for the local communities.[39] There was a provision in an agreement between the FD and village *panchayats* for undertaking tree plantation activities under the project. Although serious attention was given to meet the physical targets in the Social Forestry project, the concept of involving local people in the forest activities could not be fully realised as the forest officials were reluctant to shift their roles from 'implementers' to mere 'facilitators' of the programme.[40] Owing to the historical animosity between forest officials and villagers with regard to access to forest resources, the local communities were even more reluctant to cooperate in the project. A general apprehension among the villagers was that the FD would take over common lands that people had been using for meeting their daily requirements of fodder and fuelwood.[41] As with Social Forestry programmes in other parts of India, the control over funds remained with forest officials who, in most cases, worked in alliance with the local elites and misappropriated project funds (Khator 1989). Moreover, the procedure for the involvement of *panchayats* and the role played by them was not clearly defined (Patnaik and Singh n.d.: 17). In 1984, the *panchayats* were dissolved in J&K and no elections of local bodies took place until the end of 1990s owing to the turbulent political conditions in the state. Since the only link between the FD and village community was broken after the dissolution of *panchayats*, the functionaries of the Department decided to form Village Forest Committees (VFCs) to protect degraded forest lands (ibid.). It is important to note that unlike *panchayats*, these VFCs were not democratically elected bodies and, in most cases, had no effective powers.

The National Forest Policy of 1988 reviewed the strategy of forest management in the country and again laid an emphasis on the need for forest conservation along with meeting the basic requirements of forest dependent communities. Unlike the

---

[39] The programme had three major components: a) Farm Forestry – encouraging farmers to plant trees on their own farms by distributing free or subsidised seedlings, b) Woodlots – planted by the forest departments to meet the needs of the local populations. The woodlots were to be maintained by the people under the supervision of forest officers, and c) Community woodlots – planted by the communities themselves on common lands (Khator 1989: 57–58).

[40] Interview with Lal Chand, Director of Soil Conservation Department, Jammu (20 February, 2007).

[41] Ibid.

previous forest policies and regulations, it was held that rights and concessions of the rural people living within and around forests needed to be fully protected, and their domestic requirements of fuelwood, fodder, minor forest produce and timber should be maintained as a priority. The policy set a national goal to have a minimum of one-third of the total land area of the country under forest or tree cover. Notably, the forest policies of India (1952 and 1988) were not formally extended to the J&K state owing to its special constitutional status. Nevertheless, there were hardly any conflicts between the forest regulations of the centre and the state and all major forest management programmes initiated by the central government (such as the Social Forestry or creation of the SFCs) were also implemented in J&K.[42] The J&K state had no Forest Policy of its own until 1990 and its forests were regulated through various acts and legislative orders of the state from time to time.

A turning point in the forest management of J&K came with the formulation of the first Forest Policy in 1990 when the state realised that the regeneration of forests was not keeping pace with the timber extraction. Under this policy, the felling of trees was banned in 18 catchments and timber extraction was limited to only 5 million cubic feet annually (Kawosa 2001: 45). While the policy aimed to exploit forests sustainably and conserve village forests through community participation, it also resulted in the curtailment of concessions and grazing rights available to pastoralists and local villagers (ibid.). In order to realise the Forest Policy of 1990, the state enacted the J&K Forest (Conservation) Act of 1990, which totally banned the export of timber in unfinished or semi-finished form from the state, and also specified that no forest land could be diverted to non-forestry use. As the new policy was beginning to take roots, the state was hit by militancy and terrorism. Large areas of forests were burnt and the roadside plantations were cleared during the anti-insurgency and combing operations. The damage to the forests became a common affair and about 4 million cubic feet of SFC timber was torched by militants. Many forest officers were killed by terrorists and forestry operations came to a virtual standstill (Kawosa 2001). After the heavy deployment of military and para-military forces in the state due to militancy, it is estimated that between 1990–1999, on an average, 30% of the firewood was supplied to the army (GoJ&K 2001). This, in turn, reduced the availability of forest produce for the village poor.

To summarise, in the post-colonial period, we observed changing interpretations of 'national interest' with regard to the management and use of forest resources. In the first three decades after independence, the Indian state promoted maximum timber extraction and revenue generation following the agenda of nation-building and industrialisation. The state forest authorities adopted the colonial model of scientific forestry and continued giving primacy to commercial exploitation over subsistence use of forests. The state interventions were justified by labeling local residents as 'ecologically irresponsible' and stressing the need for scientific management by trained forest staff (Bryant 1997: 217, see also Jewitt 1995). The Indian foresters also drew upon international scientific debates and writings, particularly with regard to forest influences on climate, hydrological regimes and soil erosion, thus, provid-

---

[42] Ibid.

6.4 Conclusion

ing part of the source material required to argue their case for more strict enforcement of regulations (Saberwal 1999). We observed that in J&K, in actual practice, the silvicultural principles with regard to forest management were often violated and forests were felled unabatedly under 'uniform felling' system until the 1970s. Although there were concerns for regenerating forests but more to realise commercial interests rather than meeting either conservation goals or the daily requirements of fuelwood, fodder and minor forest produce for local communities. In fact, there was not any significant change in the relationship between forest officials and local communities, and the locals continued to be seen as 'forest destroyers'. Therefore, in J&K, similar to other parts of India, conflicts and contestations between the state and local communities continued though now under new political and economic circumstances.

I also noted that by the early 1980s, the agenda of 'forest conservation' became dominant, again under the label of 'national interest'. This emerging concern for forest conservation, however, brought two major shifts in the forest management of J&K. First, the nationalisation of timber extraction took place and control passed from private lessees to the SFC. A second and more important shift was that local communities who were earlier referred as 'forest destroyers' started being seen as potentially 'active protectors' who ought to be involved in the conservation programmes. This shift in thinking within policy circles from 'custodial' to 'social' management of forests was influenced and shaped substantially by the concerns and interests of international conservation community. As Mayers and Bass (2004: 3) argue, 'efforts were made to identify common goals, much of them geared to pinning down the components of sustainable forest management'. The scientific knowledge, apart from legitimising state intrusion in systematically managing forests and enhancing forest yield to meet commercial requirements, found another application for articulating the goals of 'forest conservation' and 'sustainable development' in the following years.

6.4 Conclusion

In this chapter, I demonstrated that forest bureaucracy used the language of 'scientific management' and 'strict regulation' for maintaining control over forest resources both in the colonial and post-colonial periods. We saw that 'scientific forestry' proved subservient, first to imperial and commercial interests of the colonial state and then to the varying national interests of the independent state in the post-colonial period. In both cases, I have argued that it played a crucial role in legitimising the interventions of forest bureaucracy and, at the same time, curtailing the rights and access of poor local communities dependent on forest resources for their sustenance. Until the 1970s, the local communities were regarded as 'illegitimate' appropriators or forest destroyers whose free access to forest resources needs to be regulated for meeting commercial as well as conservation goals. However, a subsequent shift in the national and state forest policies took place in the following years owing to the changes in international conservation and development thinking emphasising the role of local communities in forest regeneration, coupled with the rising influence of domestic grassroots environmental action demanding local control of forest resources. Consequently, the new policies now focused less on revenue generation and more on the need to protect village forests through the active involvement of local communities and giving primacy to their subsistence requirements.

The shift in the potential role of local communities in forest conservation and use figured more clearly in the central government sponsored programme of 'Joint Forest Management' which was initiated in J&K in 1992. In the following two chapters, I demonstrate how power relations determine access to and control over forest resources in the state. I assess the *joint* management of forests by involving village communities, analyse the *split* role of the state in the implementation of the programme and highlight the complex relationship between law, science and politics in forest conservation. I also illustrate the diverse ways in which law and science set the stage for various forms of politics from the macro- to the microlevel over issues of 'participation', 'decentralisation' and 'sustainable development' in the context of forest management in J&K.

References

Barton GA (2002) Empire forestry and the origins of environmentalism. Cambridge University Press, Cambridge
Beinart W, Hughes L (2007) Environment and Empire. Oxford University Press, Oxford
Blaikie P (1985) The political economy of soil erosion in developing countries. Longman, London
Bryant RL (1997) The political ecology of forestry in Burma 1824–1994. Hurst and Company, London
Chambers R (1983) Rural development: putting the last first. Longman, London
Chambers R (1987) Sustainable livelihoods, environment and development: putting poor rural people first. Paper No. 240. Institute of Development Studies, Sussex
Champion H, Osmaston FC (1962) *E.P. Stebbing's the forests of India*, vol IV. Oxford University Press, Oxford
Chhatre A (2003) The mirage of permanent boundaries: politics of forest reservation in the Western Himalayas, 1875–97. Conservation and Society 1(1):137
Cleghorn HF (1861) The forests and gardens of South India. W. H. Allen, London
Gadgil M, Guha R (1992) This fissured land: an ecological history of India. Oxford University Press, Delhi
GoJ&K (1988) Working plan for Doda Forest Division, 1978–79 to 1987–88. Jammu and Kashmir Forest Department, Srinagar
GoJ&K (2001) Digest of forest statistics. Statistics Division: Government of Jammu and Kashmir, Srinagar
GoJ&K (2005) *Digest of forest statistics*. 2005. Jammu and Kashmir Forest Department, Srinagar
Guha RC (1989) Unquiet woods. Oxford University Press, New Delhi
IFRI (1961) Hundred years of forestry 1861–1961, vol II. Indian Forest Research Institute, Dehradun
Jewitt S (1995) Europe's 'others'? Forestry policy and practices in colonial and postcolonial India. Environ Plann D Soc Space 13(1):67–90
Kawosa MA (2001) Forests of Kashmir: a vision for the future. Natraj Publishers, Delhi

References

Khator R (1989) Forests: the people and the government. National Book Organisation, New Delhi

Mayers J, Bass S (2004) Policy that works for forests and people. Earthscan, London

McDonell JC (1894) Annual report of the Forest Department, J&K State 1893–94. The Civil and Military Gazette Press, Lahore

McDonell JC (1902) Annual report of the Forest Department, J&K State 1901–02. The Civil and Military Gazette Press, Lahore

McDonell JC (1903) Annual report of the Forest Department, J&K State 1902–1903. The Civil and Military Gazette Press, Lahore

Milward RC (1911) Progress report of Forest Administration in the J&K State 1910–11. Central Jail Press, Srinagar

Ministry of Agriculture and Irrigation (1977) Report of the National Commission on Agriculture 1976. Government of India, New Delhi

Negi SS (1994) Indian forestry through the ages. Indus Publishing Company, New Delhi

Pant M (2008) Working plan for Jammu Forest Division, 1998–99 to 2007–2008. Working Plan and Research Circle, Jammu and Kashmir Forest Department, Jammu

Pathak A (2002) Laws, strategies, ideologies: legislating forests in colonial India. Oxford University Press, New Delhi

Patnaik P, Singh S (n.d.) Study on participatory Forest Management in J&K: state policy implication and human resource development. Report prepared for the Jammu and Kashmir Forest Department

Rajan SR (2006) Modernizing nature. Oxford University Press, Oxford

Rangarajan M (1996) Fencing the forest. Oxford University Press, New Delhi

Saberwal VK (1999) Pastoral politics: shepherd, bureaucrats and conservation in the Western Himalaya. Oxford University Press, Delhi

Schumacher EF (1973) Small is beautiful: a study of economics as if people mattered. Blond and Briggs, London

Shiva V (1991) Ecology and the politics of survival: conflicts over natural resources in India. Sage, New Delhi

Sivaramakrishnan K (1999) Modern forests: statemaking and environmental change in colonial Eastern India. Oxford University Press, Stanford

Sivaramakrishnan K (2003) Scientific forestry and genealogies of development in Bengal. In: Greenough P, Tsing AL (eds) Nature in the global south: environmental projects in South and Southeast Asia. Duke University Press, Durham, pp 253–287

Stebbing EP (1926) The forests of India, vol III. Bodley Head, London

Tucker R (1982) The forests of Western Himalayas: the legacy of British colonial administration. J For Hist 1982:112–123

Zutshi C (2004) Languages of belonging: Islam, regional identity and the making of Kashmir. Hurst, London

Chapter 7
Joint Management of Forests and Split Role of the State: The Politics of Forest Conservation in J&K

Abstract In this chapter, I aim to understand the politics around the National Afforestation Programme (NAP) and examine the interplay between the central Ministry of Environment and Forests (MoEF), the J&K Forest Department (FD), the State Forest Corporation (SFC), local NGOs and forest field-staff. The following questions are addressed in this chapter: what is the agenda of the NAP and how do various players associated with it interact with each other? What potential does the programme have for sustainable management of forest resources through local participation and partnership? How can the local state be understood in its divided role of both implementing the programme, and contravening the forest laws and regulations? In light of these questions, I examine the relations of power among players involved in forest management and highlight diverse forms of politics from national to local level.

Keywords Jammu and Kashmir · Joint forest management · National afforestation programme · Split role · Illegality · Conservation politics

Participatory forest management became an integral part of international discussions on environment and development from the 1970s onwards (e.g. the World Forestry Conference in Jakarta, 1976). By the late 1980s, international and national forest conservation policies started emphasising the significance of 'local participation', 'joint management' and 'decentralisation' in the successful implementation of programmes. Such policies rest on the idea that forest dependent communities are to be empowered through their involvement in decision-making (by forming forest user committees), and benefited by both wage employment opportunities (e.g. tree-plantations, fencing and other developmental activities) and increased availability of biomass for their subsistence needs. The policy makers also started to acknowledge that local communities have 'indigenous' knowledge of the forests, which can be brought to planning for the provision of their own needs. The policies based on these ideas aimed at forest protection and regeneration of degraded forest lands surrounding villages with the help of village residents. Sceptics argue that participatory forest management interventions, although significant in their own

© Springer International Publishing AG 2018
S. Gupta, *Contesting Conservation*, Advances in Asian Human-Environmental Research, https://doi.org/10.1007/978-3-319-72257-3_7

143

right, have inadequately addressed the issues of resource *access* and *control*, crucial in both forest conservation and community empowerment. A study of agenda, interests and power relations of various players involved in forest management is, therefore, necessary to understand the factors that actually influence access to and control over resources (e.g. forest produce and project funds) and, in turn, determine the outcome of forest conservation policies.

In this chapter, I aim to understand the politics around the National Afforestation Programme (NAP) and examine the interplay between the central Ministry of Environment and Forests (MoEF), the J&K Forest Department (FD), the State Forest Corporation (SFC), local NGOs and forest field-staff. The following questions are addressed in this chapter: what is the agenda of the NAP and how do various players associated with it interact with each other? What potential does the programme have for sustainable management of forest resources through local participation and partnership? How can the local state be understood in its divided role of both implementing the programme, and contravening the forest laws and regulations? Overall, I examine the relations of power among players involved in forest management and highlight diverse forms of politics from national to local level.

The structure of this chapter is as follows. In Sect. 7.1, I introduce the JFM programme started in J&K in 1992 for achieving sustainable management of forests through local participation and partnership. Following this, I discuss the objectives of the NAP initiated in 2002 for the effective implementation of the previous programme of 'joint forest management' through decentralisation. I point out that the broad range of objectives of the NAP could lead to serious problems of implementation since the project implementing agencies may fail to appropriately operationalise the various different objectives in relation to each other. This may result in different actors prioritising objectives as per their own interests and manipulating regulations according to their capacities. In Sect. 7.2, I highlight the interplay between the MoEF, the J&K FD and the SFC. I argue that, in actual practice, both central government and the State Forest Department are reluctant to devolve effective powers, often resulting in repeated re-centralisation with regard to authority, control and access to resources. Following this, in Sect. 7.3, I highlight two forms of micropolitics at village level. I discuss the dilemmas of the forest field-staff (or street level bureaucrats) and examine their split role in meeting conservation needs by restricting the access of villagers to local resources and, at the same time, transgressing the forest regulations to cater to the demands of local populations. I also examine the issues of conflict between the SFC and local contractors, and point out the fact that although the state and market actors are able to make profits by relentlessly exploiting rich forest resources, the poor struggle even to meet their subsistence needs from local forests.

My main argument in this chapter is that the 'joint forest management' under the NAP does not effectively facilitate access for the poor to local forest resources but provides a new ground for legitimising the power and intrusion of forest bureaucracy and conservation community with regard to control over resources. These interventions, however, do not go unchallenged and are reconfigured through various forms of micropolitics. The ways in which the intrusion of forest managers and their control over local resources is understood, negotiated and resisted by the village communities is the topic that I leave for the next chapter.

7.1 Joint Management of Forests: New Arenas of 'Partnership' and 'Participation'

In the last two decades, the conservation community has widely acknowledged the fact that forest conservation cannot be undertaken without the support and participation of local people, and that livelihood concerns and future development goals need to be at the centre of any viable conservation strategy that involves people (see for example, Sunderland and Campbell 2008; Barrett et al. 2005; Arnold 2002; Hulme and Murphree 2001; McShane 2003; Pimbert and Pretty 1995). Optimists maintain that the institutional governance, with wide acceptability among resource appropriators, promotes people's involvement in resource conservation, inculcates mutual sharing of responsibility and accountability, and lessens conflicts (see Vira 1999; Gadgil 1998; Zuhair 1998; Sarin 1996). Furthermore, stress is laid on the need to create 'sustainable livelihoods' in conservation programmes and on bridging the gap between state objective of forest conservation and interests of local populations (see for example, Ellis 2000; Bebbington 1999; Scoones 1998; Carney 1998; Chambers and Conway 1992).

Sceptics, on the other hand, have warned against the optimism associated with the idea of local community involvement. For example, Lele (1991) cautions against the tendency to believe that local participation and social equity automatically ensure environmental sustainability (see also Scott 2006; Gray and Moseley 2005). It is argued that local appropriators may not always enjoy full freedom to have the institutional governance of their choice (see Baland and Platteau 1996; Karlsson 1999; Reardon and Vosti 1995). Supporting this view, Mabee and Hoberg (2006) suggest that, in practice, equality in decision-making is constrained primarily by the structure of statutory authority, leaving little scope for those who are poor and marginalised. Briefly noting the claims and cautions of various scholars on sustainability and participation, I will now turn to explain and analyse the rationale, objective and policy guidelines of the Joint Forest Management Programme (JFM) in J&K.

As discussed in Chap. 6, few attempts for managing forests jointly with the local communities were made by the FD during colonial and postcolonial periods. By the late 1980s, international forest conservation policies started to advocate decentralisation and joint management of forest resources to achieve the goals of sustainability. This global discourse of participatory management of forests also found its way in the national policies on forest conservation. In 1990, the concept of local participation in forest conservation found a firm basis under the JFM when the MoEF, Government of India circulated its guidelines for the protection and development of

146   7 Joint Management of Forests and Split Role of the State: The Politics of Forest...

degraded forest areas. As discussed in the previous chapter, after the dissolution of *panchayats* in 1984, Village Forest Committees (VFCs) were formed to protect degraded forest lands to maintain the link between the Social Forestry Department and local communities. By the time of the JFM, 121 local forest committees had already been constituted in J&K under the Social Forestry programme.[1] However, until the promulgation of the JFM, there were no clear guidelines for the constitution of the VFCs, their roles or responsibilities.

In 1992, the J&K state notification on JFM entitled 'Rehabilitation of Degraded Forests and Village Plantation Rules' was issued and the procedure of constitution and functioning of the VFCs became formalised, hence attaining legal status. This new policy stressed primarily the management of forests for conservation and meeting local needs, and regarded commercial exploitation of forests and revenue generation as secondary objectives. It was held that active participation of local people is vital for regeneration, maintenance and protection of plantations for the afforestation and rehabilitation of degraded forests and wastelands (GoJ&K 1992). Although the JFM programme involved participation in the management and protection of natural forests, it has been applied only to the degraded forests and plantations raised on community lands thus far.[2] This severely limits the scope of enhancing effective local access and control of forest resources through the JFM.

Under the JFM, the FD and village community enter into an agreement to jointly protect and manage forest lands around villages by sharing responsibilities and benefits. The JFM is based on the principle that through local participation, the villagers would get better access to a number of non-timber forest products (NTFPs) and a share in timber revenue in return for their increased responsibility for forest protection. The central MoEF is the funding agency for the various activities under the JFM programme in J&K and the state FD is the main project implementing body. Apart from ensuring the active participation of different sets of actors involved, the FD is also responsible for solving disputes arising out of the management of village forest land.

Under this programme, a Village Forest Committee (VFC) was to be constituted in each forest range for protecting and managing degraded forest lands.[3] Among the various functions, its major roles include assisting the forest staff in rehabilitating the degraded forests, protecting the plantations and preventing any timber thefts,

---

[1] Interview with the Director of Soil Conservation Department, Jammu (22 January, 2007).

[2] By 2006, around 40,000 ha of forest land was being managed jointly by Forest Department and 4861 VFCs in J&K (Pai and Datta 2006).

[3] An adult male or female member of each household residing around degraded forests has a right to become a member of the VFC. It has an executive committee comprising 11 members, including two women and two representatives from Scheduled Caste/Scheduled Tribe/Backward Classes, elected annually by the General Village (Rehabilitation of Degraded Forests) Committee which constitutes all adult village residents. The Executive Committee selects one of its members as president and forester/forest guard acts as secretary. The member-secretary is required to organise at least four meetings of the Executive Committee and two meetings of the General Body in a year, and has also to coordinate various joint forest management activities under the programme (GoJ&K 1992).

7.1 Joint Management of Forests: New Arenas of 'Partnership' and 'Participation'

encroachments, grazing in enclosed area, forest fires or any other damage caused to the plantations by local communities or outsiders. In return, the committee, after approval of the Forester/Range Officer and Divisional Forest Officer, is allowed to collect grass, fodder, dry and fallen wood from the plantation site free of royalty in 'a manner which ensures sustainable yields of such produce from the area' (GoJ&K 1992). Under this programme, the village community was entitled to 25% of the final harvest of the produce of plantation in kind or sale after deducting the costs incurred by the FD to raise plantations. However, it is important to note that the Divisional Forest Officer (DFO) or Range Officer (RO) is authorised to take strong action including termination of the membership against any member who has failed or neglected to perform his function under the mentioned rules.

NGOs may serve as facilitating agencies between the local communities and the FD, and are mainly involved in activities such as disseminating information, organising awareness programmes, providing training for income generation activities and forming local self-help groups.[4] Thus, central and state governments, and NGO actors seem to converge around a position that supports forest conservation and management through the involvement of local communities. This is also a position shared widely by bi- and multilateral agencies, as well as think tanks and consultants.

The initial JFM notification of 1992 was amended in 1999 to make the programme more practical and people-oriented (GoJ&K 1999). Under the revised guidelines, the village was made the basic unit for constituting a VFC instead of the forest range. Also, the number of members in the Executive Committee was raised from 11 to 15, including at least two women, three members from Scheduled Caste/Scheduled Tribe/Backward Caste, a representative of nomads and two members from each village falling under the jurisdiction of *panchayat*, to be nominated by *sarpanch* (*panchayat* president). The VFC elections were now to be held every 3 years as per the revised notification. In order to provide a greater incentive to these communities for their involvement in forest conservation programmes, the benefits for the village communities were raised to 50% in the case of degraded forests and 75% in the case of wastelands. The state officials claimed that the village committee was now made an 'equal partner' of the FD in discharging the various functions as compared to their 'assisting' role in the previous regulation.[5]

By the late 1990s, the central government also began to encourage decentralisation policies with the aim of devolving natural resource management to local institutions. The key rationale behind this shift in management practice was that devolution would increase access of the rural poor to their surrounding forests, enhance the decision-making power of local communities, and also result in an

---

[4] In some other states of India, foreign donors have also been active supporters of the programme, most notably the World Bank and the UK's Department for International Development (DFID). These agencies have adopted the JFM as part of their overall agenda of good governance, decentralisation, poverty alleviation, sustainable development, environmental protection etc. In the state of J&K, however, this association with foreign donors is missing.

[5] Interview with the Director of Soil Conservation Department, Jammu (20 February, 2007).

148　　7 Joint Management of Forests and Split Role of the State: The Politics of Forest...

increased responsibility of villagers to manage natural resources sustainably. Optimists (for example, D'Silva and Nagnath 2002; Guha 2000; Saigal 2000; Corbridge and Jewitt 1997; Lynch and Talbot 1995) suggest that although devolution policies fail to deliver on many of our hopes, they are a better alternative than the centralised management of forests. Sceptics (such as Blaikie 2006; Upreti 2001; Bazaara 2003; Campbell et al. 2001) argue that devolution policies have transferred little or no authority to local forest users and, thus, have had no significant positive impacts on the livelihoods of the poor. It is argued that local institutions set up under devolution have often been accountable to forest departments and other government offices rather than to local people (Ribot 1998) and the possibilities for genuine comanagement have been quite limited (Baland and Platteau, Balland and Platteau 1996). Below, I explain the NAP initiated by the MoEF to further strengthen the JFM by means of devolution of funds to the local forest committees called Joint Forest Management Committees.

7.1.1 National Afforestation Programme: Facilitating JFM Through 'Decentralisation'

In 2000, taking stock of previous JFM projects, the MoEF recognised that for the effective implementation of the JFM and the active participation of local people, there is a need to reduce bottlenecks in the flow of funds from the centre to the JFMCs. Subsequently, in 2002–2003, the MoEF formulated a scheme called the 'National Afforestation Programme' with the aim of facilitating JFM programmes in the country. This scheme was to be implemented through a two-tier decentralised mechanism of Forest Development Agencies or FDAs (at the forest division level) and Joint Forest Management Committees or JFMCs (at the village level) (see Fig. 7.1). The FDAs were required to strengthen the existing JFMCs and create new ones in villages where these committees did not already exist. Notably, in contrast to previous JFM projects, under the NAP, the funds for JFM activities are given directly by the MoEF to the FDAs headed by territorial conservator of forests, bypassing state FD headquarters (especially the offices of the Principal Conservator and the Chief Conservator of Forests). The funds are then released by the FDAs to individual JFMCs. I explain the structure and composition of FDAs and JFMCs later in the section.

The J&K FD started implementing the NAP in 2003.[6] The objectives of the NAP can be broadly classified into four main categories.[7] First, the programme emphasises

---

[6] In 2003, the J&K State Forest Department got approval for initiating FDA projects in 27 divisions. An amount of 20.44 *crore* was released against the total cost of 73.76 *crore* (GoJ&K 2006). By 2005, 31 FDAs, 1070 JFMCs and 47,839 hectares of the state forest area had been covered under the FDA (ibid.). Specifically, the FDA Udhampur was sanctioned 341.91*lakhs* to treat 2250 ha under this scheme from 2002–2003 to 2006–2007 (ibid.).

[7] I have categorised the objectives into four themes. The full list of short-term and long-term objectives can be found in the 'Operational Guidelines for National Afforestation Programme' published by the J&K Forest Department, Jammu (n.d.).

7.1 Joint Management of Forests: New Arenas of 'Partnership' and 'Participation'

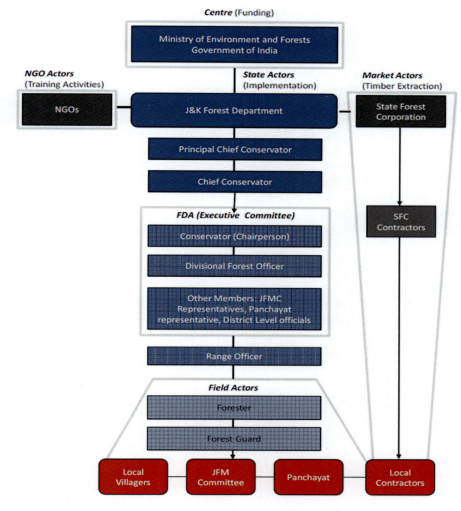

Fig. 7.1 Actors in Forest Administration and Management, J&K

the 'need for protecting and conserving natural resources' by regeneration of degraded forests, and checking land degradation and deforestation. It is held that the first step towards this is securing 'people's participation' in planning and regeneration efforts by forming village level committees so as to ensure both sustainability and equitable distribution of forest products.

Second, an emphasis is laid on 'increasing the availability of fuelwood, fodder and grass' from regenerated areas. It is held that there is a need for 'promotion of fuel saving devices' to encourage efficient use of fuelwood, 'reduce the drudgery of rural women in collecting wood' and protecting the environment. The programme also emphasises 'conservation and improvement of non-timber forest produce' (such as bamboo, cane and medicinal plants) and 'production of other non-timber products' (e.g. wax, honey, fruits and nuts) from regenerated areas.

Third, the programme specifies taking measures to 'improve the quality of life' and 'self-sustenance of people' living in and around forest areas. It proposes to 'undertake skill enhancement programmes' and 'improve the employment opportunities of the rural people' especially the disadvantaged sections such as scheduled castes, scheduled tribes and landless rural labourers. Fourth, 'development and extension of improved technologies' (such as clonal propagation, use of root trainers for raising seedlings, mycorrhizal inoculation) is aimed for increasing productive value of the forests. In addition to this, the programme also aims to 'develop agro-forestry', 'common property resources' and 'water resources' through plantation and water harvesting programmes.

The FDAs are registered as Federation of all JFMCs within forest divisions (territorial) under the Societies Registration Act.[8] The main functions of the FDAs include implementation of centrally sponsored afforestation schemes as well as beneficiary oriented activities such as agro-forestry and avenue plantations. The FDAs are also assigned tasks of 'planning financial outlays' for various activities under the programme, 'assisting JFMCs for microplanning' and 'deciding entry point activities',[9] 'formulating guidelines for utilisation or sharing of usufructs' and monitoring the activities of local NGOs. These agencies are also entrusted to take initiatives for 'marketing the local products', 'creating environmental awareness' and encouraging active participation of local villagers in the protection of forests against illicit felling of timber or forest fires. The JFMCs, on the other hand, are registered with the territorial Conservator of Forests.[10] Assisting the various forest

---

[8] The FDAs comprise a general body and an executive body. The conservator of forests is the chairperson of the general body. Its members include presidents of JFMC general bodies, not more than 50 out of which 20 should be female representatives. One non-official representative is nominated by the *panchayat*. Other members include range forest officers and DFOs. The executive body also has the conservator of forests as the chairperson, and the DFO as the member secretary. Its members include ex-officio members, District Development Officer, District level officers of Agriculture, Rural Development, Animal Husbandry, Soil Conservation, Tribal Welfare, Industries, Public Health & Engineering and Education Department, one non-official representative nominated by *panchayat* and 15 nominees from the JFMCs including a minimum of seven women. The general body is required to meet at least once a year and the executive body once every 3 months.

[9] Entry point activities include creation of community assets such as checking dams for irrigation, digging of wells for drinking water requirements, construction of sheds for schools and community use, installing energy saving or energy alternative devices etc.

[10] The general body of the JFMC includes all adult members of the village according to their willingness to participate. The meetings of the general body are chaired by the president of the committee who is elected after the voting process and holds the position for 1 year. According to the guidelines, a female member is to be elected as president at least once every 3 years. The member secretary of the executive body is the ex-officio member of the general body. The president of the general body also functions as president of the Executive body and forester as member secretary. A treasurer is appointed from the members of the JFMC by the member secretary in consultation with the president. The bank accounts are managed by the treasurer and the member secretary. One member representing the *panchayat* of the area is nominated by the member secretary. There are six other members drawn from the general body who would be nominated by the member secretary in consultation with the president of whom three are to be women.

7.1 Joint Management of Forests: New Arenas of 'Partnership' and 'Participation'

activities, the major roles of the JFMCs include 'preparation of microplans', 'choice of species to be planted', 'proposing physical and financial targets', 'suggesting entry point activities', 'developing usufruct sharing mechanisms' and 'awareness programmes' and 'fund raising activities'.

The JFM guidelines and objectives presented above indicate three key analytical issues. First, as noted in the previous chapter, there was a shift in forest policies from a focus on 'timber extraction and revenue generation' (until the late 1970s) to 'forest conservation and participation' (by the late 1980s). Above, we saw a new shift in forest management strategies based on 'partnership' with the formulation of the JFM in 1992 and 'decentralisation' with the introduction of the NAP in 2002.[11] Pointing to the risks involved in continuous policy experimentation, Mayers and Bass (2004: 136) note the emergence of conflicting signals and uncertainty, and argue that these may not help policy improvement unless the experiments are reviewed in the light of locally agreed sustainable development indicators. Sundar et al. (2001), in the context of the JFM in other parts of India, also observe that changes in the law and in official policy statements may be necessary, but they are by no means sufficient for real changes to take place. Indicating the uncertainties in which such policy changes can result, Menzies (1993), in the context of rural policies in Yunnan (China), argues that the frequent changes led to a belief amongst local communities that governmental policies may change again, taking away or further limiting the terms under which they manage forested land. To put in other words, the constantly changing guidelines and their practical implications (with regard to changes in structure and composition of the JFMCs, flow of funds, execution) create ambiguities for both agents and recipients of development concerning what they will achieve and how.

Second, from the JFM objectives (under the NAP), it can be observed that the stated goals range from 'protection of natural resources' to 'production of non-timber forest produce', or even 'initiating agro-forestry' and 'introduction of improved technologies'. Based on the range of activities to be covered under the programme, there is a wide range of responsibilities entrusted with the FD, most of which overlap with those of the JFMCs, thereby leaving scope for the FD to put the blame on village communities for the failure of projects and vice-versa. Although the JFM implies new partnerships and the NAP aims to form new mechanisms of decentralised control of forests, they fail to provide a focused set of objectives. As Sundar et al. (2001) note, the JFM objectives, while laudable, raise several difficulties for those who might wish to implement them because it is not clear how these can be understood in relation to each other. Furthermore, comanagement programmes, if poorly designed, enable a variety of actors and institutions to promote and advance their individual interests (Bene et al. 2009). Hence, there is a need to prioritise whether the primary goal of the programme is forest conservation,

---

[11] In 2010, the MoEF has again changed the JFM implementation policy by transferring project implementation powers to the *panchayats* in place of the JFMCs. The analysis of this policy shift is beyond the scope of the research on which the book is based.

availability of biomass for local communities, devolution of power, local participation or introduction of income generation activities, without which different actors may prioritise the objectives of the programme according to their own interests and assess its performance accordingly. And each objective, of course, is backed by different interests with varying forms and degrees of power.

Third, it can be noted that benefits in terms of access to non-timber forest products are promised to the local communities in return for their shared responsibility in protecting the forests. However, to avail these benefits, the villagers require a prior approval of both field-staff and the Divisional Forest Officer (DFO) which is to be granted on the basis of satisfactory performance of tasks assigned to them. Furthermore, the field-staff and the DFO can terminate the JFMC membership of a person whose performance is found 'unsatisfactory' (see also Poffenberger and Singh 1996). This reveals that, despite claims of 'equal partnership' and 'decentralisation of authority' in the programme, the forest bureaucracy retains ways for maintaining control over local communities and their access to resources. This observation also corroborates Dean's (1999: 209) argument 'rights to resources are made conditional on the basis of performance' and that very often, governmental strategies go against the interests of their target population. Supporting this view, Sundar (2001) observes how under the JFM, 'gifts' and 'subsidies' are constructed as though state control of forest resources were a timeless social fact. Similarly, Rathore (1996) suggests that conservationists and forest officials, instead of granting resource-use 'rights', prefer 'concessions' since government agencies can easily exercise control over concession grants and withdraw them at will. This reflects that under the guise of decentralisation, there is still a strong coercive element in the project outlines concerning forest management.

To summarise, in this section, I have outlined the factors that can indeed determine the effective implementation of the programme. The broad scope of the NAP, a lack of focus in the objectives, multiple and overlapping roles for the parties, constant shifts and amendments in policies, contradictory imperatives, and avenues for centralised control within decentralised structures can all play a decisive role in the weak implementation of the programme. In some cases these factors can result in ambiguities concerning the responsibilities and rights of the parties involved, lack of interest among the village communities and misuse of power by implementing agencies or actors in the name of 'partnership', 'decentralisation' and 'sustainable use' of forest resources. I provide evidence for the possibilities indicated here in the rest of this chapter and also while presenting the narratives of local communities in the next chapter. Below, I note how the programme is perceived differently by the various actors and in what ways their understandings of the NAP lead to the opening up of new spaces of cooperation as well as conflict.

7.2 Setting the Scene: Interplay Between Centre, State and Non-state Actors

The contemporary global conservation thinking gives an impulse to sustainable forest management by stimulating multi-scale and multi-stakeholder partnerships for conservation and sustainable management. These changes are raising new questions about the linkages between actors at multiple scales (for example, see Ros-Tonen et al. 2005), warrant a 'rigorous analysis of stakeholders' (Njogu 2005: 293), and require an understanding of how to achieve a consensus between stakeholders with diverging interests in forest management and planning (see Berg and Biesbrouck 2005). Li (2007) argues that community forest management is an 'assemblage' that has been created due to struggles between villagers and forest bureaucracies over the access to village forests. One of the greatest challenges facing conservation, therefore, is to be able to engage with a broader set of stakeholders and civil society at various levels (see Kumar 2006). As Greenberg and Park (1994) argue, it is not enough to focus on local cultural dynamics, rather the relationship between policy and politics in the context of environment needs to be explicitly addressed to understand relative powers at many levels of environmental and ecological analysis. Supporting this, Leach et al. (1999) maintain that just as power relations pervade the institutional dynamics of everyday resource use, they pervade any negotiation process and, therefore, different social actors have very different capacities to voice and stake their claims. It is, hence, important to examine how different actors set stages for politics according to their hierarchical powers, varied perceptions and conflicting interests.

In this section, I discuss the power and interests of central, provincial and non-state actors involved in forest conservation and exploitation in J&K, and examine the interplay between them. First, I present the perceptions of forest officials with regard to the JFM and the NAP, and examine how the FD is redefining its role under the premise of 'equal partnership' with the JFMCs. I, then, present the role of the SFC and its association with the FD. Following this, I highlight the concerns of the central MoEF regarding the implementation of the NAP and its relationship with J&K FD. I also briefly outline issues of conflict between the FD and a local NGO called Almi Khudai Khidmatgar Association (AKKA) working in the field of community forest management in the Jammu region.

7.2.1 FD and JFMCs

Since the early 1990s, various working plans and reports of the FD have emphasised the significance of forest conservation both for maintaining ecological balance in the state and providing sustainable livelihoods to forest dependent communities. The agenda of the FD, therefore, has largely coincided with that of international conservation community in preserving natural resources by involving local populations. The various senior forest officials such as the Principal Chief Conservator of Forests (PCCF), the Chief Conservator of Forests (CCF), the Conservator of Forests (CF) and the Divisional Forest Officer (DFO) held the view that since the economy of the state is primarily based on agriculture and forest related activities, the forests serve as an important source of livelihoods for the majority of its people.[12] According to these officials, the management of forests in the state should rest on the principle of 'sustainability' despite several challenges, including constantly rising human and livestock population, their increasing dependence on natural resources for sustenance and high demand of fuelwood and timber in the state, owing to harsh winter conditions. They maintained that the state is looking for ways to conserve its forests as well as provide livelihoods to its people through various forest programmes. Explaining the significance of local participation, the CF, Jammu stated:

> The agenda of forest conservation cannot be realised without the active involvement of the communities. Although afforestation schemes in the state since independence also emphasised the role of people's involvement in forestry programmes, after the 1980s, various central government initiated projects such as the Social Forestry, the Joint Forest Management and the National Afforestation Programme have made the involvement of local communities mandatory [...] We are trying to reach out to the communities for the successful implementation of the NAP by forming the JFMCs and initiating various developmental activities in the villages.[13]

On the NAP, the latest programme for decentralised forest conservation, the forest officials maintained that the local communities have been empowered by means of devolution of funds. They argued that the structure of the FDA ensures 'equal partnership' and 'participation' of forest staff and local communities at various stages of the programme. For example, the FDAs are supposed to prepare a project plan specifying the financial outlay, area to be covered and range of activities to be undertaken. These plans are then sent to the National Afforestation and Eco-Development Board (NAEB) of the MoEF for the release of funds. After the project is approved by the MoEF, the CF as chairperson of the FDA receives the first grant from the MoEF for enabling the FDAs to undertake microplanning in consultation with the JFMCs and local communities. The aim of microplanning is to link the objectives of the programme with local needs. The villagers are meant to propose entry point activities and participate in rehabilitation plan of the forests by indicating areas most suitable for various land uses, for example, afforestation, fuelwood and fodder plantations, horticulture, and development of pasture lands. According to the NAP guidelines, the next instalment of funds is linked to the satisfactory implementation of the work programme and utilisation of previous funds. The funds released by the MoEF for the implementation of the work programme are transferred

---

[12] The information presented in the following discussion is based on my interviews with various senior forest officials in Jammu from February to April, 2007. Some of the points presented here were common to the interviews with different officials and, therefore, I do not provide specific references while noting them.

[13] Interview with the CF, East Circle, Jammu (22 March, 2007).

7.2 Setting the Scene: Interplay Between Centre, State and Non-state Actors

to the account of the JFMCs (jointly managed by the JFMC chairperson and forester) after receiving receipt from the FDA about the previous utilisation of funds.

A Memorandum of Understanding (MoU) is signed by the FDAs with the JFMCs which outlines the mutual obligations and rights of both the partners. The MoU grants right to the FDAs to withdraw funding from a JFMC if its performance is found to be unsatisfactory. It may be mentioned that under this scheme, contractors or any intermediate agencies are not permitted to be engaged for execution of any of its works so that the full benefit of wages goes to the village residents. The FD maintains that the programme, besides addressing conservation concerns, also empowers local communities by generating employment and building capacities.

The forest officials state that the JFM programme has been institutionalised successfully in J&K despite the initial challenges to its implementation in the 1990s due to terrorism in the state. In Jammu region, the programme is claimed by the forest officials to be most successful in Udhampur and Nowshera forest divisions.[14] The officials associate high dependence of local people on the surrounding forests with the active 'participation' of the villagers in the programmes. 'Local participation' (although difficult to measure and quantify) is considered as the key criteria for the success of the programme by the officials.[15] It is important to note that their understanding of 'success' was based on the level of 'local participation'. For example, the DFO explained to me that when a training camp was organised at Udhampur to demonstrate various income generation activities, most of the participants who attended the camp were from Navni and Chinnora villages.

On the basis of my interviews with senior forest officials, below I note some of the claims made by them suggesting the successful implementation of the programme. According to the PCCF,[16] there is an increase in the forest cover since the JFM programme although the state has experienced a fall in forest revenue in the last two decades due to restrictions on timber extraction and trade.[17] He observed that although the programme has not successfully met the timber needs of the local communities, the availability of biomass, especially firewood and fodder, has certainly increased over the years due to establishment of closures (fencing of pasture lands and degraded forest areas). The villagers are allowed to collect grass and fuelwood from open forests and are supplied timber at concessional rates. According to him, the FD is now more open to incorporating the suggestions of local

---

[14] Interviews with the PCCF, J&K, Jammu (12 February, 2007) and the DFO, Udhampur (3 February, 2007).

[15] As stated in chapter two, due to the problem of militancy in Nowshera, I chose to examine the functioning of the NAP in Udhampur. I selected Navni and Chinnora villages (under FDA Udhampur) for my study because in these villages, the programme was considered most successful by the DFO.

[16] Interview with the PCCF, Jammu (12 February, 2007).

[17] In 1980–1981, the total forest area of J&K was 20,174 sq. kms, which increased to 20,230 sq. kms. by 2005 (GoJ&K 2005: 29). In 1980–1981, the J&K FD generated a net surplus of 2221.97 *lakh* rupees which reduced to 226.59 *lakh* rupees by 1991–1992. By 2004–2005, the FD was generating a net loss of 6901.34 *lakh* rupees which means that the gross revenue was much less than the total expenditure (ibid.: 200).

156   7 Joint Management of Forests and Split Role of the State: The Politics of Forest…

communities with regard to access to resources. For example, the PCCF informed me that on the demand of the local communities, they were allowed free access to non-timber forest produce such as anardana (*Punica granatum*)[18] which was earlier collected only by the contractors after the issuance of permits by the FD.

The forest officials view that the NAP has resulted in cordial relations between the staff and the villagers. As the CF argued, 'the forest administration approach has changed from pure policing to people friendly'.[19] He further added that the villagers are consulted in various planning issues such as the choice of species to be planted for regeneration, choice of area for entry point activities, and distribution of benefits and responsibilities. The officials maintained that the programme has helped in controlling the illegal timber trade significantly, which was earlier posing a big threat to the state forests. They also claimed that owing to the cordial relations between staff and local communities, villagers now complain to the forest authorities immediately whenever any damage to forest crops occurs due to illegal felling by timber smugglers or by the residents of neighbouring villages. It is pertinent to mention that these are the claims made by the senior forest officials. I briefly scrutinise some of these claims in Sect. 7.3 of this chapter and more fully in the next chapter, based primarily on the experiences and narratives of the village communities.

While recognising the importance of community involvement in forest protection and management, the FD also sees the forest citizens as the main destructors of state forests. Illicit tree felling by local residents, lopping and overgrazing by pastoralists (such as *bakerwals* and *gujjars*), forest fires, and encroachments on forest lands by villagers in connivance with the revenue officials are some of the issues that the FD considers as major challenges to forest regeneration. The forest officials stressed the need of *pucca* (concrete) boundary pillars for demarcating forest lands in order to prevent encroachments by the local village residents (after bribing the local revenue officials). The FD maintained that there is tremendous pressure on forests for meeting the demands of fuelwood and fodder, and that the nominal grazing fee or *kahcharai* provides incentives for overexploitation of forest resources by pastoralists and other village residents (GoJ&K 1993). I shall return to these claims while presenting the contradictory narratives of village residents and pastoralists in the next chapter.

On being questioned about the effectiveness of the NAP, the DFO, Udhampur responded that the FD takes into account the local needs while planning entry point activities.[20] She further stressed that the interests of weaker sections such as scheduled castes, scheduled tribes and women are represented in the formation of the JFMCs. She claimed that JFMC meetings are organised by foresters regularly in which the local concerns for forest protection and development are addressed.[21] By

---

[18] Wild pomegranate seeds used as spice in Indian cuisines.

[19] Interview with the CF, East Circle, Jammu (22 March, 2007).

[20] Interview with the DFO, Udhampur (3 February, 2007).

[21] The DFO informed me that there are supposed to be two meetings of the JFMCs in a year. Other senior officials were, however, not sure of the frequency of JFMC meetings in a year. It appeared to me that they had confined themselves to the administrative tasks in the head-offices and rarely made field visits.

7.2 Setting the Scene: Interplay Between Centre, State and Non-state Actors

means of various training programmes and workshops, the villagers are demonstrated the ways to develop alternative sources of income generation (such as mushroom cultivation, bee-keeping or fruit preservation) in order to reduce their dependence on forests.

These senior forest officials held the view that the NAP is at the infancy stage and issues such as employment generation for local villagers or creating markets for their local products are yet to be tested. They maintained that although some self-help groups have been formed in the villages as an attempt to make village communities self-sufficient, these committees require time to become strengthened and institutionalised. To what extent the creation of self-help groups, alternative income generation activities, local participation and other developmental activities have benefited the villagers will be assessed in the following chapter. Suffice here is to say that the FD officials subscribe to the conservation discourse that supplicates technocratic and apolitical solutions (such as creation of user committees and self-help groups) to deeply political problems of access and control of local forest resources and their multiple uses.

Recall the point made above by the senior forest officials regarding cordial relations between the FD and village communities under the NAP. However, they also provided a contradictory narrative while arguing that the implementation of the programme is challenging because 'it is very difficult to change the mindset of the field-staff and local villagers'.[22] They pointed out that on the one hand, the field-staff is unwilling to shed its power and develop friendly relations with the local communities and on the other, the local people are reluctant to trust them owing to their previous experiences with forest administration based on policing and curtailment of local access to forest resources. The CF argued that the field-staff lack motivation as well as adequate training to implement these policies which are meant to be participatory in nature.[23] The Conservator also explained that the FD is understaffed, particularly at field level, which results in unnecessary delays in project implementation. According to him, foresters and guards should preferably be local residents of the area so that a good rapport could be established between the staff and local villagers. He also suggested that there are very few NGOs or grassroots organisations in J&K to support community mobilisation activities and self-help group formation. Although NGOs such as the AKKA are working with the FD, their primary job is to train the officials and local people in income generation activities. I shall return to this point while discussing the relationship of NGOs and FD later in this section.

According to the DFO Udhampur, the villagers are less interested in forest conservation *per se* and they see the programme as just another development project. The entry point activities under the programme do provide scope for undertaking development activities (such as construction of sheds for schools, check dams for soil and water conservation, digging of wells for drinking water), but only 10% of

---

[22] Interview with the CCF, Jammu (4 March, 2007).

[23] Interview with the CF, East Circle, Jammu (22 March, 2007).

158   7   Joint Management of Forests and Split Role of the State: The Politics of Forest...

the total project funds could be spent on such activities. This prevents villagers from feeling motivated and actively participating in the programme. Another problem in project implementation, according to the DFO, is that of political rivalries at the village level. If the chairperson of the JFMC and the *sarpanch* of village *panchayat* are affiliated to different political parties, the project is likely to face problems due to intra-village conflicts and rifts. Apart from this, explaining some bureaucratic complexities in the project implementation, the DFO further informed me:

> We are completely dependent on the central government and the SFC for funds and revenue. The delay in the release of money by the Ministry is a major hindrance for project implementation in the state. When I joined the Department, I applied for funds to the Conservator of Forests for organising training camps for the villagers in my division. I was asked to produce utilisation certificate of the funds from the previous year. The earlier funds were not spent by the former DFO and hence lapsed. I could not produce the utilisation certificate and was unable to obtain the next instalment on time. As soon as I got the money, I arranged a three day camp in Udhampur which should have been organised months before […] My tenure here is just for three years. Until the time I understand the challenges in the project implementation, I will get transferred to a new department.[24]

7.2.2   SFC and FD

As noted in the previous chapter, since 1978, the responsibility of extraction of timber has been entrusted with the SFC, in return for the royalty paid to the FD. This royalty is one of the main sources of revenue for the FD. A significant change occurred in the 1990s because of the restriction on green felling in the state. The Supreme Court of India in 1996 ordered the J&K state to impose a complete ban on green felling and restrict the annual timber extraction to a maximum of 80 *lakh* cubic feet (cft.) in selected compartments.[25] This has substantially reduced the quantity of timber extraction. For instance, in 1989–1990, the SFC extracted 113.57 *lakh* cft. of timber which got reduced to 76.67 *lakh* cft. in 1998–1999 and merely 27.16 *lakh* cft. in 2004–2005 (GoJ&K 2005). The SFC was now allowed only to access dry, fallen and diseased timber in the state forests. Also, during the early 1990s when terrorism was at its peak, the timber extraction activities got hampered and militancy related fire incidents resulted in low returns for the investments made by the SFC.

The list of compartments from which timber can be extracted is made by the FD and the SFC jointly and is sent to higher authorities, the CCF of FD and the Managing Director, SFC for approval. After permission is granted by these authorities, marking is done by the FD following which the SFC contractor, the DFO and field-staff go to the area and supervise the labour for activities until the logs are carried to roads and transported to sales depots within the state.

The SFC claims that apart from transacting timber business in the state, the corporation has benefited the government largely by paying royalty which, in turn, is

---

[24] Interview with the DFO, Udhampur (3 February, 2007).

[25] Honourable Supreme Court Judgement in Writ Petition Number 202/95.

used for various forest development activities. The SFC has been supplying logs free of charge to the FD both for providing timber to the villagers at concessional rates and for meeting their firewood requirements. Moreover, the SFC maintains that they provide employment to the local villagers under their various timber extraction activities.[26] The SFC carries out timber extraction activities through selected contractors who, in turn, employ local contractors and villagers. The Corporation claims that the organisation has been contributing to conserving and developing forests of the state by removing and disposing the diseased and fallen logs in the compartments, thus maintaining the forest hygiene. The SFC also maintains a Forest Development Fund which finances development projects in the forestry sector and has also contributed to funding FD programmes in the state. I explain the activities and functioning of the SFC at the ground level later in this chapter (Sect. 7.3) but for now, it is important to note some of the areas of conflict between the FD and the SFC.

The SFC argues that although the Supreme Court had restricted the annual timber extraction to 80 *lakh* cft., the FD, in order to meet conservation needs, allows it to extract much less than the prescribed limit. For example, in 2007–2008, the SFC has been allowed to extract only 35 *lakh* cft. by the FD. Moreover, the SFC officials claimed that they have to face constant interference of the FD concerning selection of species to be extracted, quality of stocks, techniques for timber felling etc. They argued that, for all practical purposes, the SFC working remains under the check of the FD leaving little autonomy for the corporation to carry out its business. Contrary to the claims of the SFC, the DFO, Udhampur narrated:

> For the last three years, the SFC is making a loss and has failed to pay the due royalties to the FD. The establishment of the SFC has created unnecessary problems with regard to timber extraction and sale, for example, difficulties in inter-organisational coordination, duplication of work, delays in operations etc. [...] Until 1978, the FD used to sell timber and other minor timber forest products at subsidised rates but the SFC has kept their rates very high for its own profits.[27]

7.2.3 MoEF and FD

The central MoEF is the main sponsoring agency for the JFM projects under the NAP in J&K.[28] Explaining the status of the JFM in the country, the Deputy Inspector General of Forests (DIG), MoEF argued that the relationship between the FDs and local communities has improved to some extent after the JFM.[29] However, she emphasised that more time is needed to develop the capacities of the JFMCs before

---

[26] Interview with the Deputy Financial Advisor, SFC, Jammu (28 March, 2007).

[27] Interview with the DFO, Udhampur (3 February, 2007).

[28] In some other states of India (e.g. Rajasthan, West Bengal, Madhya Pradesh and Haryana), international development agencies such as the World Bank and the Japan Bank for International Cooperation are the main sponsoring agencies for the JFM.

[29] Interview with the DIG, MoEF, New Delhi (15 January, 2007).

giving complete control to the villagers to manage forests surrounding their villages. Owing to the previous experience of lack of financial powers to the village forest committees in earlier JFM projects, she argued that the MoEF in 2002 initiated the NAP for further decentralising forest management activities by directly transferring funds to the FDAs and eventually to the JFMCs, thus aiming to reduce the bottlenecks created by the state forest departments. The major constraint faced by the MoEF in implementing FDA projects is the lack of follow up action by the J&K FD which is supposed to submit regular evaluation reports to the MoEF. As the DIG, MoEF stated:

> It is generally experienced that the FDAs do not submit progress reports in time but request for the release of funds. In J&K, less than 30% of FDAs applied for release of funds in 2003–04 and no evaluation report has been submitted by the FD thus far. It is difficult for us to release grants without receiving progress reports for the previous activities [...] The problem is not with availability of funds for the project but until the J&K FD shows its serious concern, the Ministry cannot ensure effective implementation of the programme.

There is another area of tension between the central government and J&K state with regard to the issue of power and control *within* the state FD. The Indian Forest Service (IFS) officials of J&K cadre, who are not domiciles of the state, experience marginalisation within the FD, owing to the antagonistic relations between the Indian government and the state of J&K since independence.[30] The effective power is generally held by officials of the J&K State Forest Service, even though these officials are junior in terms of departmental hierarchy. Requesting anonymity, a senior IFS officer informed me:

> In J&K, there is an intense rift between the local officials and outsiders within the FD. Owing to the special status of J&K, most officers of central government, in real practice, have only de jure powers while junior forest officials from J&K who are well-connected to the politicians in Kashmir valley hold de facto powers. It is they who actually control the functioning of the FD.

7.2.4   NGOs and FD

Unlike other states of India, the presence of non-governmental development organisations is very limited in J&K, possibly due to the political instability, militancy, and cross-border conflicts in the state since independence. Currently, a local NGO called AKKA based in Jammu has marginal presence in the area of community forest management and conservation programmes. Specifically, the AKKA in association with the FD has organised training camps for the villagers on various income

---

[30] Recall that at the time of the merger of J&K princely state with the Indian Union in 1947, the J&K state was to enjoy full autonomy except in the fields of defence, foreign affairs and communications. Over the years, the autonomy promised liquidated and the central government started to intervene in all areas of governance including forests. As such, the forest officials of J&K have, often, adopted a hostile attitude towards their superior colleagues coming from the Indian Forest Service.

7.2 Setting the Scene: Interplay Between Centre, State and Non-state Actors 161

generation activities and creation of self-help groups in Udhampur.[31] While a member of the AKKA stated that the organisation's function is to train local communities in alternative income generation activities (such as mushroom cultivation, production of jams, pickles and dried fruits), it is beyond their capacities to assist them in marketing these products.[32] Arguably, lack of market opportunities renders such training activities futile and useless for meeting the livelihood needs of the villagers.

On the question of their experience of working with the FD, the members of the AKKA argued that the collaboration has not been encouraging thus far. In any joint project, the FD takes all the credit for the work done and the NGO faces unnecessary delays in receiving funds from the FD. They explained the uncertainty involved in working with the FD since, in many cases, the official gets transferred before the programme is completed, and hence, the project is left mid-way through.

Noting problems of collaboration between the FDs and NGOs in other parts of India, Sundar et al. (2001) argue that an important reason why it is hard for FDs to work with NGOs is that both organisations have very different overall agendas i.e. NGOs more interested in local community mobilisation and empowerment, and FDs primarily concerned with forest conservation (see also Levine 2002; Mahanty 2002 and Mencher 1999). Further, Sen (1999) argues that in India, there have been several instances of delays in funding by the state and stoppages of development projects due to 'whims of bureaucrats'. As Farrington and Bebbington (1993) observe, depending on their histories working with the FD, some NGOs will be more while others less reluctant to enter into partnerships with government. It is yet to be seen how the NGO sector in J&K will expand in the years to come.

On the basis of discussion in this section, it can be argued that although various actors involved in forest conservation seem to converge around the common goals of the 'sustainable use' of forests, 'local participation' and 'decentralisation of authority', there are areas of tension between them with regard to organisational autonomy, power and control of funds. Scholars such as Blaikie (2006); Oyono (2004); Bazaara (2003); Muhereza (2003); Ribot (2002); Campbell et al. (2001) and Fomete (2001), in different country contexts, argue that the main impediment to decentralised forest management is that the governments rarely devolve actual powers which, often, result in repeated *re-centralisation* in relation to power, authority and control over decision-making and access to local resources. In the narratives above, although the MoEF maintained that the objective of decentralisation is being realised through new NAP guidelines, we witnessed that there is still centralised control over funds. The middle rank officers such as the DFO also struggle to secure funds on time from the FD headquarters due to red-tapism, so typical of Indian bureaucracy. We also noted that while the state officials claimed that they ensure 'equal partnership' of local communities in the implementation of programme and devolution of funds to the JFMCs, there is scope for withdrawal of funds by the FD

---

[31] I attended one such camp organised by the AKKA for income generation activities in village Chinnora. I present my observations on it in the next chapter.

[32] Interview with a member of the AKKA, Jammu (6 February, 2007).

162   7   Joint Management of Forests and Split Role of the State: The Politics of Forest...

if the performance of committees is found unsatisfactory, again indicating the presence of centralised control within apparently decentralised structures.

We also witnessed the issues of conflict between the FD and the SFC over autonomous functioning. Apart from this, the rift between central and state forest officials in terms of de facto powers could also be seen as a significant factor in the implementation of the programme. While claiming that the programme has resulted in cordial relations between staff and local communities, we observed contradictory narrative when the senior forest officials argued that the field-staff lacks motivation to shed power and make the programme more participatory. It is also pertinent to examine the attitude of the field-staff towards senior forest officials, a point which I illustrate in the next section.

7.3   The Politics of Forest Resource Control

So far, we have observed that NAP policy regulations and laws can set the stage for politics between the various players involved. I have noted mechanisms of centralised control at different levels from the MoEF to the FD headquarters to the divisional level. How these policies unfold at field level, creating new avenues for cooperation and conflict, is the main focus of the discussion that follows in the rest of this chapter and the next. Below, I examine the micropolitics of resource access and control in the villages Navni and Chinnora, acclaimed by the DFO (Udhampur) for being the most successful example of JFM under the NAP. Specifically, I explain the functioning of contractors involved in timber extraction and the relationship between the forest guard and villagers with regard to access to local forests.

7.3.1   Navni and Chinnora: A Brief Introduction

Navni and Chinnora, the twin villages, are located in the Udhampur district of the Jammu region. These villages are quite remote, the district headquarters being approximately 90 km away.[33] The villages fall under the Navni-Dudu forest range of the Udhampur forest division (see footnote 8 in Chap. 2 for demographic composition of the two villages).

Before the land reforms in 1971 (discussed in Chap. 2), major landholdings belonged to *Thakurs* and *Brahmins* who used to employ *Gaddis*, poor Muslims and lower castes for tilling their lands but after 1971, lands were redistributed to the actual tillers. As such, *Gaddis*, Scheduled Castes and Muslims have been the major beneficiaries of the land reforms in these villages, although the upper castes have continued to maintain their privileged position in terms of status hierarchy and also

---

[33] It takes about seven to 8 h by road to reach Navni and Chinnora from Udhampur due to mountainous track and poor condition of road.

7.3 The Politics of Forest Resource Control

by diversifying their sources of income, with the exception of agriculture (e.g. government jobs, shop-keeping, construction etc.). The literacy rate is very low in these two villages i.e. 29.3% in Navni and 22.9% in Chinnora.

Agriculture is the primary occupation of the majority of the population. Paddy, maize and mustard are the major crops grown here. The people of these villages are also engaged in some handloom and handicraft activities such as blanket weaving and basket making. Some others work as pastoralists, labourers, carpenters, potters, shopkeepers, timber contractors etc. Firewood is the main source of fuel for most of the villagers. Some women are engaged in collecting and selling firewood in the village while others collect non-timber forest products and herbs such as vanafsha (*Viola odorata*), guchchi (*Morchella esculenta*), doop (*Jurinea macrocephala*), kaur (*Picorrhiza kurroa*), patis (*Aconitum heterophyllum*), kasrorh (*Diplazium frondosum*), kundian (*Mushroom rhizo*) and anardana (*Punica granatum*), which are either used for self-consumption or sold in the local market for additional incomes. Major trees found in the forests surrounding these villages include deodar (*Cedres deodara*), banj (*Quercus incana*), keekar (*Acacia eburnea*), champ (*Alnus nitida*), maru (*Quercus dilatada*), akhrot (*Juglans regia*), kathi (*Indigofera heterantha*) and kaimbal (*Lannaea grandis*). Due to harsh winter conditions in the region, the inhabitants of these villages are heavily dependent on the surrounding forests for timber and fuelwood apart from fodder and other NTFPs.

After independence, the first *panchayat* elections in Navni were held in 1978. Due to terrorism in the state from the late 1980s onwards, *panchayats* were suspended and revived only in 2001 when the political climate in the state began to change. Navni and Chinnora villages were considered to be one for revenue purposes until 2000, but after the elections in 2001, two separate *panchayats* of Navni and Chinnora were formed. Both Navni and Chinnora fall under the FDA Udhampur, and had a common JFMC from 2003 to 2007. As indicated above, the forests of Navni and Chinnora, apart from generating revenue for the state are also very crucial for meeting subsistence needs of the villagers. But who actually controls the forest resources, especially timber, and regulates its access is what I explain below.

7.3.2 Our Forests, Their Timber

In the previous section, I pointed to some issues of conflict between the SFC and the FD with regard to organisational autonomy, payment of royalties, selection of compartments, timber pricing etc. In the following discussion, I highlight the politics between the SFC and local contractors in relation to control of forest resources and their exploitation.

For the purpose of timber extraction, an agreement is made between the FD and the SFC, permits are granted and compartments for timber extraction are assigned to the SFC contractors. These SFC contractors are mainly based in urban areas,

Jammu and Udhampur. The first job of the SFC contractor is to employ a local contractor (known as a *munshi*) from a neighbouring village who, in turn, employs labourers for various timber extraction tasks such as felling trees (or *gharan*), sawing (or *charan*) and carrying logs to roads to be finally transported to the sales depots.

Contrary to the claims of the senior forest officials that the illegal timber trade is largely under control, my interviews with SFC contractors, local contractors and villagers in Navni and Chinnora suggested that it is very much prevalent, a point also covered in the international media.[34] The SFC contractors blamed local contractors for engaging in illegal activities and the destruction of forests in connivance with the field-staff.[35] These contractors argued that they work within a transparent system based on tenders and auctions, and extract timber as per regulations laid by the SFC, but it was the local contractors who were the main culprits in violating forest laws and regulations. For instance, they explained, in the 1980s, the government allowed village residents (by issuing permits) to transfer logs from their old houses to the places where they had built new constructions. This policy made vigilance of timber transportation very challenging as it became difficult to ascertain where the timber was coming from and going to. Amidst this change in government policy which was basically meant to help villagers, local contractors started purchasing timber logs from villagers and selling them in the black market, making high profits. Misusing the permits to transport timber, they also started felling deodar wood illegally from neighbouring forests and selling it to timber traders in towns. Due to such illegal activities, the government again imposed strict regulations and stopped issuing permits to the local residents for timber transportation. Following this, the timber extraction and transportation was brought completely under the supervision of the SFC contractors who held exclusive rights for this purpose.

Ramesh Chander is an SFC contractor who currently resides in Udhampur town. He was the *sarpanch* of Dudu village in 2001 and, interestingly, has also served as the JFMC chairperson in the past. He has been engaged in the timber extraction business for the last 10 years. Supporting the point made above with regard to the role of local contractors in illegal activities, he narrated:

> The local contractors adopt various strategies to exploit forest resources illegally. The local contractors make an agreement with individual landowners for tree felling and selling. Generally, the landowner bribes the field-staff and gets his permission to cut more trees than he actually has. The local contractor then cuts the trees from his land and few others from

---

[34] Newspaper reports suggest that the illegal timber trade has flourished in the last decade or so due to a boom in construction activities in the state. For example, a report in a leading British daily, The Guardian (2010), suggests that corruption hinders moves to halt the illegal trade fuelled by the construction boom and notes, 'the insurgency that has wracked the region for two decades is at a low ebb and an economic boom, in part fuelled by cash poured in by central government to win hearts and minds, has meant a voracious appetite for wood for new homes, hotels and other constructions'.

[35] Interviews with SFC contractors Krishan Lal, Jammu (22 November, 2008) and Ramesh Chander, Udhampur (20 March, 2007).

7.3 The Politics of Forest Resource Control

the nearby forest land in collaboration with the field-staff. The timber is then sold in the black market and the profits shared by the contractor, landowner and field-staff.

Om Prakash is a local contractor in Navni and has been associated with timber extraction activities since 1975. He explained that SFC contractors visit the field occasionally and, hence, the responsibility of carrying out various timber extraction activities remains with the local contractors who work under the supervision of the FD and SFC field-staff. He informed me that although he had tried to gain a direct association with the SFC, he was unsuccessful since it was difficult to get SFC contracts without good financial backing. He mentioned that the SFC sanctions timber extraction permits only to those contractors who own some property, have substantial bank savings and can deposit a fixed amount as security to the SFC (generally 2% of the project cost). Most village residents are not so affluent and, therefore, unable to become SFC contractors. They are bound to work under SFC contractors who come from outside villages.

Contrary to the argument presented by the SFC contractor above, Om Prakash argued that it is actually the SFC in connivance with their contractors which is destroying the forests. For their convenience (and also reducing transportation costs), the SFC contractors bribe the forest officials and cut the green trees in the forests surrounding villages rather than those which have been marked by the FD in distant forest compartments. He maintained that due to snowfall, trees in the upper reaches suffer damage and need to be cleared for the growth of young plants. Whilst the task of removal of fallen wood is entrusted to the SFC, the SFC officials hardly make an effort to engage in these activities in distant regions or upper reaches.[36] Not only can the fallen logs fulfil the demands of timber for villagers in Navni and Chinnora, but these can also be used for making coal, furniture or firewood. In the absence of any efforts to maintain forest hygiene, these potentially valuable woods get wasted because they are eventually burnt by pastoralists who go up the mountains during the summer to graze their cattle. He argued that if the task of forest hygiene is assigned to local contractors, it can prove to be beneficial both for the maintenance of forests as well as meeting subsistence needs of the villagers. Highlighting the indifference of SFC and FD field-staff in dealing with the problems of poor villagers, he commented:

> These are our forests, yet they [SFC] own the timber. The irony is that even if kail or any other tree is damaged in the surrounding forests due to snowfall, the field-staff let it spoil but do not allow it to be taken away by the poor for meeting their subsistence requirements. They say that it cannot be supplied to anyone because it is government property. So, most of the time, this fallen wood is neither used by the government nor the people [...] Such fallen trees even kill the small plants. This is how the rich forests of the state, which could be used for local sustenance, livelihoods and revenue generation, get destroyed due to the carelessness of forest authorities and the SFC.

---

[36] The main species in the upper reaches include deodar (*Cedres deodara*), spruce (*Picea morinda*), tos (*Also picea*), kail (*Pinus exelsea*), kheeru (*Acacia catechu*), deri (*Toona serrata*), kharari (*Celtus australis*), samel (*Bambax ceiba*), madrala (*Sorbaria tomentosa*) and padar (*Pavetta indica*).

7.3.3 Split Role of the Field-Staff: Forest Regulations vis-a-vis Local Needs

Long and Van der Ploeg (1989) rightly suggest that it is important to focus upon intervention practices as they evolve and are shaped by the struggles between various participants rather than simply on intervention models (see also Bardach 2000; Liddle 1992).[37] Lipsky (1980) argues that implementers or 'street-level bureaucrats' face an ongoing duality between being responsive to their clients' needs and ensuring policies are properly implemented. The forest field-staff, as a street-level bureaucrat, is an important link between the FD and local villagers and plays a key role in the implementation of forest programmes. Although extensions of a hierarchical chain of command, they can also act as independent agents with their decisions reflecting their own personal circumstances (Vira 1999). The discretionary power that forest guards exert in recording forest offences, deciding the amount of fines and recommending applications for timber concessions, provides them with a significant amount of power and authority over forest dependent villagers (Vasan 2006). Emphasising the need to analyse the roles and experiences of lower level government workers, Corbridge et al. (2005: 7) argue that the ways 'technologies of rule' actually work at local levels depends on the manner in which 'they are interpreted by field officials and seized upon, understood, reworked and possibly contested' by differently placed social groups. Moreover, these officials are often local residents and are involved in complex relations of sociality and reciprocity (Agrawal 2005). Since the separation between private and professional relationships of field-level staff fails in Indian rural society, their identity is shaped by both their professional role as implementers of forest policy and also by their role as villagers in rural social networks (Corbridge et al. 2005). This presents a dilemma concerning the social roles and identities that a forest guard negotiates. Such a dilemma is well expressed by a forest official in Karnataka in the following words: 'I think it is too hard to do both policing which requires terrorising villagers, and eco-development projects which requires being friendly' (Mahanty 2002: 3761). Below, I note the dilemmas of the field-staff in Navni and Chinnora villages and highlight the *split role* of the Forester and Forest Guard in the protection and conservation of forests.

The Forester and Forest Guard in Navni emphasised the need of conserving forests for maintaining ecological balance as well as meeting the biomass requirements of the local people. They argued that fuelwood and fodder security would motivate the people to participate in various forest development activities under the NAP, and maintained that the formation of decentralised bodies such as the JFMCs and devolution of funds under the FDA has positively contributed to bridging their gap with the local communities.[38]

---

[37] For a bibliography on policy process, see Mooij and de Vos (Mooij and de Vos 2003).

[38] It is pertinent to note that the Forester along with the JFMC Chairperson effectively control the project funds at the village level.

7.3 The Politics of Forest Resource Control

On the question of plantation activities under the NAP, they argued that the FD has established nurseries and planted trees such as deodar, keekar, akhrot, maru and banj on the degraded forest lands. According to them, the choice of the species is made in consultation with the JFMC considering the needs of the local people. For example, the Forester argued that deodar is very useful to melt snow and its timber is durable for construction purposes, keekar prevents soil erosion and also provides firewood and fodder, and maru and banj are both used for the supply of fuelwood. I shall deal with the issues of people's participation in selecting species and their access to local forests in the next chapter. In this section, I confine to presenting the perspective of the field-staff.

According to the Forest Guard, he is under constant pressure from senior authorities to check illegal felling. Being native to Navni, it is difficult to cater to the local needs of timber and fuelwood and, at the same time, adhere to the forest regulations that restrict local people's access to forest resources. In the following narrative, the Forest Guard, explained the dilemmas he faces with regard to the demand of timber and firewood in the village:

> The forest laws are not in consonance with the needs of villagers. In winter, people come to me every day with their demands for timber to repair their houses. They also demand fuelwood at the times of marriages, community feasts and funerals. Their needs are very genuine and I give them the best possible help even going against law.

While maintaining that forest conservation is important, the Forest Guard argued that it should not be at the cost of local requirements. He argued that while restricting access to resources (e.g. creation of enclosures to check open grazing), there should be proper provision for compensation to the affected populations. He added that although few entry point activities have been undertaken such as construction of ponds, bridges and roads, they are not sufficient in motivating the local communities to support the forest conservation programmes. The villagers still see the FD as 'encroacher' of their local resources. He stressed the fact that being local to Navni, he understands the local problems, but often other field officials are indifferent to the subsistence needs of the villagers. For instance, they even put restrictions on the collection of dry fuelwood unless they are bribed by the local residents. Explaining the challenges he faces in being accountable to both villagers and senior forest officials, the Forest Guard narrated his split role:

> I have to consider many things while helping the local people. Before letting a person collect timber or firewood from the forests, I make sure he can employ labourers to carry the log of wood or twigs as soon as possible else he might be caught by the Forester or police in the village. If caught, he may be fined heavily and I will be questioned by the higher officials [...] At times, I take the risk by allowing them to collect damaged timber when I know that their demands are genuine [...] Sometimes, I am afraid of villagers for scolding me for not helping them and at other times, I fear senior officials for helping villagers against law. I have to consider forest laws as well as local needs but senior officials take only the law into consideration. Since I am local to this place, I have to walk hand-in-hand with the people of this village as well as secure my job.

The Forest Guard explained that although the local villagers are entitled to timber at concessional rates, they need to apply for permits. Once the application is made by the villagers to the DFO, he does the necessary paper work including preparation of a report on the availability of damaged timber in the village forests, checking applicants' records regarding illegal tree felling, reporting their last consumption of timber etc. If the applicant has cordial relations with the forest staff and has participated in fencing or other forest protection activities, then he is eligible to get permits and can avail timber at concessional rates.

Apart from the highly cumbersome process, there are three main problems villagers face with regard to permits as explained by the Forest Guard. First, many villagers are too poor to pay the cost of permits, which is Rs. 350. Second, only 10–15 permits are issued in a year and applications are accepted only for 9 months to restrict the demands for timber. Also, if a villager is granted timber, he can avail it again only after a period of 3 years. Third, they are not supplied deodar (for its high market value) but only kail which is less durable. Also, the amount of timber granted is quite low when compared to their needs for constructing or repairing houses.[39] The Forest Guard explained:

> The construction of one house in a colder region requires a minimum of five trees. To meet this requirement, there is no other option than allowing the villagers to collect timber against forest laws.

The discussion above suggests the *split role* of the field-staff in simultaneously catering to the demands of local residents and senior authorities. In spite of the overly optimistic claims about 'people's participation' and 'sustainability', we can see the incompatibility between the archaic forest regulations and local requirements. The poor are denied access to good quality timber for subsistence needs and have to cope with poor quality housing, a point also raised by Sundar et al. (2001) in their study of the JFM in other parts of India. We also noted that the forest authorities maintain exclusive control over local resources in the form of permits and concessions. Menzies (in the context of rural China) also argues that the foresters retain substantial control over forest management by issuing or denying permits to the local populations (ibid, 1993). We saw that given the cumbersome legal process of getting permits for timber and restrictions imposed on open grazing and collection of fuelwood, the poor can hardly see any incentives to strictly abide by the forest laws and regulations. They may often find it more practical to negotiate with the field-staff in order to meet their basic needs, a process in which both the forest guard and village poor transgress the lines of legality. It remains to be seen which other incentives, apart from helping the poor, can motivate the field-staff to transgress forest laws. I shall return to this point while discussing illegal timber trade and rent seeking in the next chapter.

---

[39] For each permit, the applicant is entitled to receive one log of wood and 20% of the market value of the log is charged to the villagers after they obtain permits.

7.4 Conclusion

From this chapter, I draw three main conclusions. First, policy regulations and guidelines set the initial stage for politics with regard to control of funds and the activities to be undertaken as part of project implementation. This politics is shaped by factors such as the broad scope of the NAP, lack of focussed objectives, frequent amendments to regulations concerning project execution and flow of funds. I pointed to the ambiguities or uncertainties such factors can result in for the various actors in terms of their responsibilities and rights. I have argued that concessions granted to villagers in return for the responsibilities of forest conservation and regeneration are conditional upon their performance as forest protectors and law abiding forest citizens. Also, the local users need the approval of both the field-staff and senior forest officials before accessing any benefits or concessions, and their membership in the JFMCs could be terminated by the forest staff in the case of their performance being found unsatisfactory. It signals to the presence of a coercive element in the NAP, as opposed to the claims of being 'people-friendly'.

Second, although the senior forest officials claim that the NAP is based on the premise of 'participation' and 'decentralisation', I found that the forest bureaucracy is reluctant to devolve effective powers onto lower levels, resulting in repeated *re-centralisation* with reference to power, authority and control over decision-making and access to local resources. Despite efforts to reduce the bottlenecks in the transfer of funds from the centre to the JFMCs, the funds suffer delays or lapses due to failures on the part of forest officials in producing progress reports, utilisation certificates or conducting evaluation exercises. This indicates the presence of centralised control within apparently decentralised structures of the joint forest management.

Third, while the FD and the SFC exclusively control local forest resources and exploit them for commercial purposes, it is the local populations who are expected to limit their dependence on forests and perform the task of forest conservation and regeneration of degraded forest lands. The SFC contractors and the FD make profits out of valuable forest resources but the local populations are devoid of accessing resources even for meeting subsistence needs. I have also highlighted the *split role* and dilemmas of the field-staff in catering to the demands of village residents for access to forest produce as well as those of senior authorities concerning forest conservation and enforcement of forest regulations.

In closing, I presented in this chapter the arguments of optimists who maintain that participation and decentralisation lead to the sustainable management of forests along with meeting local needs. I also noted the viewpoints of sceptics who maintain that there is a strong coercive element in the forest policies and that, in actual practice, there is re-centralisation of power and resources rather than effective decentralisation. What is evidenced in my study of the NAP in J&K is that the claims of sceptics are not unfounded, especially in the analysis of the functioning of forest bureaucracy. Yet, at the village level, these policies do not always result in fixed and determined outcomes as the sceptics would want us to believe. In the next chapter, I shall demonstrate that the intrusion of forest bureaucracy and the conservation ideas that it brings are negotiated and reconfigured according to the differential power and abilities of local actors. As such, forest conservation interventions do not go uncontested – they are experienced and accepted differently by various social groups within the village community, leading to multiple forms of micropolitics of forest use and management.

References

Agrawal A (2005) Environmentality: technologies of government and political subjects. Duke University Press, London

Arnold JEM (2002) Clarifying the links between forests and poverty reduction. Int For Rev 4(3):231–233

Balland JM, Platteau JP (1996) Halting degradation of natural resources: is there a role for rural communities? Oxford University Press, Oxford

Bardach E (2000) A practical guide for policy analysis: the eightfold path to more effective problem solving. Chatham House Publishers, New York/London

Barrett CB, Lee DR, McPeak JG (2005) Institutional arrangements for rural poverty reduction and resource conservation. World Dev 33(2):193–197

Bazaara N (2003) Decentralisation, politics and environment in Uganda. Working paper, Environmental Governance in Africa series. World Resources Institute, Washington, DC

Bebbington A (1999) Capitals and capabilities: a framework for analysing peasant viability, rural livelihood and poverty. World Dev 27(12):2021–2044

Bene C, Belal E, Baba M (2009) Power struggle, dispute and alliance over local resources: analysing 'democratic' decentralization of natural resources through the lenses of Africa inland fisheries. World Dev 37(12):1935–1950

Berg J, Biesbrouck K (2005) Dealing with power imbalances in forest management: reconciling multiple legal systems in South Cameroon. In: Ros-Tonen, Dietz T (eds) African forests between nature and livelihood resources: interdisciplinary studies in conservation and forest management. The Edwin in Mellen Press, New York, pp 221–254

Blaikie P (2006) Is small really beautiful? Community based natural resource management in Malawi and Botswana. World Dev 34(11):1942–1957

Campbell BM, de Jong W, Luckert M, Mandondo A, Matose F, Nemarundwe N, Sithole B (2001) Challenges to proponents of CPR systems: despairing voices from the social forests of Zimbabwe. World Dev 29(4):589–600

Carney D (1998) Sustainable rural livelihoods. Regional Development Dialogue 8. Department for International Development, London

Chambers R, Conway G (1992) Sustainable rural livelihoods: practical concepts for the twenty-first century. IDS discussion paper 296. Institute of Development Studies, Sussex

Corbridge S, Jewitt S (1997) From forest struggles to forest citizens? Joint forest management in the unquiet woods of India's Jharkhand. Environ Plan A 29(12):2145–2164

Corbridge S, Williams G, Srivasatava M, Veron R (2005) Seeing the state. Cambridge University Press, Cambridge

D'Silva E, Nagnath B (2002) Behroonguda: a rare success story in joint forest management. Econ Polit Wkly 9:551–557

Dean M (1999) Governmentality: power and rule in modern society. Sage, London

Ellis F (2000) Rural livelihoods and diversity in developing countries. Oxford University Press, Oxford

References

171

Farrington J, Bebbington A (1993) Reluctant partners: non-governmental organisations, the state and sustainable agricultural development. Routledge, London

Fomete T (2001) The forestry taxation system and the involvement of local communities in forest management in Cameroon. Rural Development Forestry Network paper 25. Oversees Development Institute, London

Gadgil M (1998) Grassroots conservation practices: revitalising the traditions. In: Kothari A, Pathak N, Anuradha R, Taneja B (eds) Communities and conservation: natural resource management in South and Central Asia. Sage, New Delhi, pp 219–238

GoJ&K (1992) Notification No. SRO-61. 19 March, 1992. Government of Jammu and Kashmir

GoJ&K. (1993) Udhampur forest division annual report 1992–1993. Jammu and Kashmir Forest Department, Udhampur

GoJ&K (1999) Notification No. SRO-17. 12 January, 1999. Government of Jammu and Kashmir

GoJ&K (2005) Digest of forest statistics. 2005. Jammu and Kashmir Forest Department, Srinagar

GoJ&K (2006) FDA Udhampur report. Jammu and Kashmir Forest Department, Udhampur

GoJ&K (n.d.) Operational guidelines for National Afforestation Programme. Jammu and Kashmir Forest Department, Jammu

Gray L, Mosley W (2005) A geographical perspective on poverty-environment interactions. Geogr J 171(1):9–23

Greenberg J, Park T (1994) Political ecology. J Polit Ecol 2(1):1–12

Guha RC (2000) Environmentalism: a global history. Oxford University Press, New Delhi

Hulme D, Murphree M (2001) African wildlife and livelihoods: the promise and performance of community conservation. James Currey, London

Karlsson B (1999) Ecodevelopment in practice: Buxa-Tiger reserve and forest people. Econ Polit Wkly 24(30):2087–2094

Kumar C (2006) Whither 'community-based' conservation? Econ Polit Wkly 41(52):5313–5320

Leach M, Mearns R, Scoones I (1999) Environmental entitlements: dynamics and institutions in community based natural resource management. World Dev 27(2):225–247

Lele SC (1991) Sustainable development: a critical review. World Dev 19(6):607–621

Levine A (2002) Convergence or convenience? International conservation NGOs and development assistance in Tanzania. World Dev 30(6):1043–1055

Li TM (2007) Practices of assemblage and community forest management. Econ Soc 36(2):263–293

Liddle WR (1992) The politics of development policy. World Dev 20(6):793–807

Lipsky M (1980) Street-level bureaucracy: dilemmas of the individual in public services. Russell Sage Foundation, New York

Long N, Van der Ploeg JD (1989) Demythologising planned intervention. Sociol Rural 29:226–249

Lynch O, Talbot K (1995) Balancing acts: community based forest management and national law in Asia and the Pacific. World Resources Institute, Washington, DC

Mabee H, Hoberg G (2006) Equal partners? Assessing co-management of forest resources in Clayoquot Sound. Soc Nat Resour 19(10):875–888

Mahanty S (2002) NGOs, agencies and donors in participatory conservation. Econ Polit Wkly 37:3757–3765

Mayers J, Bass S (2004) Policy that works for forests and people. Earthscan, London

McShane TO (2003) Protected areas and poverty, the linkages and how to address them. Policy Matters 12:52–53

Mencher J (1999) NGOs: are they a force for change? Econ Polit Wkly 34(30):2081–2086

Menzies N (1993) Putting people back into forestry: some reflections on social and community forestry. For Soc 1(1):6–7

Mooij J, de Vos V (2003) Policy processes: an annotated bibliography on policy processes with particular emphasis on India. ODI Working Paper 221. ODI, London

Muhereza EF (2003) Commerce, kings and local government in Uganda: decentralising natural resources to consolidate the central state. Working Paper 8. Environmental Governance in Africa Series. World Resource Institute, Washington, DC

Njogu JG (2005) Beyond rhetoric: policy and institutional arrangements for partnership in community based forest biodiversity management and conservation in Kenya. In Ros-Tonen, Dietz T (eds) African forests between nature and livelihood resources: interdisciplinary studies in conservation and forest management. The Edwin in Mellen Press, New York, pp 285–316

Oyono PR (2004) Social and organisational roots of Cameroons forest management decentralisation model. Eur J Dev Res 16(1):174–191

Pai R, Datta S (eds) (2006) Measuring milestone. In: Proceedings of the National Workshop on joint Forest management (JFM). Ministry of Environment and Forest, Government of India, New Delhi

Pimbert M, Pretty J (1995) Beyond conservation ideology and the wilderness myth. Nat Resour 19(1):5–14

Poffenberger M, Singh C (1996) Communities and the state: re-establishing the balance in Indian forest policy. In: Poffenberger M, McGean B (eds) Village voices, forest choices: joint forest management in India. Oxford University Press, New Delhi, pp 56–85

Rathore BMS (1996) Joint management options for protected areas: challenges and opportunities. In: Kothari A, Singh N, Suri S (eds) People and protected areas: towards participatory conservation in India. Sage, New Delhi, pp 93–113

Reardon TA, Vosti S (1995) Links between rural poverty and environment in developing countries. World Dev 23(9):1495–1506

Ribot J (1998) Theorising access: forest profits along Senegal's charcoal commodity chain. Dev Chang 29:307–341

Ribot JC (2002) Some concepts and a proposed framework for contributions. Paper prepared to guide contributions to the international conference on decentralisation and the environment, Bellagio, Italy, 18–22 February

Ros-Tonen M, Zaal F, Dietz T (2005) Reconciling conservation goals and livelihood needs: new forest management perspectives in the twenty-first century. In: Ros-Tonen M, Dietz T (eds) African forests between nature and livelihood resources: interdisciplinary studies in conservation and forest management. The Edwin in Mellen Press, New York, pp 3–30

Saigal S (2000) Beyond experimentation: emerging issues in the institutionalisation of joint forest management. Environ Manag 2(3):269–281

Sarin M (1996) Joint forest management: the Haryana experience. Centre for Environmental Education, Ahmedabad

Scoones I (1998) Sustainable rural livelihoods: a framework for analysis. Working Paper No. 72. Institute of Development Studies, Sussex

Scott L (2006) Chronic poverty and the environment: a vulnerability perspective. CPRC. Working Paper 62. Overseas Development Institute, London

Sen S (1999) Some aspects of state-NGO relationships in India in the post-independence era. Dev Chang 30(2):327–355

Sundar N (2001) Beyond the bounds? Violence at the margins of new legal geographies. In: Peluso NL, Watts M (eds) Violent environments. Cornell University Press, Ithaca, pp 328–353

Sundar N, Jeffery R, Thin N (2001) Branching out: joint forest management in India. Oxford University Press, New Delhi

Sunderland T, Campbell B (2008) Conservation and development in tropical forest landscapes: a time to face the trade-offs? Environ Conserv 34(4):276–279

The Guardian (2010, July 14) Kashmir fears forests will disappear through 'timber smuggling'

Upreti BR (2001) Contributions of community forestry in rural social transformation. Some observations from Nepal. J Forest Livelihood 1(1):31–34

Vasan S (2006) Living with diversity. Indian Institute of Advanced Study, Shimla

Vira B (1999) Implementing joint forest management in the field: towards an understanding of the community-bureaucracy interface. In Jeffery R, Sundar N (eds) A new moral economy for India's forests? Discourses on community and participation. Sage, New Delhi, pp. 254–275

Zuhair M (1998) Country report Maldives: Asia Pacific forestry sector outlook study. Working Paper Series. APFSOS/WP30. Food and Agriculture Organisation, Rome

Chapter 8
The Micropolitics of Forest Use and Control: New Spaces for Cooperation and Conflict

Abstract In the previous chapter, I analysed the micropolitics of forest use and control by explaining the split role of the field-staff, and areas of tension between the SFC and local timber contractors. This chapter extends the discussion on micropolitics further by revealing new spaces of cooperation, conflict and contestations that have emerged under the NAP in Navni and Chinnora. The following questions are addressed in this chapter: (a) How effective has the NAP been in enhancing people's participation through decentralised management of forest resources? (b) To what extent has the programme solved the problem of access to timber, fuelwood and fodder for village residents? (c) In what ways do local power relations determine the access of villagers to forest resources? (d) How do the differential capacities of various forest users determine their behaviours in transgressing forest regulations and indulging in illegal timber harvesting? On the whole, in this chapter, I present the experiences of the village community with the forest field-staff to assess how far the agenda of partnership between the FD and the village community has been realised in actual practice.

Keywords Power · Resource access · Micropolitics · Corruption and illegality · Labour exploitation

The supporters as well as critics of participatory forest management programmes have conventionally understood these interventions in relation to issues of cooperation and conflict between forest bureaucracies and village communities. Recently, some scholars have pointed out that the local power relations and field-staff shape the outcome of conservation policies to a far greater extent than the policy debates of a hegemonic state (e.g. see Springate-Baginski and Blaikie 2007; Vasan 2006; Li 2005; Sarin et al. 2003; Saberwal 1999; Ribot 1998; Rangarajan 1996; Neumann 1992). The forest edge, as Li (2007) suggests, is a site of struggle but it is difficult to control merely by coercive mechanisms. Other commentators have warned against the unbridled faith in 'decentralisation', and argue that it may 'risk strengthening ties of patronage and further entrenching local elites' (Francis and James 2003: 327). Depending on their respective bargaining power and differential

© Springer International Publishing AG 2018
S. Gupta, *Contesting Conservation*, Advances in Asian Human-Environmental Research, https://doi.org/10.1007/978-3-319-72257-3_8

abilities, village residents devise various strategies to secure their interests amidst the opportunities and challenges created by interventions like the NAP. It is, thus, important to examine the agendas and roles of different players at the microlevel and understand how local power relations could be crucial in defining the gap between forest management policies and on-the-ground realities.

In the previous chapter, I analysed the micropolitics of forest use and control by explaining the split role of the field-staff, and areas of tension between the SFC and local timber contractors. This chapter extends the discussion on micropolitics further by revealing new spaces of cooperation, conflict and contestations that have emerged under the NAP in Navni and Chinnora. The following questions are addressed in this chapter: (a) How effective has the NAP been in enhancing people's participation through decentralised management of forest resources? (b) To what extent has the programme solved the problem of access to timber, fuelwood and fodder for village residents? (c) In what ways do local power relations determine the access of villagers to forest resources? (d) How do the differential capacities of various forest users determine their behaviours in transgressing forest regulations and indulging in illegal timber harvesting? Overall, in this chapter, I present the experiences of the village community with the forest field-staff to assess how far the agenda of partnership between the FD and the village community has been realised in actual practice.

The chapter is divided into three sections. In section one, I present the interplay between the JFMC, field-staff and *panchayat*, and highlight the areas of tension between these agencies. I focus on various forms of micropolitics concerning bottlenecks created by the FD in the devolution of funds, the domination of field-staff in decision-making, emerging hostility between the JFMC and other village residents, information asymmetries between the field officials and the JFMC members, and conflicting powers and functions of *panchayat* and the JFMC. In section two, I analyse the differential impact of forest laws and regulations on village residents. I argue that while the relatively affluent and powerful villagers are able to negotiate with the field-staff to facilitate their access by means of bribery, it is the poor who suffer the most from the increased restrictions on access to fuelwood, fodder and timber. I also illustrate how the field-staff creates opportunities for rent seeking by maintaining ambiguities between 'legal' and 'illegal' access to forest resources. Following this, in section three, I present the local understandings of the players involved in illegal timber harvesting. I maintain that although the poor, at times, indulge in illegal felling, it is merely to meet their subsistence needs. However, it is a powerful stream of actors including security forces and foresters themselves, who are involved in the relentless exploitation of forests for private gains. The principal argument in this chapter is that conservation programmes such as the NAP permeate existing relations of domination and subordination, and are reshaped by local power dynamics, resulting in differential impacts on various categories of forest users.

8.1 From Centralisation to Decentralisation: Do Blockages Disappear?

As discussed in the previous chapter, the JFMC is considered as a potential link between the village community and state. The various developmental and conservationist functions, previously performed exclusively by the FD are now to be comanaged with the JFMCs. I have shown that an important shift with regard to the flow of funds was brought under the NAP in 2002 to facilitate the effective functioning of the JFM. The stated aim was that the new programme would devolve power and authority to the JFMCs for forest protection and reduce bottlenecks created by the state forest departments in the release of funds to the JFMCs. In other words, this meant a change in the roles and powers of state and local actors with regard to control over funds as well as access to forest resources. In the following discussion, I examine the degree to which the NAP has brought a shift in the powers of the forest officials vis-a-vis local communities in Navni and Chinnora villages in relation to partnership in forest management, participation in decision-making and devolution of funds. I also highlight the symmetries (or asymmetries) with regard to information shared by the field officials with the local residents and demonstrate how the new laws and regulations set the stage for politics between various players involved.

In 2003, a common JFMC was constituted for Navni and Chinnora villages to implement the JFM project (under the NAP) funded by the central government. At the time of my fieldwork (2006–2007), the same JFMC was functioning as no elections took place in the years following its initial formation.[1] According to the JFMC members, the then Forester organised a meeting of villagers from Navni and Chinnora in order to form a village forest committee. In the meeting, he explained that under a new project of the FD, various developmental and forest regeneration activities would be undertaken jointly by the forest staff and the committee members. Out of a total of 50–60 villagers present in that meeting, the Forester asked the participants to nominate members for the proposed committee. A JFMC member from Chinnora, however, informed me that while three to four villagers had expressed their willingness to become members, the rest were simply nominated by the Forester, including Chairperson Rattan Chand (age 58), a *thakur* by caste who also happened to be the Forester's friend.[2] A list of ten names was then sent by the Forester to the FD for approval, which comprised of two *thakurs*, one Scheduled Tribe, three Scheduled Castes and four Muslims. Out of these ten members, two were women, one belonging to a Scheduled Caste and the other a Muslim. In this

---

[1] It is to be noted that the NAP guidelines suggest that the elections for the JFMCs should take place every year. However, in the villages of my study, the next round of elections took place only after 4 years. A new JFMC was constituted in November, 2007.

[2] It is important to note that *thakur* is traditionally a dominant caste group consisting mainly of landlords. Rattan Chand is an agriculturalist and a resident of Navni village.

sense, the composition of the JFMC was representative of the various different social groups in the two villages but there was disproportionate representation in terms of residence as eight members belonged to Navni and only two to Chinnora village.[3] It is pertinent to note that out of the ten committee members, eight were illiterate, the Chairperson being one of them. Having explained the formation and composition of the JFMC, I now turn to present the perspectives and experiences of the members in terms of the actual functioning of the JFMC.[4]

In my initial conversations, the committee members emphasised the importance of community participation, and argued that the villagers recognise the significance of conserving forests and their sustainable use because they are completely dependent on the surrounding forests for meeting their basic needs of timber, firewood, grass, local medicinal plants etc. They noted that village residents can protect and manage the surrounding forests better than the forest field-staff who are often not local to the village and, hence, not committed to performing their duties properly. Being unfamiliar to the issues concerning local requirements and use of forest resources, they merely restrict their job to policing the forests. Most committee members were largely unaware of the rationale behind the formation of the JFMC, a point that would become clearer in the following discussion. On the need for a village forest committee, the Chairperson, Rattan Chand, commented:

> I think the main reason for forming a village committee is that it is difficult for forest guards to take care of the whole forest beat. I also feel that the responsibility of the department to protect forests from illegal felling is now over and has become ours.

According to the Chairperson, the functions of the JFMC include village development, forest conservation and distribution of benefits and responsibilities among villagers. He stated that soon after the formation of the JFMC, the members were asked by the Forester to prepare a microplan in consultation with the villagers, specifying the various developmental and conservation activities needed in the villages. Initially, the local villagers lacked awareness of the objectives of the programme and were apprehensive owing to their experiences of curtailment of rights to use forest resources imposed by the FD in the past (as discussed in Chap. 6). They were suspicious of losing further access to the forests and, hence, showed their reluctance to participate in any joint activities with the Department. The microplan was thus made mainly by the Forester with some suggestions on entry point activities (e.g. site for the construction of bridges, ponds etc.) by the JFMC members. Explaining the various developmental and conservation activities undertaken in the last 3 years, the Chairperson informed me:

---

[3] Sansar Chand (age 42), the JFMC member from Chinnora also informed me that the disproportionate representation of the JFMC in terms of residence was mainly because the Chairperson belonged to Navni and he, along with the Forester, selected other members known to them from the same village. Some issues of inter-village conflict between Navni and Chinnora will find mention in the narratives of *sarpanch* (who belongs to Chinnora) later in this section.

[4] In order to understand the issues of cooperation and conflict between the JFMC and the FD, it was important to have repeated interactions with the members of the committee and hence the information provided below is based on several conversations with them.

8.1 From Centralisation to Decentralisation: Do Blockages Disappear?

> We have constructed two bridges, one in Navni Musral and the second in Navni Gharat. We also made four *bowlis* in Chinnora and three in Navni for securing water. For forest regeneration, we have made four closures, two in Chinnora and two in Navni. We have established two plant nurseries, one in Dhoona and other in Sira Morh. There is a watchman in both the nurseries and the committee members visit there every four days. Also, keeping in view the local demands for fodder, we allow people to cut grass from closures every three months in return for the help they offer in fencing the closures [...] We pay attention to the requirements of all the village residents as proposed by the committee members from both Navni and Chinnora villages. We also hold meetings when the Forester visits the village and take decisions collectively.

Although what the Chairperson narrated indicates the functioning of the JFMC in accordance with the prescribed guidelines, the actual picture came out more clearly in the later interactions with him, other committee members and villagers. For example, Sansar Chand, a JFMC member from Chinnora, provided a contradictory narrative with regard to the functioning of the committee in terms of decision-making powers and participation. According to him, the JFMC organised two meetings in the first few months of its formation for the preparation of microplan. The issues discussed in the meetings were related to the selection of sites for establishing closures, entry point activities and measures for preventing illegal tree felling. He informed me that after these initial meetings, the decisions were mainly taken by the Forester and Chairperson without consulting JFMC members.

Other members of the committee argued that local residents being largely unaware of the aims and objectives of the project, give the credit to the forest staff for the entry point activities undertaken and the committee members do not get due recognition for their work. They also insisted that in the absence of field visits by senior officials, they are unable to communicate to the FD their concerns over the selection of entry point activities in the village. Moreover, in the absence of any awareness building exercises for the JFMC members with regard to the overall objectives of the programme and their rights and responsibilities, they rely solely on the field officials for the purpose of information. Owing to this, they argued that the field officials are able to exploit both committee members and local villagers to serve their own interests.

After a few interactions with the Chairperson in Navni, some issues of conflict between the JFMC, field-staff and villagers also emerged on the question of control over funds, decision-making powers and roles and responsibilities of the committee. According to the Chairperson, the FD is not sensitive to the biomass requirements of the villagers and makes inappropriate choices of species to be planted for forest regeneration. For example, he mentioned that in 2006, the field officials provided keekar to the JFMC for plantation in the surrounding forests without any consultation with the committee members. The members argued that keekar yields very little fuelwood and serves no other purpose than increasing forest cover due to its fast growth. They proposed other species such as akhrot, maru and banj which are more useful for the purposes of fruits, fuelwood, fodder as well as timber. Yet, the FD did not supply these trees as they take more time to mature when compared to keekar. Also, for the purpose of tree plantation on private lands, there is reluctance on the part of field-staff in involving the committee for distribution of plants to the villagers.

As the Chairperson explained, a few months previously, the FD had provided some fruit bearing trees to be distributed among poor villagers. The Forest Guard and Forester distributed the trees according to their own discretion without consulting the JFMC regarding potential beneficiaries.

The JFMC members argued that at the time of the formation of the committee, the Forester told the members that under the programme, the committee was empowered to take decisions on its own (in consultation with the field-staff) regarding provision of timber and fuelwood to the villagers in view of their genuine needs. However, the members explained that they were never given authority to take decisions in these cases. As such, the villagers continue to approach the Forest Guard and Forester for their demands of timber and firewood. They argued that for them, the programme has been more about responsibilities than any real benefits. As the Chairperson stated:

> I go and check the closures every four days. I even fight with people in the village for the illegal collection of damaged timber from the forests. I get no rewards for it. But I have to do this, otherwise if anybody damaged a closure, the staff would put the blame on me.

Apart from decision-making, participation and responsibilities, the JFMC members also raised the issue of their control over project funds and maintained that they are hardly aware of the total funds granted by the FD for each activity. They blamed the field-staff for not discussing with them project expenses and estimates of costs for labour, transportation etc. The Chairperson complained of a delay in the approval of bills by the FD and the consequent payments to be made to the labour employed. For example, he recalled that in 2006, the committee was assigned the job of establishing a closure in Navni for which they were only provided the fencing wire but no advance payment. The committee was expected to employ the labour which was paid only after the completion of the project since the bills for labour costs were to be approved by the DFO and the entire process took more than 3 months. Explaining his lack of actual power in terms of control over funds, he stated:

> I was told by the Forester that the two of us will jointly manage all project funds. But I have no say in the utilisation of funds. It is the Forester who controls the money [...] He even kept my rubber stamp for months saying that he needs it to do all the formalities required for maintaining the records of the JFMC.

Sansar Chand, a JFMC member, reflected on his experiences with regard to benefits and associated responsibilities under the programme. Also, highlighting the problem of uncertainties regarding project costs and release of money, he narrated:

> I have not been benefited in any way other than that I am now more aware of the forest activities than the local villagers are. In fact, I had to spend from my own pocket to construct a *bowli* as an entry point activity in my ward. Neither was the amount to be spent on its construction made clear to me nor was I paid anything in advance by the Forester. So, I contacted the villagers known to me for providing labour and promised them to pay at the rate of Rs. 70 per day upon the completion of the work. The total money spent was Rs. 17,000. But, the Forester refused to pay the total amount, telling me that the Forest Department had sanctioned only Rs. 15,000 for the *bowli* [...] I came to know later from the Forest Guard that the actual money sanctioned was Rs. 25,000. So where are the benefits?

8.1 From Centralisation to Decentralisation: Do Blockages Disappear? 179

The JFMC members suggested that for effective project implementation, the government should give the power of utilising funds directly to the committee rather than to the DFO or field-staff. Although under the NAP, the money is transferred directly to the account of the JFMCs from the FDA, we noted that, in actual practice, it cannot be utilised unless approved by the DFO and the Forester. The funds remain under the effective control of the forest staff with the possibility of delays, misappropriation and leakages. Hence, the stated objective of devolution of funds to the JFMCs for effective implementation of the project seems to remain unfulfilled because of the blockages created both at the field and divisional levels.

While the devolution of funds has not resulted in reducing bottlenecks, the idea of forming local partnerships in bringing the FD closer to village communities has also been only partially realised. The JFMC members argued that instead of bridging the gap between the villagers and the FD, the programme has actually created a divide within the village community. They maintained that since the formation of the committee, fellow village residents have become hostile to them and see the committee members siding with the FD in curtailing their access to the forests. For example, if there is any case of illegal felling, a JFMC member has to accompany the field-staff to the place where the tree has been felled in order to witness the offence. However, they argued that if the offender is poor, he is punished but if he is relatively affluent, the field-staff accept bribes and allow the offender to take the timber away.

The members argued that most of the time the staff invites them to witness the offence only when they are sure that the culprit cannot pay the bribe owing to poor economic conditions. When a villager is fined, the hostility of the local villagers is directed towards the committee members instead of the field officials. Due to such cases, the villagers do not identify themselves with the committee members and blame the members both for being indifferent to their genuine requirements and acting against them. The members also complained that although the Forest Guard and the Forester benefit from bribes and salaries for doing their jobs, the committee members get a bad reputation from fellow residents for assisting the field-staff against them.

The JFMC members also pointed out that at the time of the formation of the committee, they were misinformed by the Forester with regard to the benefits that they would be entitled to. For example, a female member of the committee recalled that she was promised to be paid Rs. 500 per month for becoming a member of the committee.[5] She expressed her disappointment when no such payment was made to her.[6]

---

[5] In my conversations with the two female members of the JFMC, I found that they were not aware of the reasons for the formation of the committee, various project activities under the NAP or their responsibilities in forest protection. It appeared that although the reserved places for women in the JFMC were filled nominally, they hardly played any role in actual decision-making. The negligible role of female members in decision-making in forest programmes has also been highlighted by other scholars such as Springate-Baginski and Blaikie (2007), Agrawal (2005) and Sarin (1998).

[6] Making a similar observation, Kawosa (2001) notes that the village committee members usually consider their membership as an opportunity or stepping stone for future appointments in the Forest Department which when not achieved disappoints the members.

The Forester also promised to pay Rs. 500 per month to the watchmen employed by the JFMC for protecting enclosures. Yet, no salary was paid to them for their services. Under the NAP guidelines, however, there is no provision for cash payment to the JFMC members for their shared responsibility with the field-staff. This infers that at the time of constituting the JFMC, the Forester had misguided the villagers with such false promises in order to enrol female membership and put the burden of policing forests on the villagers.

Despite the presence of the JFMC in the village, the FD had entrusted some of the activities under the programme with the *ex-sarpanch* of Chinnora owing to her good relations with the DFO.[7] The JFMC Chairperson complained that the training camp for income generation activities organised in Chinnora in 2007 was under the supervision of *ex-sarpanch* rather than the JFMC. Explaining his marginalisation, he argued that rather than consulting him on the demands brought forward by the villagers for timber and firewood, the DFO approved the cases recommended by the *ex-sarpanch*. He also added that he was also not involved in the formation of self-help groups initiated by the *ex-sarpanch* in consultation with the DFO under the NAP. Suggesting that since the JFMC is the partner of the FD for all forest related activities, any collaboration with a third party such as the *panchayat* cannot be justified. I further elaborate on the issues of conflict between democratically elected *panchayats* and the FD promoted JFMC in the following discussion.

8.2 Panchayat and JFMC: Conflicting Powers and Functions

My interview with Basanti Gupta (age 65), the *ex-sarpanch* of Chinnora, revealed some inside facts concerning the formation of the JFMC and its functioning, disparities between the developmental activities in Navni and Chinnora, and tensions between the JFMC and the *panchayat*.[8] Although her tenure finished in 2006 after serving as *sarpanch* for 5 years, she still holds considerable power and authority in the two villages.[9] She compared the functioning of *panchayat* with the JFMC and argued that unlike the forest committee, her *panchayat* had always considered the interests of the villagers by holding regular meetings for developmental activities.

---

[7] This point has also been raised by Sundar et al. (2001: 206) in their study of the JFM in other states of India. They note that 'the creation of autonomous forest protection committees at village level is potentially in conflict with the national pressure to revitalise the panchayat'.

[8] It is to be mentioned that Basanti Gupta is *Mahajan* by caste and the majority of the villagers in Chinnora are *Gaddis*. She succeeded in winning the *panchayat* elections in 2001 against five *Gaddi* candidates which shows her popularity in the village.

[9] It is to be noted that no *panchayat* elections have taken place in J&K since 2006. Basanti Gupta is a nurse by training and the first point of contact for the villagers to get any medical treatment. As such, she was held in recognition by the villagers even before being elected as *sarpanch*.

8.2 Panchayat and JFMC: Conflicting Powers and Functions

Also blaming the JFMC for concentrating the developmental work only in a few wards of Navni, she stated:

> Out of the grants my *panchayat* received, I constructed the *panchayat* office building, water tanks, roads, playground etc. for the villagers. I have done development in all the eight wards of Chinnora unlike the JFMC which concentrated their developmental work in only a few wards of Navni, to which most of its members belong. I assigned the construction jobs to the residents of the respective wards to ensure that the work done was of the best quality. The members of the *panchayat* visited the construction sites regularly to check the progress. This way, I shared power of the *panchayat* with local residents.

She claimed that since the senior authorities approve the developmental plans without visiting the sites themselves, the community programmes, often fail to meet the actual needs of the villagers. She argued that there should be two separate JFMCs for Navni and Chinnora respectively to represent the interests of both the villages, a matter she had been pursuing actively with the DFO.[10] Also, contrary to what the JFMC members suggested, she maintained that the JFMC can work better in collaboration with the *panchayat*. She complained that the present of JFMC never sought her advice on any of the developmental or conservation matters, and argued that if they had done, she would have advised them on other developmental activities needed in the village. Further, commenting on the arbitrary process of the formation of the JFMC, she noted:

> The JFMC was formed by the Forester in 2003 without any elections […] It lacked equal representation of the people from Navni and Chinnora. The result was that the developmental activities by the JFMC were concentrated only in Navni […] I know that the JFMC Chairperson and other members are friends of the Forester – all illiterate and drunkards.

According to the *ex-sarpanch*, neither the FD nor the JFMC has taken any initiatives to create awareness about the NAP in the villages. She claimed that the JFMC Chairperson has no standing and is incapable of organising even a workshop or a training camp for the benefit of the village community. Owing to this, the DFO asked her to organise a one-day training camp for income generation activities at Chinnora in 2007.[11] As mentioned in the previous chapter, an expert from an NGO named AKKA was invited by the FD to train the local people in various forest related income generation activities such as mushroom cultivation, drying fruits, bee-keeping and honey production. This training camp did not prove beneficial, according to the *ex-sarpanch*, because the JFMC failed to take any follow-up action such as the formation of self-help groups for these income generation activities.

---

[10] After her tenure was over in 2006, she succeeded in pursuing the DFO to elect the new JFMC in 2007. However, the committee had not undertaken any projects until I conducted the second field trip to the village in November, 2008. The actual functioning of the new JFMC is beyond the scope of this study.

[11] I noted that the JFMC Chairperson was not present in the training camp and only two JFMC members attended it. This suggests that by the time of my fieldwork (2006–07), the JFMC had been surpassed by the *ex-sarpanch* and the DFO was dealing with the *ex-sarpanch* for all practical purposes. Effectively, the *ex-sarpanch* had started dominating in all matters of the FD and villagers in Navni and Chinnora even before formally taking up the position of the JFMC Chairperson of Chinnora in 2007.

Moreover, she thought that in the absence of marketing opportunities for their products, merely a one-day training camp could not succeed in motivating the villagers to start such income generation activities. Also, according to her, different departments such as rural development, horticulture, agriculture, social forestry etc. have in the past approached villagers with community based projects which have showed little tangible benefits and visible effects. This has confused the villagers and resulted in the loss of their faith and enthusiasm to participate in the governmental projects. She informed me that following the directives of the DFO, she has now formed three self-help groups in the village, two made up of females and one of males, and argued that the members of these groups have agreed to contribute money for the functioning of the group merely because of her reputation in the village.[12]

To conclude, in this section, I have argued that although there is construction of local institutions and change in the structure of flow of funds under the NAP, there is reluctance on the part of the FD to hand over effective power to the JFMC. Whilst the JFMC is representative of various social groups in the village, the members were primarily selected by the Forester. The microplan was prepared without consulting the villagers and entry point activities were decided mainly by the Forester and the JFMC Chairperson. We witnessed the domination of field-staff in the choice of tree species for regenerating forests, their distribution, and selection of sites for establishing closures. Moreover, the blockages in the transfer of money both at divisional and field levels indicate strong centralised control over funds as opposed to the decentralised management aimed under the programme. As observed, it is only after the project activity is completed and expenditure bills approved by the DFO that the money is released to the JFMC. At the field level, the effective control over funds also remains with the Forester (Secretary of the JFMC), who misappropriates them by providing wrong information to the JFMC members. This is evidenced from the lack of transparency between the Forester, the Chairperson and members of the JFMC with regard to the allocated funds for the construction works undertaken as entry point activities in the villages.

The above findings corroborate the results of some recent studies on participatory forest management in other states of India. For example, Springate-Baginski and Blaikie (2007), in their study of the JFM programme in Orissa, West Bengal and Andhra Pradesh argue that in practice, local people have very little influence on forest management plans and it is actually the FD who sets the agenda for the activities to be undertaken. Likewise, Sundar (2004) illustrates the failure of devolution of power to the participatory committees in the villages of Madhya Pradesh and argues that the committees are tightly controlled by government staff and rarely represent the actual users of the forests. Sinha (2006) maintains that the 'ruler-ruled attitude' is quite apparent among the FD officials in Jharkhand, especially the Range and the Beat officers with little participation of the villagers. Dhanagare (2000) argues that

---

[12] A self help group is formed by collecting a specified amount from each member of the group every month. Once the groups are 6 months old, they can apply for loans and start any income-generation activity.

8.2 Panchayat and JFMC: Conflicting Powers and Functions

the notion of 'joint' remains only on paper as the forest department in Uttar Pradesh still perpetuate its conventional view of forestry of working for the people than with the people. Ballabh et al. (2002) in the case of Uttaranchal and West Bengal note that the FD's control over microplans and the disposal of forest produce leave little room for 'people's participation', thus, hampering the successful implementation of the JFM projects. Referring to the functioning of the JFM under the NAP in Orissa, Sarap and Sanrangi (2009) argue that although some developmental works have been undertaken in the villages at the entry point phase, transparency has not been maintained in the use of funds and also the choice of the activities to be undertaken were decided mainly by the forester and the chairperson of the committee. Thus, on the basis of my findings from J&K and similar observations made by scholars in other states of India, I argue that empowering villagers through 'participation' (by forming village committees) and 'decentralisation' of control (by transfer of funds to the committees) does not automatically result in the altering of power relations between the FD and the villagers. Instead the diverse interests and priorities of players involved may create spaces for conflict under the cover of cooperation or comanagement of forests.

Apart from lacking power in decision-making and control over funds, we also noticed that the JFMC members experience hostility from fellow village residents due to their association with the field-staff in witnessing forest offences. While the committee members face antagonism for complaining against local villagers, we saw that such offences also open up rent seeking opportunities for the field-staff. Thus, by making the JFMC a representative of local population and a 'legitimate' body responsible for forest protection, the FD is able to shed its responsibility of forest conservation by using the committee as a vehicle to re-orient interactions of villagers with forests. At the same time, it diverts the hostility of the villagers towards the committee instead of the FD.

The JFMC members in Navni and Chinnora argued that even for the construction works undertaken as 'entry point activities', the villagers give the credit to the field-staff for bringing in money for development and do not give due recognition to the responsibility shared by the committee members. Referring to this paradoxical situation, Sundar et al. (2001: 14) argue that where the JFM fails, the FDs deflect blame towards villagers for mismanagement of already degraded commons but where it succeeds, they claim joint credit for reversing deforestation and simultaneously providing for the livelihood needs of the poor. Bhattacharya et al. (2004: 25) caution against the 'regressive effect of formal structures' (like forest committees) on community based collective action and argue that the introduction of formal structures of governance may also cause a 'schism' within social groups. In line with the above arguments, we noted in the narratives of the JFMC members that contrary to bridging the gap between villagers and the FD, the local institutions have entrenched already existing disparities and power relations within the village community.

I have also highlighted the areas of tension between the *panchayat* and the JFMC in Navni and Chinnora which suggest that local power realities are complex and, hence, even well meaning policies of decentralisation and participation unravel differently than intended by the policy makers. In reality, these policies are rarely able to secure fixed and determined outcomes; rather they become refracted and reshaped by power relations at the microlevel. We saw the conflict between the *ex-sarpanch* and the JFMC on the one hand, and emerging cooperation between the DFO and the *ex-sarpanch,* on the other. While the DFO was interested in the training workshop and formation of self-help groups to demonstrate the effective functioning of the NAP, the *ex-sarpanch*, after the end of her tenure, was trying to reconsolidate her position and power in the village by cooperating with the DFO in implementing these activities. However, the forest committee and supposed village 'beneficiaries' for whom such activities were actually targeted remained as mere spectators or passive participants in the programmes.

The issues of knowledge creation, dissemination and information are central to the effective functioning of community based natural resource management programmes. In the study of the NAP in J&K, I have shown that the lack of knowledge regarding the roles and responsibilities of different stakeholders, funds available for various activities under the programme, and lack of easy access to the senior forest officials were major sources of discontent to the committee members. Observing similar issues in state programmes, scholars reveal a considerable lack of information among village committees, for example, about the legal framework for decentralisation (see Larson 2004). Mahanty (2002) notes that in the case of forest protection in Karnataka, due to poor communication and information links with the implementing agencies, information transmission was largely informal, through personal accounts, stories and rumours. Kothari et al. (1997) argue that a basic problem in India's protected areas seems to be the absence of any form of dialogue between the government authorities and local communities resulting in misinformation about the positions, justifications and circumstances of each other. Similarly, Das (1997) notes that the interaction of villagers with the senior officials is near negligible as they visit only the forest areas and very rarely the villages. Springate-Baginski and Blaikie (2007) also report the universal tendency of the DFOs and the rangers to pay a quick visit to village leaders, organise a poorly advertised meeting and then attempt to enforce the plan. Such short visits by the government officials to the field have been referred by Chambers (1987) as 'rural development tourism'. I have explained that since the senior forest officials in J&K rarely visit the field, the laws are, often, interpreted by the field officials according to their own understandings and vested interests. In Navni and Chinnora, the committee members did not only lack information about their rights and responsibilities, they were misinformed regarding the incentives to join the JFMC. For example, the female members in the villages were misguided about the honorarium that they would be getting by becoming the JFMC members. Similarly, the committee members were also misinformed about the funds available for the construction activities under the programme.

The information asymmetries between villagers, the JFMC members and forest officials provide opportunities to the field-staff for imposing restrictions on the village residents for rent seeking. I elaborate on this point in the next section where I focus on the interactions between villagers and the field-staff. I demonstrate that even after 4 years of project implementation, the majority of the villagers lacked awareness of the aims and objectives of the NAP. I also discuss specific problems

8.3 Increased Biomass, Reduced Access

faced by ordinary villagers in terms of access to timber, fuelwood and fodder, and the differential abilities of village residents to secure their respective needs.

8.3 Increased Biomass, Reduced Access

In order to gain a nuanced understanding of 'community', Agrawal and Gibson (1999) propose to examine how the diverse interests of stakeholders can affect processes around conservation and determine their differential access. More recently, scholars such as Dressler et al. (2010), Saito-Jensen and Jensen (2010), Shackleton et al. (2010) and Springate-Baginski and Blaikie (2007) suggest that although well intentioned community forestry policies do result in collective afforestation, they may also run the risk of emboldening local elites politically and economically. Likewise, Ros-Tonen et al. (2005: 12) argue that the resource management practices are marked by power conflicts and 'exclusion from access rights are still commonplace'. Gupta (1998: 320) cautions against 'exclusions and repressions in a community' and Vasan (2006: 233) notes that diverse groups within local communities organise differently in order to contest, negotiate or even override state dictates. Likewise, Saberwal (1999) argues that the local access to and use of natural resources is influenced to a far greater extent by local power relations than by the policy regulations. Through the historical analysis of state forestry in Garhwal Himalayas, Rangan (1997) also illustrates that state ownership of forests does not result in the monolithic imposition of proprietary rights, but emerges instead as an ensemble of access and management regimes. Sundar (2001) too maintains that community involvement in conservation and management does not represent a shift away from state control; rather it is a shift to more micro-disciplinary forms of power (see also Neumann 2001). Noting the observations and cautions of these scholars, I now turn to examine the relationship between villagers and field-staff in Navni and Chinnora, and explain how forest laws and regulations have been experienced differently by village residents owing to their differential power and abilities.

My interviews with the village residents suggest that the majority of them lacked awareness about the objectives of the NAP and even the existence of the JFMC in the villages. Most of them had never participated in any of the meetings of the JFMC held in the initial months of its formation. Those few who had the knowledge of the committee stressed the fact that it was formed by the Forester without consulting the villagers. They also mentioned that a general apprehension amongst the villagers for forest conservation activities was that the FD would take away their land to establish closures. Some others even felt that if they became members of the committee, the FD staff would ask them to complain against villagers involved in illegal felling, failing which they could be punished by the forest officials. Most villagers, hence, saw the state more as a malevolent outsider than a provider for their subsistence needs, and showed little interest in participating in forest regeneration programmes of the FD. There were a few others who were aware of the JFM programme and strongly emphasised local control of forests for protection and use. For example, Majid Ahmed (age 74), a resident of Chinnora stated:

> The forest is like a mother to us. For everything, we have to go to the forests. They only help us, they never bring us harm. I believe that the forests should be protected by the local people and the department jointly [...] Forest officials get transferred and lose interest in the programmes but villagers are always here to take care of the forests. If the local people are involved by the department, they can protect the forests just as their own property.

In terms of the change in forest cover in the surrounding areas, I received mixed responses from the villagers. While some held the view that compared to 15 years ago, there is now an increase in forest cover and availability of grass and fuelwood, others considered that the forests are continuously depleting. Yet, all of them stressed that their own access to forest resources has decreased because of increasing restrictions imposed by the FD. Acknowledging the increase in pressure on the local forests due to the increasing population, they observed that the depletion of forests has been mainly due to commercial exploitation of timber and other forest produce. Moreover, they also argued that the FD is not doing enough to protect the natural loss of forests due to heavy snowfall by removing damaged timber. Instead, it has put restrictions on local populations to access forest resources even for fulfilling their basic requirements. An elderly resident of Navni, Kastoori Lal (age 78) on the issue of forest depletion due to commercial exploitation, remarked:

> When I was young, I feared thinking that if the forests continue growing this way, what will happen? Now I feel the opposite. Previously, the forests were dense and inhabited by wild animals. The presence of these wild animals prevented the local villagers from destroying the forests. But the exploitation of the forests by the Department and contractors to meet market demands has changed everything [...] In the next 50 years, there will be no forests left for the future generation.

The villagers held the view that even when the FD takes initiatives for tree plantation and forest regeneration, it is not for the benefit of local people. They explained that the FD mainly grows timber species and encloses the area for regeneration. This creates a reduced supply of fodder and fuelwood. They argued that villagers are mainly dependent on maru for fodder and firewood. Maru trees are declining due to the increasing fuelwood and fodder demands in the village and need to be replaced by fast growing species to meet the local needs. Reiterating the point made by the JFMC Chairperson (in the previous section), they argued that the FD is growing keekar which is a very fast growing species but does not yield much fuelwood. They suggested other species such as banj, champ, kaimbal and kathi which grow faster, and can also provide both firewood and fodder.[13] They suggested that walnut can also provide fuelwood as well as timber to the local villagers and can be grown both on private and forest lands. However, they are never consulted over the choice

---

[13] Poor villagers have little faith in the FD concerning the choice of species to be planted. In the past, the FDs across India have attracted public anger for the plantation of eucalyptus under the Social Forestry and Eco-development projects. Criticising this, Kumar (2002) states that by planting fast growing species, the JFM could succeed in halting forest degradation but the objectives of poverty reduction and meeting biomass requirements of the villagers are sidelined.

8.3 Increased Biomass, Reduced Access

of species or sites to be chosen for regeneration despite their knowledge and understanding of the various uses of local plant species.

When asked about the change in the attitude or behaviour of the forest staff, the poorer villagers maintained that their relationship has not changed with the field officials who continue to threaten and exploit them. Some poorer residents held the view that although the officials are inflexible in dealing with their requirements of timber and firewood, the relatively well-off residents are treated differently because they can afford to bribe the forest officials. Emphasising the problem of timber in the village, the villagers explained that in times of heavy snowfall, the field officials, instead of being responsive to their genuine demands of timber, expect them to apply for timber permits which, according to them, is a long, expensive and cumbersome process.[14] Moreover, the wood that is provided to them is of kail which is of lower quality in terms of its durability as compared to that of deodar which the villagers had been traditionally using for constructing houses. They pointed out that the forest staff provides timber to those who offer them bribes but impose all regulations and restrictions on the poor. The villagers cannot approach the DFO directly with their demands of timber because if they do so, the field-staff gets angry with them and does not recommend their case for timber concession to the DFO. In the narratives below, I illustrate that the forest officials, instead of being sensitive to the needs of the poor are actually re-enforcing the already existing disparities.

> There is no change in the relationship of forest staff with us. I have many complaints against them. The roof of my house collapsed last week and I have no choice but to sleep in my cattle shed. I am trying to get some timber to repair my roof but the Forester does not pay attention to my request. He tells me that I need a permit to get timber and the process would take a minimum of four months. Neither can I wait for four months nor can I fell any tree from the forests [...] My husband died last year, I have three little kids. I do not know whom to approach to solve my problem. (Kaushalaya Devi (age 36), Navni)

> Three years back, I needed timber for my house. I approached the Guard and he asked me to apply for permit. I told him that I could not go to the forest office at Udhampur. He asked me to pay him Rs. 3000 and after I made the payment, I got the permit at home next week. So, I have no complaints against the Department. (Zakeya Begum (age 48), Chinnora)

> There is no change in the attitude of the forest staff towards the local people. I just need one log to repair my roof but the Guard does not provide me timber [...] All forest employees are friends of the affluent. The rich get even deodar for firewood but the poor like me cannot get it even for constructing a house. Laws are only for the poor. (Rekha Kumari (age 44), Navni)

> Yes, the relationship between the forest staff and the villagers has improved in the sense that they talk courteously with us now. But they never discuss their forest activity plans with us

---

[14] It is to be noted that the actual fee for a permit is Rs. 350 but the local villagers are charged ten times more by the field officials due to their lack of awareness. We will observe this in the narratives of the villagers provided below. Those who could afford to pay Rs. 3000–3500 to the Forest Guard for a permit do so because the market value of timber is much higher, and beyond the affordability of most village residents. For example, the current market value of kail (10 cubic feet) is Rs. 14, 940 and chir (10 cubic feet) is Rs. 6260.

and make late payments for the labour we provide in the construction activities. What is worse is that the permit fee is only Rs. 350 but they charge us Rs. 3000. They say that it is their commission. (Rehmat Ullah (age 64), *ex-sarpanch*, Navni)

The Forest Department has not done anything for us. We face many problems with getting timber from the Department [...] Two years back, due to snowfall, many houses including mine were damaged. I requested the Guard many times but he told me that I needed to get a permit first. I could not apply because it was a long process and the fee was Rs. 3500 which I was not able to afford. Then, a few weeks later while crossing the river, I found a damaged log of wood which I carried and brought to my house without informing the Guard. Although I have used it to repair the roof of my house, the condition of my hut is still very dangerous as it can fall anytime. (Mool Raj (age 40), Chinnora)

Referring to the strict regulations of the FD, the villagers mentioned that, often, the field officials threaten them even for collecting dry twigs from open forest areas. The women who collect firewood and sell it in the villages for modest incomes complained that they are harassed by the field-staff on the pretext of 'illegality'. Some of them informed me that when they offer bribes, the Forest Guard even allows them to cut green branches to be used as firewood.[15] On the contrary, the Forest Guard argued that he has put restrictions on the collection of dry wood because some villagers had started supplying firewood to wood sellers outside of the villages, which is obviously not used for local consumption. The villagers exclaimed that being unclear of what is 'legal' and what is 'illegal', they are bound to succumb to the restrictions imposed by the field-staff on their access. As a result, most of them are compelled to bribe the field-staff for collecting fuelwood from the surrounding forests.

Besides the problems of timber and firewood in the village, another challenge faced by the local residents is the reduced access to fodder after the establishment of closures under various forest programmes in the villages.[16] It is mainly women and children who are involved in the task of collecting fodder and firewood for the daily consumption in the household.[17] My female respondents explained that they have to travel long distances in order to collect fodder and firewood because of the enclosures established in the surrounding forest areas. They recalled that travelling

---

[15] In the feudal times, there was a practice of *rast*, a form of informal dues paid by the local villagers for the maintenance of forests. But even now the Forest Guard collects *rast* from the villagers. It may include items for consumption such as ghee (clarified butter), maize, grains, pulses, kidney beans or even cash. According to the villagers, the Forest Guard argues that this is for the firewood he provides them. If one does not give *rast*, the Guard does not allow them to collect firewood or fodder from the forests.

[16] In the context of the JFM programmes in Andhra Pradesh, Saito-Jensen and Jensen (2010) suggest that prior to the JFM, although the forest boundaries were drawn between state-owned forests and those of villagers, they had de facto access to forest resources. The villagers used to enter the forests to collect various forest produce by avoiding the timing and locations of patrols of forest guards. However, the JFM has led to a significant redefinition of these forest boundaries by establishing closures and assigning the task of vigilance to the JFMCs. This has reduced local people's access to resources as we also see in Navni and Chinnora.

[17] As mentioned earlier, the women and children collect medicinal herbs such as vanaksha, guchhi, doop, kaur and patis from the forests and sell these to local shopkeepers for additional incomes.

8.3 Increased Biomass, Reduced Access

to distant forests in the time of heightened militancy involved high risks as the field-staff imposed restrictions on open grazing in the surrounding forests. They informed me that since the establishment of closures, the FD has never opened them for villagers to collect grass.[18]

According to the pastoralists such as *gujjars* and *bakarwals*,[19] pastoralism is no more a profitable occupation due to the growing shortage of fodder since open forests are being enclosed by the FD for forest regeneration. They argued that the livestock population in the village is declining as they have started selling their livestock and looking for alternative livelihoods. Recall the argument of senior forest officials (in the previous chapter) that an increase in livestock and human population is putting pressure on the forest resources. They blamed villagers and pastoralists for destroying the forests by overgrazing. Contrary to this, the pastoralists in Navni informed me that the livestock population has actually decreased in the last 10 years due to restrictions imposed by the FD on their access to forests.[20] Many of them have taken up agriculture to support their incomes. Others who are still engaged in livestock rearing have to buy fodder from the local residents and from the neighbouring villagers during the winter, and move to mountain tops during the summer to graze their cattle.[21] The following two narratives explain the ways in which the pastoralists are coping with the shortage of fodder in the village:

> While I manage agriculture in Navni, my brother takes the sheep to a neighbouring village in the winter and stays there for four months. He has made an arrangement for grazing with a cultivator in that village who has got several maru trees on his land [...] We are charged Rs. 1500 by him for each maru tree. For providing fodder to our cattle during these months, we spend Rs. 3000. It is an expensive affair now. (Tirth Ram (age 70), Navni)

---

[18] A few of the respondents recalled that 5 years ago, the Guard charged them a Rs. 150 fee to collect fodder from the closures. Since the FD claims to provide fodder to the villagers for free, it can be inferred that the Guard charged the villagers illegally.

[19] The *gujjars* and *bakerwals* are traditional nomadic graziers who trade in livestock and its products. Generally, *bakerwals* rear sheep and goat for wool and meat while *gujjars* rear buffaloes for milk.

[20] Sundar et al. (2001) make a similar observation in other states of India where the villagers informed them that livestock numbers had decreased since the early 1990s, corresponding to a decrease in availability of fodder. They note that this assessment by the villagers themselves contrasts with the perceptions among foresters that livestock numbers are increasing.

[21] Pastoralists in other parts of India also adopt different strategies to compensate for reduced availability of fodder in their villages. For example, Saberwal (1999: 41) mentions various methods adopted by the herders in Himachal Pradesh to cope with the shortage of fodder through local level negotiations within the herding community as well as between herding and cultivating communities. He notes that a herder may draw upon family labour to 'pay' for access to grazing. Alternatively, a herder may choose to pay for such access through a monetary transaction, or through reciprocal sharing of winter and summer grazing grounds. Similarly, Agrawal (1999) in his study of a migrant pastoral community in Rajasthan argues that shepherds negotiate with the neighbouring landowners in the villages and with state officials in order to get access to fodder from both private and common lands. Axelby (2007) also notes similar processes of negotiations in the context of Gaddi shepherds in Himachal Pradesh.

> From May to September, I go to the upper reaches of Seoi Daar because there is not enough grass in the village. We have to pay Rs. 2000 to the Forest Guard for going there.[22] In October, we move to the plains. I have some contacts with landowners in the villages of the Kathua district. They lend us their land to graze our cattle. We have to pay Rs. 5000 for 150 sheep but there is no other choice. (Mohammed Aslam (age 62), Navni)

Another important point of contention between the FD and villagers in Navni and Chinnora relates to encroachments on forest lands.[23] The practice of demarcating private lands and government lands goes far back, and some of my respondents could recall that in the pre-independence period, *bhurjis* were made to demarcate the forest lands from the private lands in Navni and Chinnora.[24] This practice continues until the present day. The forest guard measures the forest land and *patwari* (local revenue official) keeps records of the village private lands. Some senior villagers argued that the practice of encroachment on the forest lands has increased after independence because the field-staff is corrupt and it is easy for villagers to alter the boundaries after offering them bribes.[25] According to my respondents, the practice of encroachment is done more by the relatively affluent people who can bribe the field officials.[26] These encroachers sometimes involve the poorer residents (living on the nearby lands) in their encroachment activities to ensure that they do not complain to the senior authorities.[27] The poorer households are co-opted by the affluent in encroachment of forest lands by providing incentives for sharing a small portion of the total land encroached by them. Munshi (age 70), a senior resident of Chinnora and member of the JFMC explained:

> There are many examples of encroachments of the forest lands around Navni and Chinnora, but these cases are never reported. Last year, I went to complain about such encroachments to the Range Officer in Chenani. He said that they do not have employees to check this, so I should take the case to *patwari*. I did not pursue this nor did they bother to look into it [...]

---

[22] While there is no grazing fee for cows, and up to six horses, the fee is Rs. 1.15, Rs. 0.20 and Rs. 0.40 for a buffalo, sheep and goat respectively. My respondent, however, informed that the pastoralists have to pay much higher than the official fee for grazing.

[23] The FDs throughout India consider encroachment, by way of expansion of agricultural activities, as the biggest threat to the forest areas. For example, Das (1997: 57) notes that the forest officials hold the view that 'notwithstanding the people's allegations and denials', encroachment of forest land for agriculture is 'frequent and rampant'.

[24] *Bhurjis* or boundaries are conventionally made of stone and mud though the FD is now proposing to make cemented *bhurjis* in the villages to discourage the practice of encroachments.

[25] Das (1997), in the context of Rajasthan, notes that the forest guards and other field-staff generally take cash, or are bribed with milk and milk products, in exchange for which they allow the villagers to encroach on the forest land for cultivation, entry into the enclosures, and felling trees.

[26] This corroborates the findings of Sundar et al. (2001: 186) that although encroachments are a desperate response by marginalised people for subsistence, the richer households are mainly involved in this practice because 'bribing the forest staff and ploughing freshly cut forest land is expensive'.

[27] Robbins (2000) makes a similar note in the context of rural Rajasthan and argues that corruption is structurally perpetuated by co-opting the poor. The caste and class elites offer minor concessions to the poor who, in turn, surrender the ability to complain and remain complicit.

8.3 Increased Biomass, Reduced Access

Due to such insensitivity of the officers and the people, the result is that the forest land is diminishing and private land is increasing.[28]

Below I briefly note the experiences of the villagers in the construction activities sponsored by the FD under the NAP and also in the previous projects. Although the NAP emphasises income generation activities and employment provision for the villagers in return for their shared responsibility of protecting the forests, it has failed to invoke any motivation among local villagers to participate in such activities of the FD. According to the villagers, while the FD has undertaken a few construction activities such as the building of bridges, water sprouts and the fencing of forest patches for regeneration, these activities have generated little income for them as the construction works lasted for only 10–15 days in each case. Those who are employed in tree plantation (by the FD), tree felling (by timber contractors) or construction works (by the JFMC) are paid lower wages, between Rs. 50–70 per day when compared to the prescribed wages of Rs. 100 (as per the forest guidelines). The villagers, therefore, do not see these forest activities as opportunities for enhancing their incomes nor for increasing the availability of biomass. A few also indicated the reasons for their lack of faith and willingness to participate in government projects. To illustrate, in an earlier programme of watershed development in the village, the respondents informed me that under that project, activities such as the establishment of closures and tree plantation took place for a period of 5 years. But after the project was over, the closures were damaged by the villagers in order to meet their demands for firewood and open grazing. Mohammad Ashraf (age 42), a resident of Navni explained his lack of interest in governmental projects and narrated his experience of working as wage-labourer in the watershed project:

I appreciate the work the watershed project did for forest regeneration and checking water and soil erosion but we labourers suffered a lot working on this project. We were not paid for 23 months [...] What was worse was that we were not even told that the project was over. We continued waiting for our wages as we thought we would be paid after six to eight months for our labour as had always been the case. Finally, we wrote an application to the DFO to grant our wages but we were told that the project was over, so nothing could be done [...] Five years is a short period for any project to mature. As soon as the old project begins to show its effect, it ends and a new project starts. I think that the programmes should be continued for a longer period to bring effective results for the development of the village.

To summarise thus far, the discussion presented in this section raises four main issues concerning the functioning of the NAP. First, it can be noted that while there has been an increase in forest cover in the last two decades, as some respondents have suggested, the villagers' access to forest resources such as timber, fuelwood and fodder has certainly been reduced. The FD has imposed several restrictions on

---

[28] I went to see encroachments with him in a forest named *Baskar*. He showed me the land of a Muslim cultivator who had encroached nearly 30 *kanals* of forest land about 20 years ago by bribing the forester. The second case was of another Muslim agriculturalist who had encroached 40 *kanals* of forest land. Although the FD has, on paper, reclaimed the area from him, he has constructed his house on the encroached land and is still living there. He explained that this land included around 500 deri trees, which are of even better timber quality than deodar.

village residents through various mechanisms such as permits for timber concessions, providing low quality wood for housing, enclosing forest areas for regeneration, restricting open grazing, imposing grazing taxes and even prohibiting the villagers to collect dry wood from the open forest lands. This is done in the name of increasing forest cover as well as the availability of biomass for local forest users. Yet, the incentives such as timber concessions or employment generation have failed to attract villagers' interest in working with the FD. There are similar observations that can be drawn from the works of other scholars. For example, Menzies and Peluso (1991), in the case of China, note that incentives to protect forests are more effective than controls on access, but they are themselves stifled by bureaucratic procedures and may turn into another set of de facto controls imposed by the state. Edmunds and Wollenberg (2003), in the context of the JFM in India, argue that foresters retain substantial control over forest management, issuing or denying permits for a wide array of management activities. Other scholars (e.g. Peluso 2003; Peluso and Vandergeest 2001; Barber 1989) suggest that state agencies control both resources and the activities of their subjects seeking access to those resources by mapping, demarcating, reserving forest areas and administering rural territories. In J&K, we noted that the process of obtaining timber concessions remains long and complicated; villagers need to wait for several months or pay bribes to the field-staff to shorten the process. Also, contrary to the claims of senior officials (discussed in the previous chapter) that the forests are depleting as a result of an increase in the livestock population, the narratives of pastoralists and villagers in Navni and Chinnora suggest that the livestock population has, actually, decreased in the recent past and they have started selling their livestock and taking up alternative livelihoods. Due to the shortage of the availability of fodder and restrictions imposed by the FD on grazing, some of them have started purchasing fodder from the landowners within the village or from cultivators in other villages. Likewise, the access to fuelwood for the poor has declined in spite of the regeneration of surrounding forests under various projects, the NAP being the most recent.

Second, in contrast to the NAP programme guidelines, we saw that the decision-making process is mainly controlled by the FD based on their scientific expertise with little or no participation of the local forest users. For instance, the selection of species is mainly done by the forest staff at the district level using their technical forestry information. We saw that the objective of such plantations is time-bound increase in forest cover without making any assessment of the actual needs and priorities of the villagers. This finding resonates the observations and cautions of other commentators with regard to decentralised decision-making. Njogu (2005), for example, argues that despite being recognised as critical stakeholders in conservation and development projects, in practice, communities often lack the power to become definitive stakeholders. Sarin et al. (2003) suggest that instead of revitalising the indigenous knowledge of local users, the joint management system developed for supporting livelihoods and maintaining ecological services, is actually reinforcing the FD's claim to be the monopoly holder of technical forestry knowledge. Reiterating this, Randeria (2007) notes that the new regulatory regimes of forest control are based on claims to scientific knowledge while disregarding local knowledge, norms and practices in practical terms, whose revival such interventions

8.3 Increased Biomass, Reduced Access

promised. On the same lines, Saberwal (1999) argues that the FD's claims to 'scientific' expertise implicitly assert the primacy of the scientific method as the most appropriate means of acquiring a better understanding of the functioning of nature. This alarmist scientific rhetoric of FD can often serve as a means of legitimizing its own authority with regard to how forest lands are used. Agrawal (1995) points out, there is little point in giving primacy to one or the other, partly because of the difficulties of identifying a coherent set of principles that underlies any form of knowledge. Both local and science-based knowledges are socially embedded, and, thus, subject to various biases (see also Feyerabend 1993; Pickering 1992). In Navni and Chinnora, we see that the lack of opportunities to exercise choice in decision-making by villagers ultimately results in their reluctance to cooperate with the FD in spite of their indigenous knowledge and willingness to regenerate protect the surrounding forests.

Third, the forest laws and regulations are experienced *differently* by village residents owing to their differential abilities to negotiate with the field-staff. In my conversations with the villagers, I noted several references to myriad practices of rent seeking by the field-staff, signalling that corruption at the lowermost level of the FD plays a determining role in the unfair distribution of benefits. Referring to unequal access to forest resources and their distribution, Springate-Baginski and Blaikie (2007) argue that although the wealthy use the forests less for subsistence purposes than marginal farmers and the landless, they are often in a position to gain more from the new opportunities which participatory forest management offers. Likewise, Sarin et al. (2003) note that instead of enhancing multidimensional space for local forest management, the JFM interventions empower the FD, often in alliance with village elites, to reassert its control over local land and forest resources. Collins (2008) argues that the powerful groups receive facilitation by the field officials for private gain in return for providing security to exploit environmental opportunities but the poor experience marginalisation. Platteau (2009: 60) states that in situations of 'power asymmetries' within communities, there is a considerable risk of local elites distorting information in a strategic manner and 'opportunistically capturing' a substantial portion of benefits. Owing to such unequal or clientelist conditions, the local elites are able to capture the local state, a process referred by Veron et al. (2006) as 'decentralised corruption'. In Navni and Chinnora, we noted that while the poor face problems in accessing timber and fodder despite their genuine requirements, the relatively affluent village residents benefit easily from both timber concessions and fuelwood. They also benefit extensively by encroaching forest lands but the ability of the poor to do the same is severely restricted due to their weak bargaining capacities vis-à-vis forest officials.

Fourth, information asymmetries between the FD and villagers provide an opportunity for the field-staff to bolster their authority over forest resources as well as manipulate forest laws for private gains. We noted that the villagers were largely unaware of the programme as well as the actual rules and regulations governing their access, due to which they were often harassed by the field-staff for their supposed 'illegal' access to forest resources. They were also paid less and irregularly due to their lack of information regarding minimum wages for construction activities. Wardell and Lund (2006) observe that people's ignorance of their rights provides scope for local authorities to define current practices as illegal. In the narratives of the villagers, I have demonstrated that, on the one hand, the forest staff compromises with forest laws and regulations to meet the requirements of relatively affluent and on the other, it uses them against the poor even for collecting dry wood under the false charges of 'illegality'. This suggests that, in practice, forest laws and policies are often bent to serve the powerful actors but can be used to work against the interests of the poor. Arguably, rather than forest laws and regulations defining the outcome of conservation policies, it is the local power dynamics which actually shape the conforming and transgressing behaviors of various forest users, thereby, determining the success or failure of conservation programmes. This again suggests that power is dispersed and fluid, and that forest conservation outcomes are shaped by complex processes of negotiations and contestations at the ground level rather than simply determined by the rationality from the 'above'.

8.4 Illegal Timber Felling: What If Fence Eats the Grass?

In the previous chapter, I noted that the senior forest officials believe that the illegal felling and timber trade has been controlled significantly since the initiation of the programme. The local villagers, as they argued, are more willing to cooperate in forest protection because of the efforts made by the FD to enhance community participation. It was also claimed that the JFMCs, now recognised by the FD as comanagers, assist the field-staff by complaining immediately if any damage to forests (such as illegal felling, forest fires and encroachments) has occurred in the village. I also discussed the involvement of the SFC and local contractors in illegal timber extraction, and the role of the field-staff in transgressing forest regulations while dealing with the timber and firewood needs of local villagers. In this section, I extend the discussion on these points further by presenting local understandings and the experiences of village residents (and field-staff) to assess the actual situation regarding illegal timber felling and trade in Navni and Chinnora, and the various players involved therein.

Although initially the field-staff, including the Forester and the Forest Guard, shared the viewpoint of senior officers that illegal timber extraction has been under complete control for the last few years, it was after a few meetings that they narrated the inside story of illegal felling and the timber trade, and their own limitations in controlling these activities. The Forester in Navni pointed out that militancy and turbulent political conditions in the 1990s had proved beneficial for timber smugglers across the state as forest patrolling became difficult.[29] He further explained

---

[29] Newspaper reports suggest that the unrest in Kashmir during the summer of 2010 has immensely benefitted timber smugglers in the state. As noted by a press agency, 'timber smugglers have largely benefitted from the restricted movement of forest guards and other officials during the course of unrest in Kashmir. The illegal felling of trees has shown an upward trend and has become so widespread that during the last 3 months alone, forest officials have seized more than 5000 cubic feet of timber from smugglers in different parts of the Valley' (Ganae 2010).

8.4 Illegal Timber Felling: What If Fence Eats the Grass?

that while the local poor also resort to illegal felling in times of crises such as damage to their houses due to snowfall or at the time of communal feasts for want of firewood; it is the Border Security Forces (BSF) and police, deployed since the early 1990s to curb militancy, that have actually damaged the forests extensively. These agencies are involved in illegal felling and smuggling out logs of wood using their official conveyances. In other instances, they even make furniture from deodar wood in the local furniture shops of Navni and transport it to their houses outside the state. The Forester stressed that the local villagers cannot indulge in this business because there are some 20 check-posts of the FD and the police on the way from Navni to Udhampur. Explaining this paradoxical situation, he argued that the local people complain to the field-staff about imposing restrictions on their access while turning a blind eye to the 'illegal' activities of paramilitary and security forces. Although he reported such cases to the senior officials of the FD, they have been reluctant to take any disciplinary actions against the illegal practices of security forces. Moreover, to register a court case for illegal felling, a witness is needed. No villager, he explained, would come forward to complain against the BSF for fear of receiving threats from them. Explaining the dilemma faced by the field-staff in taking action against the BSF, the Forest Guard of Navni added:

> In Mantalai, a few months back, BSF personnel felled 15 deodar trees. The Forest Guard of Mantalai complained about this to the DFO. The DFO sent a letter to the Deputy Inspector General, BSF complaining about the illegal felling by the BSF personnel in the region. After this, the Guard started receiving threats until he apologised and presented ten kilograms of *ghee* [clarified butter] to the BSF personnel in the village [...] In my forest range also, they illegally fell firewood and timber [...] I am afraid of the BSF because it will start snowing next month and they will clear a forest patch, and I will have to face antagonism from the local villagers.

The Forest Guard and Forester in Navni argued that they are considerate about the needs of the local people when their requirements are genuine and allow them to collect fallen wood from the nearby forests. Yet, they blamed the local police who have been extorting money from the village residents until recently in forest related offences. They argued that, in the past, the police officials used to harass the villagers by accusing them with false charges of tree felling. The police would ask for bribes even from the people who were given permits for timber concessions by the FD. However, in the recent years, due to changes in regulations related to forest offences, if the police confiscate logs from a villager, they have to take the case to the DFO and the charges can be ascertained only after the DFO's consent. This change in procedure, according to them, has saved the innocent from the hands of corrupt police.

Some JFMC members held the view that the illegal timber trade has certainly fallen in the last 10 years or so but has not been stopped completely. Apart from the BSF, the members indicated the forest field-staff for facilitating the continued green felling in the area. As mentioned in section one, the JFMC is required to assist the field-staff in witnessing such offences but, often, the field officials accept bribes from the offenders and let them go free. Moreover, members argued that the staff also violates the forest rules by charging nearly ten times more than the actual fee

Fig. 8.1 A ring peeled pine in the closure of Navni

Fig. 8.2 A burnt deodar trunk in the closure of Chinnora

for providing permits for timber concessions. The members argued that, in many cases, the forest guards get the application sanctioned from the DFO for one log but allow the applicants to cut and carry more. While the payment of the sanctioned timber is made to the Forest Department, the field-staff benefits from the payments made to them by the applicants for the additional trees cut.

Interestingly, the JFMC members informed me that the field-staff themselves have devised several strategies for green felling. For example, they do this by peeling the trunk in ring form or burning the bottom part of the trunk, causing the trees to fall naturally in the next few months (see Figs. 8.1 and 8.2). The field officials make informal agreements with the village residents and local contractors, and apply these felling techniques mainly to the enclosed areas and conceal their malpractices. The field-staff then justifies such premature felling on the pretext of forest fires or high wind velocity. The JFMC members believe that by doing this, the field-staff is able to earn money as well as maintain cordial relations with the public.

8.4 Illegal Timber Felling: What If Fence Eats the Grass?

As indicated earlier, the village poor have also adopted strategies to meet their requirements for fuelwood and timber by either illegal felling of trees or collecting the damaged and fallen timber after bribing the field-staff. According to a senior resident of Chinnora, in response to the strict regulations, poor villagers sometimes collect the damaged wood from the forests without informing the field-staff, and hide the stolen logs (underground) in their houses to be used in times of need. Obviously, they do so because the forest laws and regulations work against their subsistence needs. Below, I present some narratives pointing out the viewpoints of village residents on the various different players involved in illegal felling:

> Most of our forests are being destroyed by the security forces [...] The BSF gets funds from the Indian government to buy coal and kerosene oil. They pocket this money and, instead, cut the trees from the surrounding forests for firewood, taking our share away. (Mohammed Ashraf (age 55), Navni)

> I believe that the SFC has caused much harm to the surrounding forests. The contractors do not cut the trees marked in the compartments because they are difficult to access but fell green ones in the areas nearby for their convenience. Alhough they are allowed only to cut champ by the Department, at times they even cut maru, banj and deodar and intermix the wood logs [...] Some local villagers also indulge in green felling and sell the logs to the local contractors for making money. (Rehmat Ullah (age 62), *ex-sarpanch*, Chinnora)

> These days, there is a big problem of getting fuelwood and timber even for our basic needs. Previously, we used to collect dry wood from the surrounding forests but now we are told by the Guard that it is illegal because the forest is no longer open for free access. Since we need fuelwood, there is no other way than to bribe the Guard or buy firewood from local woodsellers. (Mohammed Aslam (age 40), Navni)

To summarise, in this section, I have demonstrated that there are multiple actors involved in the illegal felling of trees, including the BSF, forest staff, police, timber contractors and village residents. While indulging in illegal practices, we observed that it is the already powerful actors who largely succeed in forest exploitation and destruction. I have shown that the paramilitary force, the BSF is able to transgress the forest laws due to their position of privilege and power whereas the poorer village residents bear the brunt of nature conservation costs. The field-staff experience dilemmas in fulfilling their expected duties of forest protection and meeting the timber and firewood demands of their fellow village residents. Corroborating this point, Vasan (2006) argues that the field-staff also feels helpless in checking illegal activities for 'fear of physical danger'. We witnessed that the Forest Guard himself suggests various strategies of green felling to village residents who, in turn, bribe him for the illegal harvesting of timber. Referring to very similar tactics adopted by log traders in Thailand, Dheeraprasart (2005) highlights the role of the Forest Industry Organisation in illegal timber felling and explains that the timber traders employ the villagers to inflict a premature death of the trees by cutting a thin ring around the narks and sometimes applying chemicals in a large quantity on selected trees. In the context of South East Asia, Dauvergne (1997) also reports that in exchange for gifts, money or security, implementers, often ignore or even assist illegal harvesters and timber smugglers. More recent work of Rana and Chhatre

(2016) suggests collusion among traders, large landowners, and local forest officials with regard to the sale of trees from private lands as well as public forests in Himachal Pradesh, thereby posing monitoring and enforcement challenges to the state agencies. Another work of Chhatre and Saberwal (2006) illustrate the case of Great Himalayan National Park in which the forest authorities after passing the park regulation with regard to ban on grazing and plant collection bypassed it years later for 'unofficially' permitting grazing as well as collecting timber in order not to alienate potential vote-banks. What gets reflected from this discussion above is that it is ultimately, the management of contradictory forces between development and conservation that shapes the outcome of nature conservation regulations.

The poor village residents also resort to illegal practices mainly to meet their daily requirements of fuelwood or timber in the times of emergency, although with greater probability of being harassed by the enforcement officials. Rossi (2004) cautions that although the policies are strategically reworked by various actors, according to their own powers and interests, those in weaker bargaining positions are unlikely to prevail their interests over the actors dominantly placed. Scholars such as Bloomer (2009), Chatterjee (2004) and Peluso (1993), in different contexts, have argued that poverty and lack of alternatives are the main reasons for the indulgence of poor in illegal activities. Wiggins et al. (2004) suggest that resource dependent communities may ignore or deliberately contravene policy regulations to defend their interests or livelihoods. In J&K, I have argued that it is the poor who are considered to be the major destructors of forest resources by the FD. Rather than acting against the powerful players who commit the most serious violations of forest laws, the field as well as the senior officials remain complacent and put the burden of forest conservation and regeneration on the poor by imposing restrictions on their access.

I have shown that timber concessions 'through proper channels' are difficult to avail for the poorer residents but easily available for those who can afford to bribe the field-staff. Robbins (2000: 428) aptly notes, where existing patterns of social power are strong, 'state-defined common property rules, founded on principles of equity, are fashioned into inequitable de facto rules, cemented in local social capital and trust'. In the absence of effective formal mechanisms for delivering services to the poor, MacGaffey and Bazenguissa-Ganga (2000) argue that people rely either on the trust of personal relationships or create their own system of values by operating in the 'second economy'. Likewise, Nordstrom (2004) suggests that 'shadow economies' function through negotiated rules, laws and mutual trust among the participants. Brown and Ekoko (2001) maintain that unlike conventional polarisation of local villagers versus external loggers, there are multiple actors who regulate the use, protection and exploitation of forests and determine the outcome of forest management policies. I have demonstrated that forest conservation and protection in J&K is not simply a matter of conflict or cooperation between village residents and the FD but an issue of power relations between multiple actors including timber contractors, paramilitary forces and police. Ironically, the actors who are meant to either fence the forests or provide security and protection are the key players in transgressing forest laws and regulations.

8.5 Conclusion

From this chapter, I draw three conclusions. First, while the NAP promises a rich array of benefits for local populations, from representative decision-making to increased availability of biomass, it has not resulted in bringing any effective change in the relationship of villagers with the FD in terms of control or access to local forest resources. I have shown the dominant position of the field-staff in selecting the members of the committee, preparation of microplans, selection of entry point activities, choice of species for forest regeneration, and control over funds for various activities under the programme. I have also argued that despite the claim of devolution of funds to local communities for facilitating the NAP, the release of money is strictly controlled by the forest bureaucracy at both divisional and field levels. In practice, the FD, in the guise of decentralisation and people's participation, is still able to retain centralised control of forest use and management.

Second, there are emerging areas of conflict, cooperation and contestations between the FD, JFMC members, *panchayat* and the villagers. Contrary to the claims of forest officials that the programme is successfully bridging the gap between the villagers and the FD, we saw that it is actually creating a divide between the committee members and other villagers, and between democratically elected rural local bodies and the FD sponsored JFMCs. By making the JFMC a representative of the local population in forest protection, the field-staff tries to shed its responsibility of forest conservation on to the JFMC and also divert the hostility of the villagers towards the committee. I have noted that the DFO and the *ex-sarpanch* collaborate with each other to pursue their respective agendas while sidelining the existing JFMC which was entrusted with the implementation of the various project activities. Owing to these new spaces of cooperation, conflict and contestations, I argue that conservation interventions like the NAP, often get refracted and reshaped by power relations at the microlevel.

Third, the majority of the respondents suggested that there has been an increase in the forest cover in the last ten or so years but their own access to forest resources such as timber, fuelwood and fodder has reduced. Since the forest regulations are incompatible with the subsistence needs of village residents, we saw that the villagers have no choice but to bribe the field-staff or resort to 'illegal' practices for accessing forest resources. The village residents have differential abilities to negotiate with the field-staff owing to their status determined by various factors including wealth and power. While the relatively affluent succeed in obtaining timber concessions and fuelwood easily or even in encroaching forest land, the poor are harassed for the collection of dry twigs or damaged timber. Moreover, the blame of forest destruction is put on the poorer residents and pastoralists but the powerful actors such as the BSF, police and timber contractors who indulge in extensive forest exploitation 'illegally', go scot-free.

From the above discussion it can be argued that both law and science are being used as a means to legitimise the control of forest bureaucracy, to serve the interests of the powerful and affluent, and at the same time, to restrict the access of poorer residents to forest resources. A general response from the villagers was that they are confused regarding what is 'legal' and what is not. The forest laws and regulations keep changing and villagers are, often, not informed about the rights which affect their access to resources. Since the senior forest officials in J&K rarely visit the field, the villagers are dependent on the field officials to understand the new regulations. The forest laws are, often, interpreted by the field officials according to their own understandings and vested interests for creating rent seeking opportunities. Furthermore, we observe that scientific knowledge is privileged over indigenous or local understandings of forest use and management even under the NAP. Apart from reinforcing the expert authority of the FD by means of selection of species to increase forest cover, we also see that the field-staff themselves disseminate strategies for causing the premature death of trees.

Overall, the case study of forest use and control in J&K explained the shift in management strategies from 'communal' to 'custodial' to 'joint'. Beginning from the colonial period to post-colonial times, we saw a continuous decline in the power of local forest users to control local resources. While the scientific rationality succeeded in limiting the access of local users to forest resources, it proved subservient first to imperial and then to varying national interests of forest exploitation and conservation. In the last two decades, the science of 'sustainability' has changed the focus of forest use from commercial exploitation to forest regeneration through local participation and decentralisation, based on the assumption of convergence of interests of forest managers, the conservation community and forest dependent populations. I have demonstrated that the broad objectives of the NAP, the split role of the state in forest protection and exploitation, information asymmetries between the FD and villagers, lack of actual devolution of power and funds to forest citizens, differential abilities of village residents to access forest resources and extensive destruction by both timber contractors and security forces, result in various forms of politics from macro- to microlevels. I argue that owing to the differential power and diverse interests of players involved in forest use and control, the forest conservation policies are negotiated, manipulated and reworked in very different ways than originally envisaged by policy makers. While the FD bureaucracy, timber contractors, security forces and local elites are all able to secure their respective interests, it is the poor who are compelled to limit their forest use and bear the cost of forest conservation and regeneration.

References

Agrawal A (1995) Dismantling the divide between indigenous and scientific knowledge. Dev Chang 26:413–439

Agrawal A (1999) Greener pastures: politics, markets and community among a migrant pastoral people. Duke University Press, Chapel Hill

Agrawal A (2005) Environmentality: technologies of government and political subjects. Duke University Press, London

References

Agrawal A, Gibson C (1999) Enchantment and disenchantment: the role of community in natural resource conservation. World Dev 27(4):629–649

Axelby R (2007) It takes two hands to clap: how *gaddi* shepherd in the Indian Himalayas negotiate access to grazing. J Agrar Chang 7(1):35–75

Ballabh V, Balooni K, Dave S (2002) Why local resource management institutions decline: a comparative analysis of van (forest) panchayats and forest protection committees in India. World Dev 30(12):2153–2167

Barber VV (1989) The state, the environment, and development: the genesis and transformation of social forestry policy in new order Indonesia. PhD thesis. University of California, Berkeley

Bhattacharyya D, Jayal NG, Mohapatra B, Pye S (2004) Introduction. In: Bhattacharyya D, Jayal NG, Mohapatra B, Pye S (eds) Interrogating social capital: the Indian experience. Sage, New Delhi, pp 15–34

Bloomer J (2009) Using a political ecology framework to examine extra-legal livelihood strategies: a Lesotho-based case study of cultivation of and trade in cannabis. J Polit Ecol 16:49–69

Brown K, Ekoko E (2001) Forest encounters: synergy among agents of forest change in Southern Cameroon. Soc Nat Resour 14(4):269–290

Chambers R (1987) Sustainable livelihoods, environment and development: putting poor rural people first, Paper No. 240. Institute of Development Studies, Sussex

Chatterjee P (2004) The politics of the governed: reflections on popular politics in most of the world. Columbia University Press, New York

Chhatre A, Saberwal V (2006) Democratising nature: politics, conservation and development in India. Oxford University Press, New Delhi

Collins TW (2008) The political ecology of hazard vulnerability: marginalisation, facilitation and the production of differential risks to urban wildfires in Arizona's White Mountains. J Polit Ecol 15:21–43

Das P (1997) Kailadevi wildlife sanctuary Rajasthan: prospects for joint management. In: Kothari A, Vania F, Das P, Christopher K, Jha S (eds) Building bridges for conservation: towards joint management of protected areas in India. Indian Institute of Public Administration, New Delhi

Dauvergne P (1997) Shadows in the forest: Japan and the politics of timber in South East Asia. The Massachusetts Institute of Technology Press, Cambridge, MA

Dhanagare DN (2000) Joint Forest Management in Uttar Pradesh: people, panchayats and women. Econ Polit Wkly 35(37):3315–3324

Dheeraprasart V (2005) After the logging ban: politics of forest management in Thailand. Foundation for Ecological Recovery, Bangkok

Dressler W, Buscher B, Schoon M, Brockington D, Hayes T, Kull C, McCarthy J, Shrestha K (2010) From hope to crisis and back again? A critical history of the global CBNRM narrative. Environ Conserv 37(1):5–15

Edmunds D, Wollenberg E (2003) Local forest management: the impacts of devolution policies. Earthscan, London

Feyerabend P (1993) Against method. Verso, London

Francis P, James R (2003) Balancing rural poverty reduction and citizen participation: the contradiction of Uganda's decentralisation program. World Dev 31(2):325–337

Ganae N (2010) Unrest in Kashmir proves juicy for jungle smugglers. Agence India Press. http://agenceindiapress.com/. Accessed 15 Nov 2010

Gupta A (1998) Postcolonial developments in the making of modern India. Duke University Press, Durham

Kawosa MA (2001) Forests of Kashmir: a vision for the future. Natraj Publishers, Delhi

Kothari A, Vania V, Das P, Christopher K, Jha S (1997) Building bridges for conservation: towards joint management of protected areas in India. Indian Institute of Public Administration, New Delhi

Kumar S (2002) Does participation in common pool resource management help poor? A social cost-benefit analysis of joint forest management in India. World Dev 30(5):1421–1438

Larson A (2004) Formal decentralisation and the imperative of decentralisation 'from below': a case-study of natural resource management in Nicaragua. Eur J Dev Res 16(1):55–70

Li TM (2005) Beyond "the state" and failed schemes. Am Anthropol 107(3):383–394

Li TM (2007) Practices of assemblage and community forest management. Econ Soc 36(2):263–293

MacGaffey J, Bazenguissa R (2000) Congo-Paris: transnational traders on the margins of the law. International African Institute in association with James Currey and Indiana University Press, London

Mahanty S (2002) NGOs, agencies and donors in participatory conservation. Econ Polit Wkly 37:3757–3765

Menzies N, Peluso NL (1991) Rights of access to upland forest resources in Southwest China. J World Forest Resour Manag 6:1–20

Neumann RP (1992) Political ecology of wildlife conservation in the Mt Meru area of northeast Tanzania. Land Degrad Rehabil 3:85–98

Neumann R (2001) Disciplining peasants in Tanzania: from state violence to self-surveillance in wildlife conservation. In: Peluso NL, Watts M (eds) Violent Environments. Cornell University Press, Ithaca, pp 305–327

Njogu JG (2005) Beyond rhetoric: policy and institutional arrangements for partnership in community based forest biodiversity management and conservation in Kenya. In: Ros-Tonen, Dietz T (eds) African forests between nature and livelihood resources: interdisciplinary studies in conservation and forest management. The Edwin in Mellen Press, New York, pp 285–316

Nordstrom C (2004) Shadows of war: violence, power and international profiteering in the twenty-first century. University of California Press, Berkeley

Peluso NL (1993) Coercing conservation: the politics of state resource control. In: Lipshutz R, Conca K (eds) The state and social power in global environmental politics. Columbia University Press, New York, pp 199–218

Peluso NL (2003) The politics of specificity and generalisation in conservation matters. Conserv Soc 1(1):61–64

Peluso NL, Vandergeest P (2001) Geneologies of the political forests and customary rights in Indonesia, Malaysia and Thailand. J Asian Stud 61(3):761–812

Pickering A (1992) Science as practice and culture. Chicago University Press, Chicago

Platteau J (2009) Information distortion, elite capture and task complexity in decentralised development. In: Ahmad E, Brosio G (eds) Does decentralisation enhance service delivery and poverty reduction? Edward Elgar, Cheltenham, pp 23–72

Rana P, Chhatre A (2016) Rules and exceptions: regulatory challenges to private tree-felling in Northern India. World Dev 77(c):143–153

Randeria S (2007) Global designs and local lifeworlds: colonial legacies of conservation, disenfranchisement and environmental governance in post-colonial India. Interventions: Int J Postcolonial Stud 9(1):12–30

Rangan H (1997) Property versus control: the state and forest management in the Indian Himalayas. Dev Chang 28(1):71–94

Rangarajan M (1996) Fencing the forest. Oxford University Press, New Delhi

Ribot J (1998) Theorising access: forest profits along Senegal's charcoal commodity chain. Dev Chang 29:307–341

Robbins P (2000) The rotten institution: corruption in natural resource management. Polit Geogr 19(4):423–443

Rossi B (2004) Revisiting Foucauldian approaches: power dynamics in development projects. J Dev Stud 40(6):1–29

Ros-Tonen M, Zaal F, Dietz T (2005) Reconciling conservation goals and livelihood needs: new forest management perspectives in the twenty-first century. In: Ros-Tonen, Dietz T (eds) African forests between nature and livelihood resources: interdisciplinary studies in conservation and forest management. The Edwin in Mellen Press, New York, pp 3–30

Saberwal VK (1999) Pastoral politics: shepherd, bureaucrats and conservation in the Western Himalaya. Oxford University Press, Delhi

References

Saito-Jensen M, Jensen CB (2010) Rearranging social space: boundary-making and boundary-work in a joint forest management project, Andhra Pradesh, India. Conserv Soc 8(3):196–208

Sarap K, Sanrangi TK (2009) Malfunctioning of forest institutions in Orissa. Econ Polit Wkly XLIV(37):18–22

Sarin M (1998) Who is gaining? Who is losing? Gender and equity concerns in joint Forest management. In: Working paper by the gender and equity sub-group, National Support Group for JFM. Society for Promotion of Wastelands Development, New Delhi

Sarin M, Singh NM, Sundar N, Bhogal RK (2003) Devolution as a threat to democractic decision making in forestry? Findings from three states in India, ODI working paper 197. Overseas Development Institute, London

Shackleton CM, Wills T, Brown K, Polunin NVC (2010) Editorial: reflecting on the next generation of models for community-based natural resources management. Environ Conserv 37(1):1–4

Sinha H (2006) People and forests. Concept Publishing Company, Delhi

Springate-Baginski O, Blaikie P (2007) Forests, people and power: the political ecology of reform in South Asia. Earthscan, London

Sundar N (2001) Beyond the bounds? Violence at the margins of new legal geographies. In: Peluso NL, Watts M (eds) Violent environments. Cornell University Press, Ithaca, pp 328–353

Sundar N (2004) Devolution, joint forest management and the transformation of social capital. In: Bhattacharyya D, Jayal N, Mohapatra B, Pye S (eds) Interrogating social capital: the Indian experience. Sage, New Delhi, pp 203–232

Sundar N, Jeffery R, Thin N (2001) Branching out: joint forest management in India. Oxford University Press, New Delhi

Vasan S (2006) Living with diversity. Indian Institute of Advanced Study, Shimla

Veron R, Williams G, Corbridge S, Srivastava M (2006) Decentralised corruption or corrupt decentralisation? Community monitoring of poverty alleviation schemes in Eastern India. World Dev 34(11):1922–1941

Wardell DA, Lund C (2006) Governing access to trees in Northern Ghana: micropolitics and the rents of non-enforcement. World Dev 34(11):1887–1906

Wiggins S, Marfo K, Anchirinah V (2004) Protecting the forest or people? Environmental policies and livelihoods in the forest margins of Southern Ghana. World Dev 32(11):1939–1955

Chapter 9
On Conservation Politics: Cooperation, Conflicts and Contestations

Abstract In this chapter, I have concluded that power is not concentrated in a single site but is dispersed and fluid, and that nature conservation interventions rather than resulting in fixed and determined outcomes are accepted, resisted and reconfigured by various actors at multiple levels. In the case of conflict regions like J&K, these conservation interventions may also coincide with the ongoing struggles between the state and citizens on the question of legitimacy to rule as well as with the attempts of the historically powerful actors to dominate the poor and marginalised. Although there are valid arguments in favour of conserving both *chirus* and forests, it is the poor that bear the brunt of nature conservation costs.

Keywords Contested conservation · Jammu and Kashmir · Power · Split role · Delegated illegality

In this book I have discussed the politics of wildlife and forest conservation by exploring two recent interventions in J&K. Following an integrative multi-level and multi-actor framework, the study has examined the process of the banning of *shahtoosh* trade and the implementation of the NAP, and analysed the power, agendas and interests of the various stakeholders involved. In the case of ban on *shahtoosh*, apart from discussing the myths, legends and realities concerning the origin of *shahtoosh*, shawl production process as well as the growth and development of the shawl industry, I highlighted the areas of conflict and tension between conservationist groups, central government and the J&K state in banning the trade of *shahtoosh*. I also analysed the various forms of micropolitics that emerged in response to the imposition of the ban in the state. In the case of forest conservation and management, I traced the systems of management and control of forests in J&K from early colonial times, and have presented the interplay between various state and non-state actors associated with the implementation of joint forest management under the NAP. I also highlighted the areas of cooperation, conflict and contestations at the village level with regard to managing, controlling and accessing forest resources.

The analysis of the two case-studies began with examining how the historical factors and circumstances influenced both the shaping of the conservation policies

---

© Springer International Publishing AG 2018
S. Gupta, *Contesting Conservation*, Advances in Asian Human-Environmental
Research, https://doi.org/10.1007/978-3-319-72257-3_9

and the determining of their outcomes. The study then examined how law, science and politics played a central role in defining access to and control over resources. By taking into account the agenda and priorities of the global conservation community, internal divisions and complexities within the state, and the differentiated concerns of social groups within the affected communities, the book has illustrated the various forms of contestations that take place when global conservation policies are transmitted to the local worlds. The narratives of the local communities were presented in the book to demonstrate how various historical, political, scientific, legal and social forces interact with each other, and how they are challenged, compromised and reconfigured.

I have engaged with diverse viewpoints on nature conservation in understanding how conservation policies meet local realities. The mainstream literature advocates global control over local natural resources for their sustainable use and management. I examined arguments in this literature ranging from the urgent need for global natural resource management (e.g. IUCN 1980; WCED 1987), to realising the goals of sustainable management of natural resources along with the sustainable livelihoods of dependent populations (e.g. Bebbington 1999; Scoones 1998; Chambers and Conway 1992), to natural resource management through local community involvement (e.g. Hulme and Murphree 2001; Gadgil 1998; Shepherd 1998; Adams 1996) and devising effective mechanisms for participation through local partnerships and decentralisation (e.g. D'Silva and Nagnath 2002; Saigal 2000; Lynch and Talbot 1995). In contrast, the alternative viewpoint shows scepticism towards the global conservation agenda and considers global management of local resources as a new form of imperialism (Calvert and Calvert 1999; Guha and Martinez-Alier 1997), often with coercive and violent force to control the local populations and their resources (see Peluso and Watts 2001; Lohmann 1996; Peluso 1993; Neumann 1992). Some scholars even caution against the dangers of repeated centralisation of actual powers within decentralised natural resource management practices (see Sarap and Sanrangi 2009; Sinha 2006; Sundar 2004; Ballabh et al. 2002; Poffenberger and Singh 1996). My observations during the fieldwork revealed that the deterministic interpretations of conservation interventions fail to provide convincing accounts of the complex and diverse processes through which the policies meet the local realities. My analysis, thus, goes beyond the claims of both 'sustainable resource management' and 'coercive conservation' schools. I have argued that power relations are at the heart of sustainable resource management and use, and that while the global conservation agenda can be seen as coercive in nature, it does not always result in straightforward compliance by the state and local actors. My findings, therefore, resonate with the arguments of scholars (such as Gupta 2009; Singh 2008; Li 2007; Rossi 2004; Leach et al. 1999) who have emphasised that global interventions rather than transforming the local realities, in positive or negative ways, are resisted, reinterpreted and reshaped in the process of implementation, resulting in such interventions taking very different forms than intended in the policy rhetoric. My analysis suggests that rather than the conservation policies shaping the local realities, or the 'state' dictating the 'community', it is the powerful stream of actors located at multiple sites which actually determines the outcomes of nature conservation interventions by undergoing through the processes of contestations and negotiations.

In the following discussion, I revisit the case-studies and reflect on the broad conclusions of this research. I begin by analysing the nature of power in conservation interventions. Following this, I highlight the spaces of cooperation, conflict and contestations between different actors in terms of: (a) relationship between law, science and politics, (b) partnerships and collaborations, (c) split role of the state, (d) different impacts and differential abilities, (e) rationality of rule, (f) illegal trade and (g) limited space for protest. I raise the issue of accountability and outline the main implications of the study for policy-making in the areas of wildlife and forest conservation in general, and specifically in the context of the conflict-ridden state of J&K.

9.1 Power as Dispersed and Fluid

I have demonstrated that power rather than located at one site is dispersed and fluid, and that when conservation policies meet local realities, differently placed actors engage in contestations at multiple sites to secure their respective interests. This means that the sole capacity of conservation interventions to change local realities can be questioned. There is no doubt that in recent times, the influence of transnational actors has considerably increased in defining the relationship of nation-states and local populations with their resources (Gupta 1998), and that the global conservation community has adopted a coercive attitude in addressing the nature conservation issues (Peluso 1993). It is also argued that the transnational conservation organisations, using moralistic constructions, present nature conservation as global necessity (Singh 2008). I have shown that despite the initial resistance of the J&K state to the banning of *shahtoosh*, the will of the international community prevailed in making the trade 'illegal'. Also, the issue was dealt largely along conservation lines with little consideration to the livelihood concerns of the local populations. The global conservation thinking on sustainable use of forests through local participation and decentralisation also became a guiding force to govern forest resources in J&K.

The other side of the narrative is that the state and local populations rather than acting as passive recipients are consistently engaged in resisting, negotiating and reshaping these conservation interventions (Gupta and Sinha 2008; Rossi 2004; Moore 2000; Li 1999). I argue that although the conservation policies from 'above' certainly play a significant role in defining the access of these actors to their resources, the conforming and transgressing behaviours of the local state and affected populations is actually determined and shaped more by the local power relations and circumstances. In other words, when the recipients of policies and interventions are faced with discourses external to their language, culture and society, the relative distance from the sources of rationality increases the room for manoeuvre available to them (Rossi 2004: 26). Referring to the dispersed and fluid

208   9 On Conservation Politics: Cooperation, Conflicts and Contestations

nature of power, I have maintained that the affected communities, located at distance from the centres of policy formulation, exercise their power by finding ways to transgress regulations imposed from 'above'. However, I have also demonstrated that it is primarily the relatively powerful actors within the affected communities who are able to manipulate policies to suit their requirements, priorities and interests.

9.2 Between Cooperation and Conflict: Spaces for Contestation

When power rather than being located at one site is dispersed and fluid, it is likely to move beyond cooperation and conflict into the domain of contestation and negotiation. In practice, resource governance is seen as a matter of contested locales that involve negotiations between state and local communities (Rose 1999; see also Rangan 1995). Salskov-Iversen et al. (2000) maintain that the 'local' is a site where transnational discourses are challenged, translated, negotiated and internalised into numerous political re-positionings. Resistance to nature conservation, Li (2005: 391) suggests, involves not simply rejection but the creation of something new, finding allies and repositioning in relation to the various powers confronted. Moreover, global interventions can also be seen to articulate with deeper histories of government attempts to regulate and discipline landscapes and livelihoods, and generate politics of contingency and contestation (Moore 2000). Therefore, as Gupta and Sinha (2008: 290) argue, there is a need to approach local development contexts as spaces of negotiation in which the power of globally dominant discourses is received, accepted, amended and rejected, leading to more diverse outcomes than the deterministic interpretations allow. Below, I discuss the spaces of cooperation, conflict and contestations through which the two conservation interventions unfolded in J&K.

9.2.1 Science, Law and Politics

I have argued that science and law played a crucial role in setting the stages for politics between various actors involved in wildlife and forest conservation efforts. While both science and law served in legitimising the rationality of the conservationist community and its intrusion in controlling local resources in J&K, we found that they also played a significant role in serving the interests of powerful actors both within the state and resource dependent populations. As discussed in Chap. 4, science and scientists were enrolled in legitimising the imposition of the ban with the view of urgent need for protecting the Tibetan antelopes and reversing the declining trend of its population size. Yet, little efforts were made to explore the

9.2 Between Cooperation and Conflict: Spaces for Contestation

possibility of *chiru* farming. In Chap. 6, I explained how scientific forestry played a crucial role in legitimising the intrusion of the forest bureaucracy in J&K for the systematic management of forests. However, the scientific management principles were compromised in favour of war-time needs and serving the commercial interests of the colonial as well as post-colonial state. In the years to come, a new application of science was promulgated by the international conservation community under the label of 'sustainable management' of forests. Chapter 8 argued that the scientific expertise, apart from legitimising the control of the professional community in protecting the forest resources, was also used by the forest bureaucracy to maintain its dominant position in terms of selection of species for forest regeneration. I have demonstrated that while science played an authoritarian role in controlling local resources (Saberwal and Rangarajan 2003; Greenough and Tsing 2003; Yearley 1996) and maintaining its superior position in decision-making (Blaikie 2006; Saberwal 2003; Agrawal 1995), it surrendered to the will of powerful actors who manipulate its application and use it to suit their own priorities and interests. Such application of science and its subservience to commercial and vested interests results in shaping the outcomes of conservation interventions very different from the claims made by the global conservation community with regard to sustainable management and use of natural resources.

Apart from science, the law served as another platform for the contending parties in both imposing the regulations and violating them. I have shown the paradoxical relationship between the centre and the state owing to the special constitutional status of J&K, posing a situation of direct conflict between the conservationists, central government and the J&K state. As discussed in Chap. 4, separate central and state wildlife laws, ambiguities with regard to the understandings of *chiru* hair and wool, as well as the origin of *shahtoosh* wool became significant points of contestation for the parties involved. This resulted in the reluctance on the part of the state agencies to respond to the pressures of the international conservation groups and central government, and to delay the imposition of the ban in J&K. In the case of the NAP, I have argued that the broad objectives of the programme, and policy incoherence provide ample space for politics between the forest managers and local populations over the sharing of responsibilities, distribution of forest produce, and evaluating the success or performance of programmes according to their own priorities (cf. Bene et al. 2009; Sundar et al. 2001). I have maintained that although the forest bureaucracy placed an emphasis on decentralised management, there are avenues for centralised control over resources (Poffenberger and Singh 1996; Menzies 1993). For example, the FD retained the power to terminate the JFMC membership in case the members fail to perform their responsibilities. Moreover, it was noted that the DFO and the field-staff retain the authority in terms of timber concession grants to the villagers, depending on the applicant's performance in abiding by the forest regulations and assisting the field-staff in forest conservation. There is also a need for prior approvals by the local communities for availing benefits under the programme, determined mainly by the field-staff or the DFO. Irrespective of the legal rights available to the forest citizens, the boundaries between legality and illegality are often (mis)interpreted by the field-staff for rent seeking opportunities. The misuse of the law is also observed when the field-staff curtails the rights of the poorer villagers in the name of illegality but transgresses the regulations to benefit the relatively affluent village residents.

9.2.2 Uneasy Partnerships and Collaborations

Mainstream conservation and development policies often assume cooperation as the underlying feature of sustainability but overlook issues of conflicts and contestations in natural resource governance. While the NAP envisages an unproblematic relationship based on cooperation between the JFMC and the FD, in actual practice, we see problems of control over funds and decision-making between the MoEF, the CF, the DFO, field-staff and the JFMC Chairperson. As discussed in Chaps. 7 and 8, there are bottlenecks in the release of funds to the state FD by the MoEF, to the DFO by the CF and to the JFMC by the forester. I have presented contradictory imperatives and claims of the senior officials and the field-staff in terms of problems of effective implementation of the projects, and between the SFC and the FD in terms of timber harvesting and revenues. In Navni and Chinnora, I have shown conflicts between the JFMC and the *panchayat* on the one hand, and emerging cooperation between the DFO and *ex-sarpanch* as well as between the forest field-staff and local elites. Although it is claimed that the JFMC is a representative of villagers, we witnessed the emerging hostility between JFMC members and ordinary villagers over issues of forest policing and protection.

In the case of *shahtoosh*, I have discussed the exploitative relationship between manufacturers and artisans (Chap. 3) and the reluctance on the part of manufacturers to respond to the demands of workers' associations for increase in wages (Chap. 4). However, we witnessed that a cooperative relationship started taking shape, although for a brief period, between manufacturers and other categories of *shahtoosh* workers, with manufacturers joining the previously formed associations of the weavers and motivating their subordinate workers to cooperate in order to challenge and resist the ban. We saw different tactics adopted by the manufacturers to resist the ban such as the false promises made by them to increase the wages and abandon the use of machines, the display of spinning wheels in demonstrations to publicise the loss of traditional occupations, and later corrupting the politicians and enforcement officials in order to continue both the use of machines for dehairing and spinning as well as the illegal trade of *shahtoosh*. I argue that by collaborating with each other and engaging in a protest movement with subordinate workers against the ban, the manufacturers could both publicly obscure their exploitative relations with their employees and also succeed in continuing to use machines and the illegal production of shawls.

9.2.3 Split Role of the State

My research also substantiates the argument that the 'state' rather than being a unified or monolithic entity is heterogeneous in nature and marked with internal complexities and divisions (cf. Corbridge et al. 2005; Vira 1999; Gupta 1995; Midgal 1988). While giving an appearance of the unity of command, I have explained that the various state actors pursue their own agendas and interests, contributing to the politics of contingency in natural resource management. The internal divisions within the state may result in unanticipated consequences of conservation interventions. Taking into account the heterogeneity of the state, I suggest that state actors create situations where there is neither effective implementation of conservation regulations nor complete disregard to the laws concerning these regulations. This, in turn, lead to perverse incentives such as rent seeking opportunities for the local officials (Wardell and Lund 2006) and formation of nexus between political actors, enforcement agencies and the local elites (Bloomer 2009; Veron et al. 2006). Corroborating *these* arguments, this book has demonstrated that the state, in practice, has performed a dual role in enforcing forest and wildlife regulations as well as contravening them. I have referred to such dual performances of the state and its constituents as *split role*, critical in understanding the incongruence between the stated objectives and actual results of nature conservation interventions. It is also argued that by weak enforcement of the banning of *shahtoosh*, the J&K state is able to show resistance to the imposition of global agendas and interventions. In the case of the NAP, various layers of state actors contravene the stated aim of the new policy and resist effective devolution of power. Furthermore, the enforcement officials in both cases perform a divided role of law enforcement along with allowing, ignoring or even participating in 'illegal' trade in natural resources (cf. Dauvergne 1997). In line with the arguments of scholars (such as Vasan 2006; Corbridge et al. 2005; Mahanty 2002; Vira 1999) in relation to local state performances in different parts of India, this research shows that the field-staff in J&K faces a challenging situation of acting as forest police as well as maintaining friendly relations with the fellow village residents.

I have discussed in Chap. 4 that the J&K state continued to delay the imposition of the ban on *shahtoosh* but, having lost the legal battle (and succumbing to international pressures), ultimately submitted to the international will. I have also shown that the imposition of the ban in 2002 resulted in party politics with the ruling party, NC (National Conference), having little leverage after the decision of the J&K High Court. The main opposition party of that time, the PDP (People's Democratic Party) demanded the lifting up of the ban and criticised the party in power for its inability to resist the will of the central government and for neglecting the issues of livelihoods as well as cultural heritage. However, when the PDP came to power in 2003, it did not take the issues of alternative livelihoods and rehabilitation seriously but instead resorted to weak enforcement of the ban which mainly helped the manufacturers and traders. While the ban had a direct bearing on the livelihoods of poor skilled workers whose voice and concerns were rarely addressed in the legal battle,

I have explained that the strong nexus between politicians, manufacturers and traders succeeded not only in persuading the state to delay the imposition of the ban but also in allowing illegal trade to continue. The enforcement officials as well as politicians saw this as another opportunity for rent seeking in the already corrupt system of governance in J&K.

In the case of implementation of the NAP, I have argued that the state FD, while welcoming project funds coming from the centre, has been reluctant in devolving actual powers to the local communities. At the field level, as explained in Chap. 7 and illustrated in Chap. 8, the forest field-staff plays a crucial role in determining the access of various actors in the village to the forest resources, including timber, fuelwood and fodder. While there are strict regulations in relation to forest use and management to restrict the access of villagers, the field-staff could be seen transgressing the law in various ways: by misappropriating the project funds for NAP activities; by allowing the village residents to collect forest resources 'illegally' in the times of need; by charging higher fees for timber concession permits; by suggesting tactics for illegal timber harvesting in lieu of bribery; by allowing relatively powerful villagers to encroach forest lands; and by turning a blind eye to the destruction of forests by para-military forces. I have argued that the state appears to align with the rationality of the conservation community in terms of promoting participation, partnership and decentralisation in forest management but various state actors, especially the field-staff, are able to secure their respective interests amidst changing discourses and policies coming from 'above'.

I have maintained that the power of the conservation community is confronted with the power of both the local state actors and differently positioned social groups within resource dependent populations that, in turn, shape the outcome of nature conservation policies. The split role of the state is decisive in the complex process of transmission of conservation policies to the local worlds. The state actors in the process of policy implementation create avenues whereby resource dependent populations cooperate, negotiate and contest for access to and use of resources. In the case of J&K, I have shown that these state practices generated the politics of contingency in relation to natural resource management, again suggesting the dispersed and fluid nature of power.

9.2.4 Differential Impact, Differential Abilities

Since communities are not homogenous, the impact of conservation interventions is likely to be differential on different social groups within the affected communities. The differential impact of nature conservation policies also leads to different kinds of conforming and resisting behaviours of differently positioned *shahtoosh* workers and village residents. While it is important not to characterise less powerful actors as passive, Rossi (2004: 26) argues that there is a difference between framing the terms of reference and being able to manipulate dominant orders of discourse subversively. Indeed, the lack of access to resources results less from people's lack of formally defined rights, but their incapacity to make claims 'stick' against those of more powerful actors (Leach et al. 1999: 241, see also Ramirez 1999). These policies even run the risk of emboldening local elites who capture the substantial portion of benefits as well as are able to access forest produce illegally (Dressler et al. 2010; Springate-Baginski and Blaikie 2007; Collins 2008).

I have argued that the ban has affected poor artisans more severely in comparison to the manufacturers, traders and agents. As observed in Chap. 5, various strategies are adopted by the manufacturers and their agents to exploit the poor artisans: by deducting wages; by supplying work to rural artisans to minimise production costs; and by supplying wool instead of cash to control labour. I also explained that raw wool traders created artificial shortages to increase the price of raw wool, and using dehairing and spinning machines to reduce labour costs. While the powerful actors have devised various ways to compensate for the losses after the ban, the poor skilled workers have been unsuccessful in securing their interests. They either work for their manufacturers to produce *shahtoosh* illegally for reduced wages or if they try to compensate for the losses by resorting to tactics, such as adulteration of wool, they do not succeed because their employers are able to detect the adulterated threads. I have also explained that while males have suffered from their loss of incomes and social status, females have additionally experienced a decline in their decision-making power and recognition in the family.

In the case of the NAP, as discussed in Chap. 8, although the forest cover has increased (as claimed by forest officials and villagers) in the last few years, the villagers' access to forest produce has reduced. However, again we saw differential access of villagers to forest resources and differential abilities to transgress the laws in connivance with the forest officials. While the relatively affluent are able to access timber and fuelwood easily by bribing the field-staff, it is the poorer who suffer the most from the increased restrictions on forest use. Based on these observations, I have argued that the new conservation interventions, in fact, have seeped into the existing relations of domination and subordination within the local communities along with entrenching the gap between enforcement officials and the poorer citizens in J&K. In both cases, I have demonstrated that the state has served the interests of the powerful actors and local elites rather than the poor and marginalised, who are left to bear the cost of nature conservation and protection.

9.2.5 Rationality of Rule

In this book I have shown that while conservation policies create an appearance of agreement about environmental management, they do not necessarily change the belief systems of the affected populations (cf. Singh 2008). This research has pointed to the limitations of the power of conservation community by demonstrating that it is actually the rationality of the powerful stream at the microlevel which determines the behaviour of the local communities. As I illustrated in the case of *shahtoosh* ban, the knowledge of the unsavoury origin of *shahtoosh* wool is limited to the powerful players. The unpopularity of the ban and the persistence of myths relating to the origin of *shahtoosh* even after the imposition of the ban suggest that the global rationality is understood, accepted and shaped by the local actors according to their own interests. We saw that the *shahtoosh* workers challenge the genuineness of the premise on which the ban is based. It is the perception of powerful manufacturers and traders which dominate the local belief systems and govern their behaviours in conforming or transgressing the boundaries of legality.

In the case of the NAP, I also demonstrated that there is a lack of awareness among the villagers about the objectives of the programme, and the role and functioning of the JFMC. Here, the behaviour of the people related to access of forest resources is largely determined by the field-staff. I have noted several information asymmetries among the JFMC members and villagers with regard to what is 'legal' and what is 'illegal'. By maintaining such uncertainties, I have argued, the field-staff is able to exploit the poor for rent seeking. Moreover, the JFMC members who are claimed to be equal partners under the NAP know little about the funds available for project activities. In fact, they have been misguided by the field-staff both on the funds available for the entry point activities as well as the incentives to join the programme. This suggests that conservation discourses can 'project an appearance of unified potency even while being effectively divided and weak in practice' (Singh 2008: 14). From both cases, we can conclude that it is the rationality of local powerful actors which largely rules rather than the legal and scientific rationality of the conservation community.

9.2.6   On Illegality

All negotiation processes reflect prevailing power relations and it could be argued that if powerful groups do not achieve their desired outcome through open negotiation, they are likely to achieve their ends through other means (Leach et al. 1999: 241). In this book, I have argued that the ban has officially been imposed on the state but the production and trade of *shahtoosh* still continues (although in lesser volume), owing to a strong nexus between manufacturers, politicians and enforcement officials. The poor also resort to illegal production of *shahtoosh*, however, I explained that their indulgence cannot be explained only as a result of poor economic conditions (Chatterjee 2004), absence of alternative livelihoods (Brown and Marks 2007; Wiggins et al. 2004; Peluso 1993) or a 'coping strategy' (Bloomer 2009). Non-compliance decisions, as Leon (1994) suggests, have elements of both self and collective interests, and the relationship between the two varies according to the type of illegal behaviour. Factors such as poverty and loss of livelihoods certainly play a significant role in the transgression of law by the poor artisans but I suggest that their indulgence in illegality is primarily determined by the manufacturers who control the production of *shahtoosh* and delegate the now illegal tasks of processing to their entrusted subordinate workers. I have termed this process as *delegated illegality*. Although I do not suggest that the poor skilled workers are compelled by their employers to undertake these jobs, their participation could be seen as an obligation on their part to take the orders of their employers for whom they have been working for generations. In the case of forest conservation, I have explained that while the poor villagers resort to illegal timber felling, it is more to meet their subsistence needs. Apart from the local village residents, I have argued that it is the powerful stream including the forest field-staff, police, military forces and timber contractors who carry out illegal timber harvesting for profit making. I have also noted that on some occasions, the forest field-staff has assisted the relatively affluent village residents to access timber illegally. Oddly enough, those who are authorised to protect forest citizens and their resources are the ones who are playing a significant role in forest destruction.

9.2.7 Limited Space for Protest

While I have maintained that nature conservation interventions in J&K cannot be understood simply as environmental imperialism or a new form of colonialism pursued through coercion, the conservation initiatives can certainly serve as means for the central as well as provincial states to assert their authority over environment and resource dependent populations. The environmental conservation is rarely seen by states as a goal in itself, but rather as means to achieve political as well as economic ends (Bryant and Bailey 1997). The conservation interventions can also serve as moral justifications to strengthen state military capacity (Peluso 1993) and legitimise state violence against dissenting groups (Neumann 2001; Kothari et al. 1995; Ghimire 1994). As I explained in Chap. 4, the ongoing militant separatist movement in J&K limited the space of association and protest for *shahtoosh* workers, with the state using its violent powers to crush the demonstrations against the ban on the pretext of 'anti-national' activity. My research has suggested that in regions affected by violence, nature conservation policies collide with the ongoing political struggles between the state, militant groups and the wider civil society over legitimacy to rule. I have argued that the shrunken space for protest in Kashmir was crucial to sidelining the issues of the alternative livelihoods of the affected populations. Moreover, the Kashmiri shawl hawkers often face harassment from police agencies outside the state where they are seen as suspected terrorists. This suggests that the political climate of the state largely shapes the manner in which nature conservation interventions are experienced by the affected communities as well as the way in which the state responds to the local resistance.

9.3 Who Is Accountable?

This book has demonstrated that there are valid ethical arguments in favour of nature conservation but it is the poor who bear the burden of conservation costs. While issues of environmental protection are on top of the agenda for international conservation agencies, neither the state nor any other agency is accountable for addressing the grievances of the affected populations. In Chap. 4, I have argued that while the issue of *chiru* conservation was addressed at length in the legal battle, the poorer shawl workers remained almost voiceless. Apart from meetings and discussions of the expert committee constituted to explore the prospects of *chiru* breeding, and observations made by the J&K High Court concerning adequate compensation to the artisans in Kashmir, no practical efforts were made to provide alternative livelihoods, or compensate the artisans. While no substantial efforts were made to experiment with *chiru* breeding, the logistic problems of getting the consent of all signatories of the CITES in domesticating *chirus* for commercial purposes became an impediment in the possible collaborative efforts required for the purpose. The Chinese authorities maintained that exploring the prospects of *chiru* breeding for commercial purposes is beyond their control (due to the international ban) even if the habitat of *chiru* falls within the Chinese territory. I explained in Chap. 4 that the only initiative for the rehabilitation of *shahtoosh* workers has been taken by the animal rights group, WTI. This programme can be seen at best as tokenism owing to its limited coverage and scope. While the international agencies did not take the responsibility of adequately addressing the issue of rehabilitation of affected communities, the central government passed on the task of rehabilitating the shawl workers to the state government. As I have argued, the state government remained complacent in providing alternative livelihoods to the shawl workers. In addition to this, the state initiatives to promote the shawl industry and to help artisans proved futile due to a mismatch between the actual requirements of the artisans and the mandates of the various state agencies.

In the case of forest conservation, I have discussed that while there is an apparent consensus in the policy rhetoric on the issues of forest conservation through local participation and decentralisation, little attention has been paid to the subsistence needs of the poorer forest citizens. I have explained that both the colonial and post-colonial state considered the local communities as forest destroyers and curtailed their rights to access forest produce. In recent times, the local communities have been left on their own to cope with the reduced access, either bearing the costs of forest conservation or, at best, transgressing the law to meet their subsistence requirements. I argue that unless a joint effort on the part of international conservation agencies and national and provincial governments is made to undertake the issues of rehabilitation, and unless the subsistence needs of the poor are adequately met, the nature conservation policies are unlikely to realise the goals of sustainable resource management or sustainable livelihoods.

9.4 Practical Implications

Whilst this book does not aim to generate a list of recommendations for policy-making, my findings nonetheless point towards a number of practical implications. This research illustrates that a blanket ban not only results in the loss of the livelihoods of affected populations, it may also become difficult to overturn it when the population of the endangered species begins to rise. Therefore, strategic time bound restrictions can result in sustaining the species as well as the livelihoods of the dependent populations. There is a need to run a simultaneous programme for rehabilitation when conservation measures are imposed. This can be done by a collaborative effort on the part of international agencies, the central government, provincial state and NGOs to compensate for the losses or through the formation of a support body which could take adequate steps in this regard. The frontier location of the state, the incidence of the militancy as well as the high rate of unemployment necessitate special efforts to address the issues of livelihoods if we are to prevent nature conservation interventions fuelling terrorism in the Valley. I argue that since the *shahtoosh* workers were concentrated in one region i.e. Srinagar, a timely effort to rehabilitate artisans could have saved the affected population from the miseries resulting from the ban.

As discussed in Chap. 7, the broad objectives of the NAP create ambiguities regarding what is to be achieved through the programme. While conservation of natural resources, income generation for the poor, fulfillment of subsistence requirements of the villagers, introduction of scientific technologies for forest regeneration and increase in biomass can be equally important, there is a need to prioritise and limit the targets to make the policy work practically given the limited time period of the projects. The short duration of the projects and constant shifts in policies may result in a lack of motivation among the villagers to participate in the programmes. The lack of awareness regarding the project activities as well as forest regulations and inaccessibility of senior officials increase the chances of their exploitation at the hands of field-staff. As argued earlier, the forest officials not only legitimise their control over the forest resources on the pretext of scientific rationality, they use and abuse the law to secure their own interests. The choice of species for forest regeneration are often not the ones required by the forest dependent communities and the ambiguities in terms of what is 'legal' and what is not increase their vulnerability to harassment by the enforcement officials. I suggest that science and law, rather than serving the interests of the forest bureaucracy, police and local elites, should be used in the service of the poor and marginalised communities in order to realise sustainable ecological practices. Moreover, the overlap of roles and functions of the JFMC and the forest field-staff in forest protection and development activities results in blame games. Effective forest conservation policies should be based on the idea of increasing the access of the poor to the forest resources rather than putting the costs of regeneration of the degraded forest lands on them, especially when the reasons for forest degradation are very different from the presumed linkages between poverty and environmental destruction as claimed in mainstream literature.

9.5 Conclusion

In this book, I have demonstrated that power is not concentrated in a single site but is dispersed and fluid, and that nature conservation interventions rather than resulting in fixed and determined outcomes are accepted, resisted and reconfigured by various actors at multiple levels. Although there are valid arguments in favour of conserving both *chirus* and forests, it is the poor that bear the brunt of nature conservation costs. While the relatively powerful actors are able to manipulate the laws and regulations depending on their differential power and abilities, the poor suffer from the loss of traditional occupations or reduced access to forest resources. In the case of conflict regions like J&K, I argue that these conservation interventions may also coincide with the ongoing struggles between the state and citizens on the question of legitimacy to rule as well as with the attempts of the historically powerful actors to dominate the poor and marginalised. Although much intellectual energy has been spent on advocating or challenging global control over local resources, little effort has been made to emphasise the need of wider responsibility for recompensing the losses of the resource dependent populations. If the demands for saving *chirus* from extinction or regenerating degraded forest areas are pressing, equally pressing, I insist, are the issues of the livelihoods of the poor artisans and the subsistence requirements of the forest dependent populations who are somehow trying to survive in the already fragile economy of J&K. There is no reason to deny the need and significance of global concern for conserving forest and wildlife resources; however, I suggest, global resource control without matching accountability to the affected communities is unlikely to meet the goals of sustainable resource management.

References

Adams W (1996) Future nature: a vision for conservation. Earthscan, London

Agrawal A (1995) Dismantling the divide between indigenous and scientific knowledge. Dev Chang 26:413–439

Ballabh V, Balooni K, Dave S (2002) Why local resource management institutions decline: a comparative analysis of van (forest) panchayats and forest protection committees in India. World Dev 30(12):2153–2167

Bebbington A (1999) Capitals and capabilities: a framework for analysing peasant viability, rural livelihood and poverty. World Dev 27(12):2021–2044

Bene C, Belal E, Baba M (2009) Power struggle, dispute and alliance over local resources: Analysing 'democratic' decentralization of natural resources through the lenses of Africa inland fisheries. World Dev 37(12):1935–1950

Blaikie P (2006) Is small really beautiful? Community based natural resource management in Malawi and Botswana. World Dev 34(11):1942–1957

Bloomer J (2009) Using a political ecology framework to examine extra-legal livelihood strategies: a Lesotho-based case study of cultivation of and trade in cannabis. J Polit Ecol 16:49–69

References

219

Brown T, Marks S (2007) Livelihoods, hunting and the gamemeat trade in northern Zambia. In: Davies G, Brown D (eds) Bushmeat and livelihoods: wildlife management and poverty reduction. Blackwell, Oxford, pp 92–106

Bryant R, Bailey S (1997) Third world political ecology. Routledge, London

Calvert P, Calvert S (1999) The south, the north and the environment. Pinter, London

Chambers R, Conway G (1992) Sustainable rural livelihoods: practical concepts for the twenty-first century, IDS discussion paper 296. Institute of Development Studies, Sussex

Chatterjee P (2004) The politics of the governed: reflections on popular politics in most of the world. Columbia University Press, New York

Collins TW (2008) The political ecology of hazard vulnerability: marginalisation, facilitation and the production of differential risks to urban wildfires in Arizona's White Mountains. J Polit Ecol 15:21–43

Corbridge S, Williams G, Srivasatava M, Veron R (2005) Seeing the state. Cambridge University Press, Cambridge

D'Silva E, Nagnath B (2002) Behroonguda: a rare success story in joint Forest management. Econ Polit Wkly 9:551–557

Dauvergne P (1997) Shadows in the forest: Japan and the politics of timber in South East Asia. The Massachusetts Institute of Technology Press, Cambridge, MA

Dressler W, Buscher B, Schoon M, Brockington D, Hayes T, Kull C, McCarthy J, Shrestha K (2010) From hope to crisis and back again? A critical history of the global CBNRM narrative. Environ Conserv 37(1):5–15

Gadgil M (1998) Grassroots conservation practices: Revitalising the traditions. In: Kothari A, Pathak N, Anuradha R, Taneja B (eds) Communities and conservation: natural resource management in south and Central Asia. Sage, New Delhi, pp 219–238

Ghimire KB (1994) Parks and people: livelihood issues in national parks management in Thailand and Madagascar. Dev Chang 25:195–229

Greenough P, Tsing AL (2003) Introduction. In: Greenough P, Tsing AL (eds) Nature in the global south. Duke University Press, Durham, pp 1–28

Guha RC, Martinez-Alier J (1997) Varieties of environmentalism: essays north and south. Earthscan, London

Gupta A (1995) Blurred boundaries: the discourse of corruption, the culture of politics and the imagined state. Am Ethnol 22(2):375–402

Gupta A (1998) Postcolonial developments in the making of modern India. Duke University Press, Durham

Gupta Saurabh (2009) The politics of development in rural Rajasthan, India: evidence from water conservation and watershed development programmes since the early 1990s. PhD thesis. University of London, London

Gupta S, Sinha S (2008) Beyond 'dispositif' and 'depoliticisation': spaces of civil society in water conservation in rural Rajasthan. In: Lahiri-Dutt K, Wasson R (eds) Water first: issues and challenges for nations and communities in South Asia. Sage, New Delhi, pp 271–294

Hulme D, Murphree M (2001) African wildlife and livelihoods: the promise and performance of community conservation. James Currey, London

IUCN (1980) World conservation strategy: living resource conservation for sustainable development. http://data.iucn.org/dbtw-wpd/edocs/WCS-004.pdf. Accessed 20 Feb 2011

Kothari A, Suri S, Singh N (1995) People and protected areas: rethinking conservation in India. Ecologist 25:188–194

Leach M, Mearns R, Scoones I (1999) Environmental entitlements: dynamics and institutions in community based natural resource management. World Dev 27(2):225–247

Leon M (1994) Avoidance strategies and governmental rigidity: the case of the small-scale shrimp fishery in two Mexican communities. J Polit Ecol 1:67–81

Li TM (1999) Compromising power: development, culture and rule in Indonesia. Cult Anthropol 40(3):277–309

Li TM (2005) Beyond "the state" and failed schemes. Am Anthropol 107(3):383–394

Li TM (2007) Practices of assemblage and community forest management. Econ Soc 36(2):263–293

Lohmann L (1996) Freedom to plant: Indonesia and Thailand in a globalising pulp and paper industry. In: Parnwell M, Bryant R (eds) Environmental change in South East Asia. Routledge, London, pp 23–48

Lynch O, Talbot K (1995) Balancing acts: community based forest management and national law in Asia and the Pacific. World Resources Institute, Washington, DC

Mahanty S (2002) NGOs, agencies and donors in participatory conservation. Econ Polit Wkly 37:3757–3765

Menzies N (1993) Putting people back into forestry: some reflections on social and community forestry. For Soc 1(1):6–7

Midgal J (1988) Strong societies and weak states: state-society relations and state capabilities in the third world. Princeton University Press, Princeton

Moore DS (2000) The crucible of cultural politics: reworking "development" in Zimbabwe's eastern highlands. Am Ethnol 26(3):654–689

Neumann RP (1992) Political ecology of wildlife conservation in the Mt Meru area of northeast Tanzania. Land Degrad Rehabil 3:85–98

Neumann R (2001) Disciplining peasants in Tanzania: from state violence to self-surveillance in wildlife conservation. In: Peluso NL, Watts M (eds) Violent Environments. Cornell University Press, Ithaca, pp 305–327

Peluso NL (1993) Coercing conservation: the politics of state resource control. In: Lipshutz R, Conca K (eds) The state and social power in global environmental politics. Columbia University Press, New York, pp 199–218

Peluso NL, Watts M (2001) Introduction. In: Peluso NL, Watts M (eds) Violent environments. Cornell University Press, Ithaca/New York, pp 1–30

Poffenberger M, Singh C (1996) Communities and the state: re-establishing the balance in Indian forest policy. In: Poffenberger M, McGean B (eds) Village voices, forest choices: joint forest management in India. Oxford University Press, New Delhi, pp 56–85

Ramirez R (1999) Stakeholder analysis and conflict management. In: Buckles D (ed) Cultivating peace: conflict and collaboration in natural resource management. World Bank Institute, Ottawa, pp 101–126

Rangan H (1995) Contested boundaries: state policies, forest classifications, and deforestation in the Garhwal Himalayas. Antipode 27(4):343–362

Rose N (1999) Powers of freedom: reframing political thought. Cambridge University Press, Cambridge

Rossi B (2004) Revisiting Foucauldian approaches: power dynamics in development projects. J Dev Stud 40(6):1–29

Saberwal V (2003) Conservation by state fiat. In: Saberwal V, Rangarajan M (eds) Battles over nature: science and politics of conservation. Permanent Black, New Delhi, pp 240–266

Saberwal V, Rangarajan M (2003) Introduction. In: Saberwal V, Rangarajan M (eds) Battles over nature: science and politics of conservation. Permanent Black, New Delhi, pp 1–30

Saigal S (2000) Beyond experimentation: emerging issues in the institutionalisation of joint forest management. Environ Manag 2(3):269–281

Salskov-Iversen D, Hansen H, Bislev S (2000) Governmentality, globalisation and local practice: transformations of a hegemonic discourse. Alternatives 25(2):183–222

Sarap K, Sanrangi TK (2009) Malfunctioning of forest institutions in Orissa. Econ Polit Wkly XLIV(37):18–22

Scoones I (1998) Sustainable rural livelihoods: a framework for analysis, Working paper no. 72. Institute of Development Studies, Sussex

Shepherd A (1998) Sustainable rural development. Macmillan, Basingstoke

Singh S (2008) Contesting moralities: the politics of wildlife trade in Laos. J Polit Ecol 15:1–20

Sinha H (2006) People and forests. Concept Publishing Company, Delhi

Springate-Baginski O, Blaikie P (2007) Forests, people and power: the political ecology of reform in South Asia. Earthscan, London

References

Sundar N (2004) Devolution, joint forest management and the transformation of social capital. In: Bhattacharyya D, Jayal N, Mohapatra B, Pye S (eds) Interrogating social capital: the Indian experience. Sage, New Delhi, pp 203–232

Sundar N, Jeffery R, Thin N (2001) Branching out: joint forest management in India. Oxford University Press, New Delhi

Vasan S (2006) Living with diversity. Indian Institute of Advanced Study, Shimla

Veron R, Williams G, Corbridge S, Srivastava M (2006) Decentralised corruption or corrupt decentralisation? Community monitoring of poverty alleviation schemes in Eastern India. World Dev 34(11):1922–1941

Vira B (1999) Implementing joint forest management in the field: towards an understanding of the community-bureaucracy interface. In: Jeffery R, Sundar N (eds) A new moral economy for India's forests? Discourses on community and participation. Sage, New Delhi, pp 254–275

Wardell DA, Lund C (2006) Governing access to trees in Northern Ghana: micropolitics and the rents of non-enforcement. World Dev 34(11):1887–1906

WCED (1987) Our common future. Oxford University Press, Oxford

Wiggins S, Marfo K, Anchirinah V (2004) Protecting the forest or people? Environmental policies and livelihoods in the forest margins of southern Ghana. World Dev 32(11):1939–1955

Yearley S (1996) Sociology, environmentalism, globalization. Sage, London

Bibliography

Adams W (1996) Future nature: a vision for conservation. Earthscan, London

Adams WM (2001) Green development: environment and sustainability in the third world. Routledge, London

Agrawal A (1995) Dismantling the divide between indigenous and scientific knowledge. Dev Chang 26:413–439

Agrawal A (1999) Greener pastures: politics, markets and community among a migrant pastoral people. Duke University Press, Chapel Hill

Agrawal A (2005) Environmentality: technologies of government and political subjects. Duke University Press, London

Agrawal A, Gibson C (1999) Enchantment and disenchantment: the role of community in natural resource conservation. World Dev 27(4):629–649

Ahad A (1987) Kashmir to Frankfurt: a study of arts and crafts. Rima Publishing House, New Delhi

Ahmed M (2004) The politics of pashmina: the Champas of Eastern Ladakh. Nomadic Peoples 8(2):89–106

Ahuja B (2006) The unique and eloquent legacy of Kashmir handmade pashmina. Wildlife Trust of India, New Delhi

Ames F (1986) The Kashmir shawl. Antique Collectors Club, Suffolk

Arjunan M, Holmes C, Puyravaud JP, Davidar P (2006) Do development initiative influence local attitudes towards conservation? A case-study from the Kalakad- Mundanthurai Tiger Reserve, India. J Environ Manag 79:188–197

Arnold JEM (2002) Clarifying the links between forests and poverty reduction. Int For Rev 4(3):231–233

Ashley C, Carney C (1999) Sustainable livelihoods: lessons from early experience. Department for International Development, London

Axelby R (2007) It takes two hands to clap: how *gaddi* shepherd in the Indian Himalayas negotiate access to grazing. J Agrar Chang 7(1):35–75

Ballabh V, Balooni K, Dave S (2002) Why local resource management institutions decline: a comparative analysis of van (forest) panchayats and forest protection committees in India. World Dev 30(12):2153–2167

Balland JM, Platteau JP (1996) Halting degradation of natural resources: is there a role for rural communities? Oxford University Press, Oxford

Bamzai PNK (1962) A history of Kashmir. Metropolitan Book Co. Private Ltd, Delhi

Bamzai PNK (1980) Kashmir and Central Asia. Light and Life Publishers, New Delhi

Barakat S, Chard M, Jacoby T, Lume W (2002) The composite approach: research design in the context of war and armed conflict. Third World Q 23(5):991–1003

© Springer International Publishing AG 2018
S. Gupta, *Contesting Conservation*, Advances in Asian Human-Environmental Research, https://doi.org/10.1007/978-3-319-72257-3

Barber VV (1989) The state, the environment, and development: the genesis and transformation of social forestry policy in new order Indonesia. PhD thesis. University of California, Berkeley

Bardach E (2000) A practical guide for policy analysis: the eightfold path to more effective problem solving. Chatham House Publishers, New York/London

Barraclough S (1997) Rural development and the environment: towards ecologically and socially sustainable development in rural areas. United Nations Research Institute for Social Development, Geneva

Barrett CB, Lee DR, McPeak JG (2005) Institutional arrangements for rural poverty reduction and resource conservation. World Dev 33(2):193–197

Barton GA (2002) Empire forestry and the origins of environmentalism. Cambridge University Press, Cambridge

Baumann P, Sinha S (2000) Panchayati raj institutions and natural resource management. Overseas Development Institute, London

Baviskar A (2001) Written on the body, written on the land: violence and environmemtal struggles in Central India. In: Peluso NL, Watts M (eds) Violent environment. Cornell University Press, Ithaca/New York, pp 354–379

Baviskar A (2002) The politics of the city. Seminar 516:41–47

Baviskar A (ed) (2007) Waterscapes: the cultural politics of a natural resource. Orient Longman, New Delhi

Bazaara N (2003) Decentralisation, politics and environment in Uganda. In: Working Paper, Environmental governance in Africa series. World Resources Institute, Wahington, DC

Bebbington A (1999) Capitals and capabilities: a framework for analysing peasant viability, rural livelihood and poverty. World Dev 27(12):2021–2044

Beinart W, Hughes L (2007) Environment and empire. Oxford University Press, Oxford

Bene C, Belal E, Baba M (2009) Power struggle, dispute and alliance over local resources: analysing 'democratic' decentralization of natural resources through the lenses of Africa inland fisheries. World Dev 37(12):1935–1950

Berg J, Biesbrouck K (2005) Dealing with power imbalances in forest management: reconciling multiple legal systems in South Cameroon. In: Ros-Tonen, Dietz T (eds) African forests between nature and livelihood resources: interdisciplinary studies in conservation and forest management. The Edwin in Mellen Press, New York, pp 221–254

Bhatt S (2004) Kashmir ecology and environment: new concerns and strategies. APH Publishers, New Delhi

Bhattacharyya D, Jayal NG, Mohapatra B, Pye S (2004) Introduction. In: Bhattacharyya D, Jayal NG, Mohapatra B, Pye S (eds) Interrogating social capital: the Indian experience. Sage, New Delhi, pp 15–34

Blaikie P (1985) The political economy of soil erosion in developing countries. Longman, London

Blaikie P (2006) Is small really beautiful? Community based natural resource management in Malawi and Botswana. World Dev 34(11):1942–1957

Blaikie P, Brookfield H (1987) Land degradation and society. Methuen, London

Bloomer J (2009) Using a political ecology framework to examine extra-legal livelihood strategies: a Lesotho-based case study of cultivation of and trade in cannabis. J Polit Ecol 16:49–69

Bose S (1997) The challenge in Kashmir: democracy, self-determination and a just peace. Sage, New Delhi

Brandis D (1876) Report of the Proceedings on the forest conference of 1875. Indian Forester 2(2), October

Brown K, Ekoko E (2001) Forest encounters: synergy among agents of forest change in Southern Cameroon. Soc Nat Resour 14(4):269–290

Brown T, Marks S (2007) Livelihoods, hunting and the gamemeat trade in northern Zambia. In: Davies G, Brown D (eds) Bushmeat and livelihoods: wildlife management and poverty reduction. Blackwell, Oxford, pp 92–106

Bryant RL (1997) The political ecology of forestry in Burma 1824–1994. Hurst and Company, London

Bibliography

Bryant RL (2005) Nongovernmental organisations in environmental struggles: politics and the making of moral capital. Yale University Press, London

Bryant R, Bailey S (1997) Third world political ecology. Routledge, London

Buttel FH (1992) Environmentalisation: origins, processes and implications for rural social change. Rural Sociol 57:127–146

Calvert P, Calvert S (1999) The south, the north and the environment. Pinter, London

Campbell BM, de Jong W, Luckert M, Mandondo A, Matose F, Nemarundwe N, Sithole B (2001) Challenges to proponents of CPR systems: despairing voices from the social forests of Zimbabwe. World Dev 29(4):589–600

Campese J, Borrini-Feyeraband G, de Cordova M, Guigner A, Oviedo G (2007) Just conservation? What can human rights do for conservation and vice-versa?! Policy Matters 15:6–9

Carney D (1998) Sustainable rural livelihoods. In: Regional development dialogue 8. Department for International Development, London

Carswell G, Hussain G, McDowell K, Wolmer W (1997) Sustainable livelihoods: a conceptual approach. IDS. Mimeo, Brighton

Cederlof G, Sivaramakrishnan K (2006) Ecological nationalisms: nature, livelihoods and identities in South Asia. University of Washington Press, Seattle

Census of India (2011) Series 2, Jammu and Kashmir provisional population totals, Paper 2 of 2001, rural urban distribution of population. Registrar General and Census Commissioner of India, New Delhi

Cernea MM, Schmidt-Soltau K (2006) Poverty risks and national parks: policy issues on conservation and development. World Dev 34(10):1868–1830

Chadha SK (1991) Kashmir: ecology and environment. Mittal Publications, New Delhi

Chambers R (1983) Rural development: putting the last first. Longman, London

Chambers R (1987) Sustainable livelihoods, environment and development: putting poor rural people first, Paper No. 240. Institute of Development Studies, Sussex

Chambers R (1988) Sustainable rural livelihoods: a strategy for people, environment and development. In: Conroy C, Litvinoff M (eds) The greening of aid: sustainable livelihoods in practice. Earthscan, London, pp 1–46

Chambers R, Conway G (1992) Sustainable rural livelihoods: practical concepts for the twenty-first century, IDS Discussion Paper 296. Institute of Development Studies, Sussex

Chambers R, Saxena NC, Shah T (1989) To the hands of the poor: water and trees. Immediate Technology Publications, London

Champion H, Osmaston FC (1962) E.P. Stebbing's the forests of India, vol IV. Oxford University Press, Oxford

Chaplin S (1999) Cities, sewers and poverty: India's politics of sanitation. Environ Urban 11(1):145–158

Chatterjee P (2004) The politics of the governed: reflections on popular politics in most of the world. Columbia University Press, New York

Chaulk KG, Robertson GJ, Collins BT, Montevecchi WA, Turner B (2005) Evidence of recent population increases in common eiders breeding in Labrador. J Wildl Manag 69:805–809

Chhatre A, Saberwal V (2006) Democratising nature: politics, conservation and development in India. Oxford University Press, New Delhi

Chundawat RS, Talwar R (1995) A report on the survey of Tibetan antelope (Pantholops hodgsoni), in the Ladakh region of Jammu and Kashmir. Ministry of Environment and Forests, New Delhi

CITES (1999) Conservation of and control of trade in the Tibetan antelope. www.cites.org/eng/res/11/11-08R13.shtml. Accessed 1 Feb 2006

CITES China (2004) Letter by CITES Management Authority of China to the CITES Management Authority of India

Cleghorn HF (1861) The forests and gardens of South India. W. H. Allen, London

Collins TW (2008) The political ecology of hazard vulnerability: marginalisation, facilitation and the production of differential risks to urban wildfires in Arizona's White Mountains. J Polit Ecol 15:21–43

Corbridge S, Jewitt S (1997) From forest struggles to forest citizens? Joint forest management in the unquiet woods of India's Jharkhand. Environ Plan A 29(12):2145–2164

Corbridge S, Williams G, Srivasatava M, Veron R (2005) Seeing the state. Cambridge University Press, Cambridge

CTA (2007) Wildlife poaching still rampant in Tibet. Central Tibetan Administration

D'Silva E, Nagnath B (2002) Behroonguda: a rare success story in joint forest management. Econ Polit Wkly February 9 37:551–557

Dame J, Nüsser M (2008) Development perspectives in Ladakh, India. Geographische Rundschau International Edition 4(4):20–27

Dame J, Nüsser M (2011) Food security in High Mountain regions: agricultural production and the impact of food subsidies in Ladakh, Northern India. Food Sec 3:179–194

Das P (1997) Kailadevi wildlife sanctuary Rajasthan: prospects for joint management. In: Kothari A, Vania F, Das P, Christopher K, Jha S (eds) Building bridges for conservation: towards joint management of protected areas in India. Indian Institute of Public Administration, New Delhi

Dauvergne P (1997) Shadows in the forest: Japan and the politics of timber in South East Asia. The Massachusetts Institute of Technology Press, Cambridge, MA

Dean M (1999) Governmentality: power and rule in modern society. Sage, London

Des Chene M (1997) Locating the past. In: Gupta A, Ferguson J (eds) Anthropological locations: boundaries and grounds of a field science. University of California Press, Berkeley, pp 66–85

Dhanagare DN (2000) Joint Forest Management in Uttar Pradesh: people, panchayats and women. Econ Poitical Wkly 35(37):3315–3324

Dhar DN (2001) Dynamics of political change in Kashmir. Kanishka Publishers, New Delhi

Dhar DN (2004) Kashmir: land and its management. Kanishka Publishers, New Delhi

Dheeraprasart V (2005) After the logging ban: politics of forest management in Thailand. Foundation for Ecological Recovery, Bangkok

Dietz AJ (1996) Entitlements to natural resources: contours of political environmental geography. International Books, Utrecht

Dodds D (1998) Lobster in the rain forest: the political ecology of miskito wage labor and agricultural deforestation. J Polit Ecol 5:83–108

Dressler W, Buscher B, Schoon M, Brockington D, Hayes T, Kull C, McCarthy J, Shrestha K (2010) From hope to crisis and back again? A critical history of the global CBNRM narrative. Environ Conserv 37(1):5–15

Dryzek J (2005) Deliberative democracy in divided societies: alternatives to agonism and analgesia. Polit Theo 33:218–242

Edmunds D, Wollenberg E (2003) Local forest management: the impacts of devolution policies. Earthscan, London

Ellis F (2000) Rural livelihoods and diversity in developing countries. Oxford University Press, Oxford

Escobar A (1996) Constructing nature: elements of a post-structuralist political ecology. In: Peet R, Watts M (eds) Liberation ecologies. Routledge, London, pp 46–68

Esser D (2008) How local is urban governance in fragile states? Theory and practice of capital city politics in Sierra Leone and Afghanistan. PhD thesis. London School of Economics and Political Science, UK

Farrington J, Bebbington A (1993) Reluctant partners: non-governmental organisations, the state and sustainable agricultural development. Routledge, London

Feng Z (1999) Status and conservation of Tibetan antelope in China. In: Zhen RD (ed) The future of Tibetan antelope. Proceedings of an International Workshop on Conservation and Control of Trade in Tibetan antelope, October 12–13, 1999, Xining, Qinghai, Beijing, pp 27–28

Feyerabend P (1993) Against method. Verso, London

Fomete T (2001) The forestry taxation system and the involvement of local communities in forest management in Cameroon. In: Rural development forestry network paper 25. Oversees Development Institute, London

Bibliography

227

Forsyth T (2003) Critical political ecology: the politics of environmental science. Routledge, London

Forsyth T, Walker A (2008) Forest guardians, forest destroyers: the politics of environmental knowledge in northern Thailand. University of Washington Press, Seattle

Fox JL, Bardsen BJ (2005) Density of Tibetan antelope, Tibetan wild ass and Tibetan gazelle in relation to human presence across the Chang Tang Nature Reserve of Tibet, China. Acta Zool Sin 51:586–597

Francis P, James R (2003) Balancing rural poverty reduction and citizen participation: the contradiction of Uganda's decentralisation program. World Dev 31(2):325–337

FSI (1999) State of Forest report. Forest Survey of India, Dehradun

Gadgil M (1998) Grassroots conservation practices: Revitalising the traditions. In: Kothari A, Pathak N, Anuradha R, Taneja B (eds) Communities and conservation: natural resource management in South and Central Asia. Sage, New Delhi, pp 219–238

Gadgil M, Guha R (1992) This fissured land: an ecological history of India. Oxford University Press, Delhi

Ganae N (2010) Unrest in Kashmir proves juicy for jungle smugglers. Agence India Press. 10 November, 2010. http://agenceindiapress.com/. Accessed 15 Nov 2010

Ganhar JN (1979) The wildlife of Ladakh. Srinagar Press, Srinagar

Ganjoo SK (1998) Kashmir: history and politics. Commonwealth Publishers, New Delhi

Ghimire KB (1994) Parks and people: livelihood issues in national parks management in Thailand and Madagascar. Dev Chang 25:195–229

Go J&K (1992) Notification No. SRO-61. 19 March, 1992. Government of Jammu and Kashmir

Go J&K (1993) Udhampur forest division annual report 1992–1993. Jammu and Kashmir Forest Department, Udhampur

Go J&K (2006) FDA Udhampur report. Jammu and Kashmir Forest Department, Udhampur

GoJ&K (1988) Working plan for Doda Forest division, 1978–79 to 1987–88. Jammu and Kashmir Forest Department, Srinagar

GoJ&K (1999) Notification No. SRO-17. 12 January, 1999. Government of Jammu and Kashmir

GoJ&K (2001a) Digest of statistics 2000–01. Directorate of Economics and Statistics/Planning and Development Department, Jammu

GoJ&K (2001b) Digest of forest statistics. Statistics Division: Government of Jammu and Kashmir, Srinagar

GoJ&K (2005) Digest of forest statistics. 2005. Jammu and Kashmir Forest Department, Srinagar

GoJ&K (n.d.) Operational guidelines for National Afforestation Programme. Jammu and Kashmir Forest Department, Jammu

Goldman M (1996) Eco-governmentality and other transnational practices of a "green" World Bank. In: Peet R, Watts M (eds) Liberation ecologies. Routledge, New York, pp 166–192

Gray L, Mosley W (2005) A geographical perspective on poverty-environment interactions. Geogr J 171(1):9–23

Greenberg J, Park T (1994) Political ecology. J Polit Ecol 2(1):1–12

Greenough P, Tsing AL (2003) Introduction. In: Greenough P, Tsing AL (eds) Nature in the global south. Duke University Press, Durham, pp 1–28

Guha RC (1990) The unquiet woods: ecological change and peasant resistance in the Himalayas. University of California Press, Berkeley

Guha RC (2000) Environmentalism: a global history. Oxford University Press, New Delhi

Guha RC (2003) The authoritarian biologist and the arrogance of anti-humanism: wildlife conservation in the third world. In: Saberwal V, Rangarajan M (eds) Battles over nature: science and politics of conservation. Permanent Black, New Delhi, pp 139–157

Guha RC, Martinez-Alier J (1997) Varieties of environmentalism: essays North and South. Earthscan, London

Gupta A (1995) Blurred boundaries: the discourse of corruption, the culture of politics and the imagined state. Am Ethnol 22(2):375–402

Gupta A (1998) Postcolonial developments in the making of modern India. Duke University Press, Durham

Gupta S (2009) The politics of development in rural Rajasthan, India: evidence from water conservation and watershed development programmes since the early 1990s. PhD thesis. University of London, UK

Gupta S (2013) Democratic transition in Jammu and Kashmir: lessons from recent nature conservation interventions. In: Arora V, Jayaram N (eds) Roots and routes of democracy in the Himalayas. Routledge, New Delhi

Gupta S (2014) Worlds apart? Challenges of multi-agency partnership in participatory watershed development in Rajasthan, India. Dev Stud Res 1(1):100–112

Gupta S, Sinha S (2008) Beyond 'dispositif' and 'depoliticisation': spaces of civil society in water conservation in rural Rajasthan. In: Lahiri-Dutt K, Wasson R (eds) Water first: issues and challenges for nations and communities in South Asia. Sage, New Delhi, pp 271–294

Hareven T (2002) The silk weavers of Kyoto. University of California Press, London

Harriss-White B (2003) India working essays on society and economy. Cambridge University Press, Cambridge

Hassnain FM (1992) The beautiful Kashmir valley. Rima Publishing House, New Delhi

Hempel L (1999) Conceptual and analytical challenges in building sustainable communities. In: Mazmanian D, Kraft M (eds) Towards sustainable communities: transition and transformations in environmental policy. The MIT Press, Cambridge, MA, pp 33–62

High Court of Jammu and Kashmir (2000) Proceedings of the Public Interest Litigation PIL (CWP) No. 293/98. May, 2000

High Court of Jammu and Kashmir (2003) Verdict of the division bench of the J&K High Court on the Writ Petition filed by the Wildlife Protection Society of India

Hill E (2010) Worker identity, agency and economic development: Women's empowerment in the Indian informal economy. Routledge, London

Homer-Dixon T (1999) Environment scarcity and violence. Princeton University Press, Princeton

Hovland I (2005) What do you call the heathen these days? The policy field and other matters of the heart in the Norwegian Mission Society. Paper presented at Workshop on Problems and possibilities in multi-sited ethnography, Brighton, University of Sussex, UK, 27–28 June 2005

Huber T (2003) The chase and the dharma: the legal protection of wild animals in pre-modern Tibet. In: Knight J (ed) Wildlife in Asia: cultural perspectives. Routledge, London, pp 36–55

Hulme D, Murphree M (1999) Policy arena: communities, wildlife and the "new conservation" in Africa. J Int Dev 11(2):277–285

Hulme D, Murphree M (2001) African wildlife and livelihoods: the promise and performance of community conservation. James Currey, London

IFAW and WTI (2001) Wrap up the trade: an international campaign to save the endangered Tibetan antelope. International Fund for Animal Welfare and Wildlife Trust of India, New Delhi

IFAW and WTI (2003) Beyond the ban: a census of shahtoosh in the Kashmir valley. International Fund for Animal Welfare and Wildlife Trust of India, New Delhi

IFRI (1961) Hundred years of forestry 1861–1961, vol II. Indian Forest Research Institute, Dehradun

Irwin J (1973) The Kashmir shawl. Her Majesty's Stationary Office, London

IUCN (1980) World Conservation Strategy: Living resource conservation for sustainable development. http://data.iucn.org/dbtw-wpd/edocs/WCS-004.pdf. Accessed 20 Feb 2011

IUCN (2016) IUCN red list of threatened species. www.iucnredlist.org/details/15967/0. Accessed 1 Aug 2017

Jammu and Kashmir (1925) State progress report on forest administration, J&K State 1924–25. The Kashmir Merchantile Press, Srinagar

Jammu and Kashmir State (1911) Progress report of the forest administration in J&K 1909–10. The Civil and Military Gazette Press, Lahore

Jammu and Kashmir State (1921) Progress report of forest administration in the J&K State 1918–19. The Civil and Military Gazette Press, Lahore

Bibliography 229

Jammu and Kashmir State (1923) Progress report on forest administration in the J&K State 1920–21. The Civil and Military Gazette Press, Lahore

Jan A (2005) Protest movements in J&K. Zeba Publications, Srinagar

Jayakar P (1959) Cotton Jamdanis of Tanda and Benaras. Lalit Kala, Delhi

Jayal NG (2001) Balancing political and ecological values. Environ Politics 10(1):65–88

Jha PS (2003) The origins of a dispute: Kashmir 1947. Oxford University Press, New Delhi

Jha UM, Misra DC (2006) Economics of silk weavers. Sunrise Publications, New Delhi

Karlsson B (1999) Ecodevelopment in practice: Buxa-Tiger reserve and forest people. Econ Polit Wkly 24(30):2087–2094

Kawosa MA (2001) Forests of Kashmir: a vision for the future. Natraj Publishers, Delhi

Khan RS (2014) Pakistan's demand for shahtoosh shawls threatens rare Tibetan Antelope. http://www.theguardian.com/environment/2014/jan/17/shahtoosh-shawl-rate-tibetan-antelope-kashmir. Accessed 25 July 2014

Khator R (1989) Forests: the people and the government. National Book Organisation, New Delhi

KIN (2006) Naya Kashmir, Legal Document No 81 (Extract), Kashmir Information Network. http://www.kashmir-information.com/LegalDocs/81.html. Accessed 10 Apr 2010

Korf B (2004) Wars, livelihoods and vulnerability in Sri Lanka. Dev Chang 35(2):275–295

Kothari A, Suri S, Singh N (1995) People and protected areas: rethinking conservation in India. Ecologist 25:188–194

Kothari A, Vania V, Das P, Christopher K, Jha S (1997) Building bridges for conservation: towards joint management of protected areas in India. Indian Institute of Public Administration, New Delhi

Krishna G (2014) The shahtoosh conundrum. http://www.business-standard.com/article/opinion/geetanjali-krishna-the-shahtoosh-conundrum-114071801445_1.html. Accessed 25 July 2014

Kumar S (2002) Does participation in common pool resource management help poor? A social cost-benefit analysis of joint forest management in India. World Dev 30(5):1421–1438

Kumar C (2006) Whither 'community-based' conservation? Econ Polit Wkly 41(52):5313–5320

Kumar A, Wright B (1997) Fashioned for extinction: an expose of the shahtoosh trade. Wildlife Protection Society of India, New Delhi

Larson A (2004) Formal decentralisation and the imperative of decentralisation 'from below': a case-study of natural resource management in Nicaragua. Eur J Dev Res 16(1):55–70

Lawrence W (1895) The valley of Kashmir. Henry Frowde, London

Le Billon P (2005) Corruption, reconstruction and oil governance in Iraq. Third World Q 26(4):685–703

Leach M, Mearns R, Scoones I (1999) Environmental entitlements: dynamics and institutions in community based natural resource management. World Dev 27(2):225–247

Lele SC (1991) Sustainable development: a critical review. World Dev 19(6):607–621

Lele SC (1993) Sustainability: a plural, multidimensional approach. Working paper. Oakland, CA, USA: Pacific Institute for Studies in Development, Environment and Security

Leon M (1994) Avoidance strategies and governmental rigidity: the case of the small-scale shrimp fishery in two Mexican communities. J Polit Ecol 1:67–81

Leslie D Jr, Schaller G (2008) Pantholops hodgsonii. Mamm Species 817:1–13

Levine A (2002) Convergence or convenience? International conservation NGOs and development assistance in Tanzania. World Dev 30(6):1043–1055

Lewis D, Mosse D (2006) Development brokers and translators : the ethnography of aid and agencies. Kumarian Press, Bloomfield

Li TM (1999) Compromising power: development, culture and rule in Indonesia. Cult Anthropol 40(3):277–309

Li TM (2003) Situating resource struggles: concepts for empirical analysis. Econ Polit Wkly 29:5120–5128

Li TM (2005) Beyond "the state" and failed schemes. Am Anthropol 107(3):383–394

Li TM (2007) Practices of assemblage and community forest management. Econ Soc 36(2):263–293

Liddle WR (1992) The politics of development policy. World Dev 20(6):793–807

Lipsky M (1980) Street-level bureaucracy: dilemmas of the individual in public services. Russell Sage Foundation, New York

Liu W (2009) Tibetan Antelope (in Chinese). China Forestry Publishing House, Beijing

Lohmann L (1996) Freedom to plant: Indonesia and Thailand in a globalising pulp and paper industry. In: Parnwell M, Bryant R (eds) Environmental change in South East Asia. Routledge, London, pp 23–48

Long N, Van der Ploeg JD (1989) Demythologising planned intervention. Sociol Rural 29:226–249

Lynch O, Talbot K (1995) Balancing acts: community based forest management and national law in Asia and the Pacific. World Resources Institute, Washington, DC

Mabee H, Hoberg G (2006) Equal partners? Assessing co-management of forest resources in Clayoquot sound. Soc Nat Resour 19(10):875–888

MacGaffey J (1991) The real economy of Zaire: the contribution of smuggling and other unofficial activities to national wealth. James Currey, London

MacGaffey J, Bazenguissa R (2000) Congo-Paris: transnational traders on the margins of the law. International African Institute in association with James Currey and Indiana University Press, London

Mahanty S (2002) NGOs, agencies and donors in participatory conservation. Econ Polit Wkly 37:3757–3765

Maran T (2003) European mink. Setting of goal for conservation and the Estonian case-study. Galemys 15:11. www.secem.es/galemys/pdf

Marcus G (1995) Ethnography in/of the world system: the emergence of multi-sited ethnography. Annu Rev Anthropol 24:95–117

Markowitz L (2001) Finding the field: notes on the ethnography of NGOs. Hum Organ 60:40–46

Martinez-Alier J (2003) The environmentalism of the poor: a study of ecological conflicts and valuation. Edward Elgar, Cheltenham

Mawdsley E (2004) India's middle classes and the environment. Dev Chang 35(1):79–103

Mayers J, Bass S (2004) Policy that works for forests and people. Earthscan, London

McDonell JC (1894) Annual report of the Forest Department, J&K State 1893–94. The Civil and Military Gazette Press, Lahore

McDonell JC (1902) Annual report of the Forest Department, J&K State 1901–02. The Civil and Military Gazette Press, Lahore

McDonell JC (1903) Annual report of the Forest Department, J&K State 1902–1903. The Civil and Military Gazette Press, Lahore

McNeill D, Lichtenstein G, Renaudeau M (2009) International policies and national legislation concerning vicuna conservation and exploitation. In: Gordon I (ed) The vicuña: the theory and practice of community based wildlife management. Springer, New York, pp 63–79

McShane TO (2003) Protected areas and poverty, the linkages and how to address them. Policy Matters 12:52–53

Mehta L, Leach M, Newell P, Scoones I, Sivaramakrishnan K, Way S (1999) Exploring understandings of institutions and uncertainty: new directions in natural resource management. In: IDS discussion paper 372. Institute of Development Studies, Sussex

Mencher J (1999) NGOs: are they a force for change? Econ Polit Wkly 34(30):2081–2086

Menon V (n.d.) Letter by Vivek Menon, wildlife Trust of India to Shashi Bhushan, director, Ministry of Textiles, Government of India, New Delhi

Menzies N (1993) Putting people back into forestry: some reflections on social and community forestry. For Soc 1(1):6–7

Menzies N, Peluso NL (1991) Rights of access to upland forest resources in Southwest China. J World Forest Resour Manag 6:1–20

Middleton N, O'Keefe P, Moyo S (1993) The tears of the crocodile: from Rio to reality in the developing world. Pluto, London

Midgal J (1988) Strong societies and weak states: state-society relations and state capabilities in the third world. Princeton University Press, Princeton

Milward RC (1911) Progress report of Forest Administration in the J&K State 1910–11. Central Jail Press, Srinagar

Ministry of Agriculture and Irrigation (1977) Report of the National Commission on agriculture 1976. Government of India, New Delhi

Ministry of Food and Agriculture (1952) National forest policy. Government of India, New Delhi. www.forests.ap.nic.in/forests%20policy-1952.htm. Accessed 24 Mar 2008

Ministry of Textiles (2005a) Minutes of the meeting of the expert group set up to look into the issues relating to shahtoosh. Udhyog Bhavan, New Delhi

Ministry of Textiles (2005b) Draft report of the expert group on the issues related to shahtoosh. Udhyog Bhavan, New Delhi

Misri ML, Bhatt MS (1994) Poverty, planning and economic change in Jammu & Kashmir. Vikas Publishing House, Delhi

MoEF (n.d.) Letter by S.P. Prabhu, the Minister of Environment and Forests, Government of India to the J&K Chief Minister, Farooq Abdullah

Mohamad M (1996) The Malay handloom weavers. Institute of Southeast Asian Studies, Singapore

Mollinga P (2008) Water, politics and development: framing a political sociology of water resources management. Water Altern 1(1):7–23

Mollinga P (2010) Boundary concepts for interdisciplinary analysis of irrigation water management in South Asia. In: Working Paper Series 64. ZEF, Bonn

Mooij J, de Vos V (2003) Policy processes: an annotated bibliography on policy processes with particular emphasis on India, ODI Working Paper 221. ODI, London

Moorcroft W (1820) Excerpt from a letter from Moorcroft to Mr. Metcalfe, the East India Company's resident at Delhi. India Office, London

Moorcroft W, Trebeck G (1841) Travels in the Himalayan provinces of Hindustan and the Punjab in Ladakh and Kashmir. John Murray, London

Moore DS (1993) Contesting terrain in Zimbabwe's eastern highlands: political ecology, ethnography and peasant resource struggles. Econ Geogr 69:380–401

Moore DS (1996) Marxism, culture and political ecology: environmental struggles in Zimbabwe's eastern highlands. In: Peet R, Watts M (eds) Liberation ecologies: Environment, development, social movements. London, Routledge, pp 125–147

Moore DS (2000) The crucible of cultural politics: reworking "development" in Zimbabwe's eastern highlands. Am Ethnol 26(3):654–689

Mosse D (2003) The rule of water: statecraft, ecology and collective action in South India. Oxford University Press, Oxford

Mosse D (2005) Cultivating development: an ethnography of aid policy and practice. Pluto Press, London

Muhereza EF (2003) Commerce, kings and local government in Uganda: Decentralising natural resources to consolidate the central state. In: Working Paper 8. Environmental Governance in Africa Series. World Resource Institute, Washington, DC

Negi SS (1994) Indian forestry through the ages. Indus Publishing Company, New Delhi

Neumann RP (1992) Political ecology of wildlife conservation in the Mt Meru area of northeast Tanzania. Land Degrad Rehabil 3:85–98

Neumann R (2001) Disciplining peasants in Tanzania: from state violence to self-surveillance in wildlife conservation. In: Peluso NL, Watts M (eds) Violent Environments. Cornell University Press, Ithaca, pp 305–327

Njogu JG (2005) Beyond rhetoric: policy and institutional arrangements for partnership in community based forest biodiversity management and conservation in Kenya. In: Ros-Tonen, Dietz T (eds) African forests between nature and livelihood resources: interdisciplinary studies in conservation and forest management. The Edwin in Mellen Press, New York, pp 285–316

Nordstrom C (2004) Shadows of war: violence, power and international profiteering in the twenty-first century. University of California Press, Berkeley

Norgaard RB (1994) Development betrayed: the end of progress and a co-evolutionary revisioning of the future. Routledge, London

Omara-Ojungu PH (1992) Resource management in developing countries. Longman Scientific, Essex

Oyono PR (2004) Social and organisational roots of Cameroons forest management decentralisation model. Eur J Dev Res 16(1):174–191

Pai R, Datta S (eds) (2006) Measuring milestone: proceedings of the National Workshop on Joint Forest Management (JFM). Ministry of Environment and Forest, Government of India, New Delhi

Painter M, Durham W (eds) (1995) The social causes of environmental destruction in Latin America. University of Michigan Press, Ann Arbor

Panikar KN (1953) Gulab Singh: foundation of Kashmir state. Sage, New Delhi

Pant M (2008) Working plan for Jammu Forest division, 1998–99 to 2007–2008. Working Plan and Research Circle, Jammu and Kashmir Forest Department, Jammu

Pathak A (2002) Laws, strategies, ideologies: legislating forests in colonial India. Oxford University Press, New Delhi

Patnaik P, Singh S (n.d.) Study on participatory Forest Management in J&K: state policy implication and human resource development. Report prepared for the Jammu and Kashmir Forest Department

Peluso NL (1993) Coercing conservation: the politics of state resource control. In: Lipshutz R, Conca K (eds) The state and social power in global environmental politics. Columbia University Press, New York, pp 199–218

Peluso NL (2003) The politics of specificity and generalisation in conservation matters. Conserv Soc 1(1):61–64

Peluso NL, Vandergeest P (2001) Geneologies of the political forests and customary rights in Indonesia, Malaysia and Thailand. J Asian Stud 61(3):761–812

Peluso NL, Watts M (2001) Introduction. In: Peluso NL, Watts M (eds) Violent environments. Cornell University Press, Ithaca/New York, pp 1–30

Pickering A (1992) Science as practice and culture. Chicago University Press, Chicago

Pimbert M, Pretty J (1995) Beyond conservation ideology and the wilderness myth. Nat Resour 19(1):5–14

Platteau J (2009) Information distortion, elite capture and task complexity in decentralised development. In: Ahmad E, Brosio G (eds) Does decentralisation enhance service delivery and poverty reduction? Edward Elgar, Cheltenham, pp 23–72

Poffenberger M, Singh C (1996) Communities and the state: re-establishing the balance in Indian forest policy. In: Poffenberger M, McGean B (eds) Village voices, forest choices: joint forest management in India. Oxford University Press, New Delhi, pp 56–85

Popham P (1998) These animals are dying out. And all because the lady loves shahtoosh. The Independent. 20 June 1998. www.independent.co.uk/news/these-animals-are-dying-out-and-all-because-the-lady-loves-shahtoosh-1166083.html. Accessed 8 June 2010

Prakash S (2000) Political economy of Kashmir since 1947. Econ Polit Wkly 35(24):2051–2060

Punjabi R (1990) Panchayati raj in Kashmir: yesterday, today and tomorrow. In: Matthew G (ed) Panchayati Raj in Jammu & Kashmir. Concept Publishing Company, New Delhi, pp 37–49

Puri B (1993) Kashmir towards insurgency. Rekha Printers Private Limited, New Delhi

Puri B (1999) Jammu and Kashmir regional autonomy. Mehra Offset Press, New Delhi

Raina T (1972) An anthology of modern Kashmiri verse 1930–1960. Sangam Press, Poona

Rajan SR (2006) Modernizing nature. Oxford University Press, Oxford

Ramasubramanian R (2004) Can environmental security bring peace to Jammu and Kashmir. Article 1407. Institute of Peace and Conflict Studies, New Delhi

Ramaswamy V (2006) Textiles and weavers in South India. Oxford University Press, New Delhi

Ramirez R (1999) Stakeholder analysis and conflict management. In: Buckles D (ed) Cultivating peace: conflict and collaboration in natural resource management. World Bank Institute, Ottawa, pp 101–126

Ramnath M (2002) The impact of ban on timber felling. Econ Political Wkly November 30:4774–4776

Bibliography

Rana P, Chhatre A (2016) Rules and exceptions: regulatory challenges to private tree-felling in Northern India. World Dev 77(c):143–153

Randeria S (2007) Global designs and local lifeworlds: colonial legacies of conservation, disenfranchisement and environmental governance in post-colonial India. Interventions: Int J Postcolonial Stud 9(1):12–30

Rangan H (1995) Contested boundaries: state policies, forest classifications, and deforestation in the Garhwal Himalayas. Antipode 27(4):343–362

Rangan H (1997) Property versus control: the state and forest management in the Indian Himalayas. Dev Chang 28(1):71–94

Rangan H (2004) From Chipko to Uttaranchal: development, environment, and ocial protest in the Garhwal Himalayas, India. In: Peet R, Watts M (eds) Liberation ecologies. Routledge, London, pp 205–226

Rangarajan M (1996) Fencing the forest. Oxford University Press, New Delhi

Rathore BMS (1996) Joint management options for protected areas: challenges and opportunities. In: Kothari A, Singh N, Suri S (eds) People and protected areas: towards participatory conservation in India. Sage, New Delhi, pp 93–113

Rawling C (1905) The great plateau. Edward Arnold, London

Reardon TA, Vosti S (1995) Links between rural poverty and environment in developing countries. World Dev 23(9):1495–1506

Redclift M (1987) Sustainable development: exploring the contradictions. Routledge, London

Rehman S, Jafri N (2006) Kashmiri shawl from Jamavar to paisley. Mapin Publishing House Ltd., Ahmedabad

Riazuddin A (1988) History of handicrafts, Pakistan-India. National Hijra Council, Islamabad

Ribot J (1998) Theorising access: Forest profits along Senegal's charcoal commodity chain. Dev Chang 29:307–341

Ribot JC (2002) Some concepts and a proposed framework for contributions. Paper prepared to guide contributions to the international conference on decentralisation and the environment, Bellagio, Italy. 18–22 February

Rizvi J (1999) Trans-Himalayan caravans: merchant princes and peasant traders in Ladakh. Oxford University Press, Delhi

Robbins P (2000) The rotten institution: corruption in natural resource management. Polit Geogr 19(4):423–443

Robbins P (2004) Political ecology: a critical introduction. Blackwell, Oxford

Rocheleau D, Thomas-Slayter B, Wangari E (eds) (1996) Feminist political ecology. Routledge, London

Rose N (1999) Powers of freedom: reframing political thought. Cambridge University Press, Cambridge

Rossi B (2004) Revisiting Foucauldian approaches: power dynamics in development projects. J Dev Stud 40(6):1–29

Ros-Tonen M, Zaal F, Dietz T (2005) Reconciling conservation goals and livelihood needs: new forest management perspectives in the twenty-first century. In: Ros-Tonen, Dietz T (eds) African forests between nature and livelihood resources: interdisciplinary studies in conservation and forest management. The Edwin in Mellen Press, New York, pp 3–30

Saberwal VK (1999) Pastoral politics: shepherd, bureaucrats and conservation in the Western Himalaya. Oxford University Press, Delhi

Saberwal V (2003) Conservation by state fiat. In: Saberwal V, Rangarajan M (eds) Battles over nature: science and politics of conservation. Permanent Black, New Delhi, pp 240–266

Saberwal V, Rangarajan M (2003) Introduction. In: Saberwal V, Rangarajan M (eds) Battles over nature: science and politics of conservation. Permanent Black, New Delhi, pp 1–30

Sachs W (ed) (1993) Global ecology: a new arena of political conflict. Zed Books, London

Saigal S (2000) Beyond experimentation: emerging issues in the institutionalisation of joint Forest management. Environ Manag 2(3):269–281

Saito-Jensen M, Jensen CB (2010) Rearranging social space: boundary-making and boundary-work in a joint forest management project, Andhra Pradesh, India. Conserv Soc 8(3):196–208

Salskov-Iversen D, Hansen H, Bislev S (2000) Governmentality, globalisation and local practice: transformations of a hegemonic discourse. Alternatives 25(2):183–222

Sarap K, Sanrangi TK (2009) Malfunctioning of forest institutions in Orissa. Econ Polit Wkly XLIV(37):18–22

Sarin M (1996) Joint forest management: the Haryana experience. Centre for Environmental Education, Ahmedabad

Sarin M (1998) Who is gaining? Who is losing? Gender and equity concerns in joint Forest management. In: Working Paper by the gender and equity sub-group, National Support Group for JFM. Society for Promotion of Wastelands Development, New Delhi

Sarin M, Singh NM, Sundar N, Bhogal RK (2003) Devolution as a threat to democractic decision making in forestry? Findings from three states in India, ODI Working Paper 197. Overseas Development Institute, London

Schaller GB (1993) Tibet's remote Chang Tang. Natl Geogr 184(2):62–87

Schaller GB (1997) Tibet's hidden wilderness: Wildlife and nomads of the Chang Tang Reserve. Harry N Abrams, New York

Schaller GB (1998) Wildlife of the Tibetan Steppe. The University of Chicago Press, Chicago

Schaller GB (2003) Letter by George Schaller to the Wildlife Trust of India, New Delhi

Schonsberg BE (1853) India and Kashmir. Hurst & Blackett Publishers, London

Schumacher EF (1973) Small is beautiful: a study of economics as if people mattered. Blond and Briggs, London

Schwabach A, Qinghua L (2007) Measures to protect the Tibetan antelope under the CITES framework. Thomas Jefferson Law Review 29. www.papers.ssrn.com/sol3/papers.cfm

Scoones I (1998) Sustainable rural livelihoods: a framework for analysis, Working paper no. 72. Institute of Development Studies, Sussex

Scott J (1985) Weapons of the weak: everyday forms of peasant resistance. Yale University Press, New Haven

Scott L (2006) Chronic poverty and the environment: a vulnerability perspective, CPRC. Working Paper 62. Overseas Development Institute, London

Sen S (1999) Some aspects of state-NGO relationships in India in the post-independence era. Dev Chang 30(2):327–355

Shackleton CM, Wills T, Brown K, Polunin NVC (2010) Editorial: reflecting on the next generation of models for community-based natural resources management. Environ Conserv 37(1):1–4

Shafi M (1990) Revival of a democratic tradition. In: Matthew G (ed) Panchayati raj in Jammu & Kashmir. Concept Publishing Company, New Delhi, pp 31–36

Shah MA (1992) Export marketing of Kashmir handicrafts. Ashish Publishing House, New Delhi

Sharma YR (2002) Politics dynamics of Jammu & Kashmir. Radha Krishan Anand & Co., Jammu

Sharma SP (2004) Kashmir's political leadership: sheikh Mohammed Abdullah and his legacy. RBSA Publishers, Jaipur

Sharma KS, Bakshi SR (1995) Economic life of Kashmir. Anmol Publications, New Delhi

Shepherd A (1998) Sustainable rural development. Macmillan, Basingstoke

Shiva V (1991) Ecology and the politics of survival: conflicts over natural resources in India. Sage, New Delhi

Showeb M (1994) Silk handloom industry of Varanasi: a study of socio-economic problems of weavers. Ganga Kaveri Publishing House, Varanasi

Simmel G (1950) The sociology of George Simmel (Trans. and ed, Kurt Wolff). The Free Press. New York

Singh S (2008) Contesting moralities: the politics of wildlife trade in Laos. J Polit Ecol 15:1–20

Sinha H (2006) People and forests. Concept Publishing Company, Delhi

Sivaramakrishnan K (1999) Modern forests: Statemaking and environmental change in colonial eastern India. Oxford University Press, Stanford

Bibliography

Sivaramakrishnan K (2003) Scientific forestry and genealogies of development in Bengal. In: Greenough P, Tsing AL (eds) Nature in the global south: environmental projects in south and Southeast Asia. Duke University Press, Durham, pp 253–287

Springate-Baginski O, Blaikie P (2007) Forests, people and power: the political ecology of reform in South Asia. Earthscan, London

Stebbing EP (1926) The forests of India, vol III. Bodley Head, London

Sterndale (1884) Natural history of the mammalia of India and Ceylon. Thacker, Calcutta

Stonich S (1993) "I am destroying the land!": the political ecology of poverty and environmental destruction in Honduras. Westview Press, Bolder

Stott P, Sullivan S (eds) (2000) Political ecology: science, myth and power. Arnold, London

Strauss L (1986) The romance of the cashmere shawl. Mapin Publishing Pvt. Ltd., Ahmedabad

Sundar N (2001) Beyond the bounds? Violence at the margins of new legal geographies. In: Peluso NL, Watts M (eds) Violent environments. Cornell University Press, Ithaca, pp 328–353

Sundar N (2004) Devolution, joint forest management and the transformation of social capital. In: Bhattacharyya D, Jayal N, Mohapatra B, Pye S (eds) Interrogating social capital: the Indian experience. Sage, New Delhi, pp 203–232

Sundar N, Jeffery R, Thin N (2001) Branching out: joint forest management in India. Oxford University Press, New Delhi

Sunderland T, Campbell B (2008) Conservation and development in tropical forest landscapes: a time to face the trade-offs? Environ Conserv 34(4):276–279

Sunita (2006) Politics of state autonomy and regional identity: Jammu and Kashmir. Kalpaz Publications, Delhi

Supreme Court of India (1996) Judgement in Writ Petition 202/95. 12 December, 1996. www.moefrolko.org/scjudgement.htm. Accessed 15 Apr 2008

The Guardian (2010) Kashmir fears forests will disappear through 'timber smuggling'

Tibet Information (2006) Guarding the homes of Tibetan antelopes. http://www.tibetinfor.com/tibetzt-en/antelope/menu.htm. Accessed 10 Jan 2009

Tisdell C (1999) Conditions for sustainable development: weak and strong. In: Dragum AK, Tisdell C (eds) Globalisation and the impact of trade liberalisation. Edward Elgar, Cheltenham, pp 23–36

Traffic India (n.d.) Shawls of shame. Traffic India, New Delhi

Tripp A (1997) Changing the rules: the politics of liberalisation and the urban informal economy in Tanzania. University of California Press, Berkeley

Tsing A (2005) Friction: an ethnography of global connection. Princeton University Press, Princeton

Tucker R (1982) The forests of western Himalayas: the legacy of British colonial administration. J For Hist 1982:112–123

Upreti BR (2001) Contributions of community forestry in rural social transformation. Some observations from Nepal. J Forestry Livelihoods 1(1):31–34

Utting P (1993) Trees, people and power: social dimensions of deforestation and forest protection in Central America. Earthscan, London

Vasan S (2006) Living with diversity. Indian Institute of Advanced Study, Shimla

Venkatesan S (2009) Craft matters: artisans, development and the Indian nation. Orient Black Swan, New Delhi

Verma T (1994) Karkhanas under the Mughals: from Akbar to Aurangzeb. Pragati Publications, Delhi

Veron R, Williams G, Corbridge S, Srivastava M (2006) Decentralised corruption or corrupt decentralisation? Community monitoring of poverty alleviation schemes in Eastern India. World Dev 34(11):1922–1941

Vira B (1999) Implementing joint forest management in the field: towards an understanding of the community-bureaucracy interface. In: Jeffery R, Sundar N (eds) A new moral economy for India's forests? Discourses on community and participation. Sage, New Delhi, pp 254–275

Wakefield W (1879) The happy valley. Searle and Rivington, London

Walton O (2009) Negotiating war and the liberal peace: National NGOs, legitimacy and the politics of peace-building in Sri Lanka. PhD thesis. University of London, UK

Wardell DA, Lund C (2006) Governing access to trees in northern Ghana: micropolitics and the rents of non-enforcement. World Dev 34(11):1887–1906

Warikoo K (1989) Central Asia and Kashmir: a study in the context of Anglo-Russian rivalry. Gyan Publications, Delhi

Watts M (2000) Political ecology. In: Barners T, Scheppard E (eds) A companion to economic geography. Blackwell, Oxford, pp 257–275

WCED (1987) Our common future. Oxford University Press, Oxford

Wells M, Brandon K (1992) People and parks: linking protected area management with local communities. World Bank, Washington, DC

Widmalm S (2002) Kashmir in comparative perspective. Routledge, London

Wiggins S, Marfo K, Anchirinah V (2004) Protecting the forest or people? Environmental policies and livelihoods in the forest margins of southern Ghana. World Dev 32(11):1939–1955

Wildlife Department, J&K (2005) Interim Survey Report of Tibetan Antelope in D.B.O. (Karakuram Nubra Wildlife Sanctuary) & Changchenmo valley (Changthang Cold Desert Wildlife Sanctuary). August–September, 2005

Wilson HH (1841) Travels in Himalayan provinces of Hindustan and the Panjab from 1819 to 1825. Himalayan Gazetteer, Calcutta

WPSI (2005) Letter by Belinda Wright, Director, Wildlife Protection Society of India to the Director, Ministry of Textiles, Government of India. New Delhi

WWF (2006) The slaughter of Tibetan antelope in Tibet's Chang Tang National Nature Reserve continues. In: WWF China-Tibet program. WWF, Beijing

Xi Z, Wang L (2004) Tracking down Tibetan antelopes. Foreign languages press, Beijing

Xinhua (2011) Tibetan antelope population hits 200,000. 21 January, 2011. http://news.xinhuanet.com/english2010/china/2011-01/21/c_13701284.htm. Accessed 3 Feb 2011

Yearley S (1996) Sociology, environmentalism, globalization. Sage, London

Younghusband FE (1909) Kashmir. Adam & Charles Black, London

Zuhair M (1998) Country report Maldives: Asia Pacific forestry sector outlook study, Working paper series. APFSOS/WP30. Food and Agriculture Organisation, Rome

Zutshi C (2004) Languages of belonging: Islam, regional identity and the making of Kashmir. Hurst, London

Index

A
Adams, W., 11
Adams, W.M., 9
Agrawal, A., 12, 72, 102, 179, 185, 189, 193
Ahad, A., 40, 43, 52, 53, 58
Ahuja, B., 43
Alternative livelihoods, 3, 85, 110, 113–118,
    189, 192, 211, 214–216
Ames, F., 51, 52, 54
Axelby, R., 189

B
Bailey, S., 8, 12, 84
Ballabh, V., 183
Bamzai, P.N.K., 27, 42
Barakat, S., 32
Bass, S., 139, 151
Bazaara, N., 161
Bazenguissa, R., 88, 109, 198
Bebbington, A., 10, 161
Bhatt, M.S., 27
Blaikie, P., 161, 179, 182, 184, 185, 193
Bloomer, J., 81, 110, 198
Bose, S., 28
Brown, K., 198
Bryant, R., 8, 127
Bryant, R.L., 84, 124
Buttel, F.H., 12

C
Calvert, P., 11

Calvert, S., 11
Campbell, B.M, 161
Carney, D., 10
Cederlof, G., 13
Chambers, R., 10, 11, 184
Chatterjee, P., 110, 198
Chhatre, A., 76, 130, 198
Chiru, 3, 4, 16, 34, 37–39, 41–43, 60, 64–71,
    73–79, 85, 89, 90, 97, 110, 118, 119,
    209, 216, 218
Chiru farming, 73–78, 90, 209
Collins, T.W., 193
Colonial period, 5, 14, 16, 23, 122–124, 135,
    136, 200
Conflict and conservation, 23
Conservation politics, 143–170, 205
Contested conservation, 15
Conway, G., 10
Corbridge, S., 166
Corruption, 11, 15, 27, 81, 90, 97, 116, 190, 193
Corruption and illegality, 11, 15, 90, 97

D
Das, P., 184, 190
Dauvergne, P., 197
Dean, M., 12
Delegated illegality, 16, 110, 118, 214
de Vos, V., 166
Dhanagare, D.N., 182
Dhar, D.N., 27
Dheeraprasart, V., 197
Dressler, W., 185

© Springer International Publishing AG 2018
S. Gupta, *Contesting Conservation*, Advances in Asian Human-Environmental
Research, https://doi.org/10.1007/978-3-319-72257-3

E
Edmunds, D., 84, 192
Ekoko, E., 198
Ellis, F., 10
Environmental politics, 2, 15

F
Farrington, J., 161
Feng, Z., 39
Fomete, T., 161
Forest history, 23, 122, 123
Forest management, 2, 5, 6, 12, 14, 16, 28, 31, 34, 119, 121, 143–145, 148, 151–154, 160, 168, 169, 173–175, 182, 192, 193, 198, 205, 212

G
Ganhar, J.N., 40
Ganjoo, S.K., 26
Gibson, C., 102, 185
Guha, R., 129
Guha, R.C., 76, 131
Gupta, A., 11, 13, 72, 185, 208

H
Hareven, T., 106, 107
Harriss-White, B., 112
Hill, E., 113
Hoberg, G., 145
Homer-Dixon, T., 88
Huber, T., 41

I
Illegality, 11, 15, 16, 66, 82, 85–87, 100, 101, 107, 109–111, 113, 118
Irwin, J., 52

J
Jafri, N., 43, 52
Jammu and Kashmir, 1, 15, 17, 23, 69
Jayakar, P., 52
Jensen, C.B., 185, 188
Jha, P.S., 26
Jha, U.M., 116, 117
Joint forest management (JFM), 2, 6, 14, 31, 122, 140, 144–146, 148, 154, 169, 205

K
Kashmir shawls, 25, 55, 79, 81
Kawosa, M.A., 135, 179
Khator, R., 129
Korf, B., 89
Kothari, A., 184
Kumar, S., 186

L
Labour exploitation, 106–111
Leach, M., 103, 153
Lele, S.C., 10, 76, 145
Leon, M., 110, 214
Leslie, D., 39
Li, T.M., 13, 153, 173, 208
Lipsky, M., 166
Liu, W., 39
Long, N., 166
Lund, C., 81, 193

M
Mabee, H., 145
MacGaffey, J., 87, 88, 109, 198
Mahanty, S., 184
Maran, T., 74
Marcus, G., 29
Martinez-Alier, J., 13
Mayers, J., 139, 151
McDonell, J.C., 125, 128
Menzies, N., 151, 192
Micropolitics, 95, 173
Midgal, J., 80
Militancy, 14, 16, 25, 28, 29, 32, 33, 64, 66, 68, 87–89, 108, 113, 138, 155, 158, 160, 189, 194, 217
Milward, R.C., 127
Misra, D.C., 109, 116, 117
Misri, M.L., 27
Mohamad, M., 103, 104, 113
Mooij, J., 166
Moorcroft, W., 54, 56
Moore, D.S., 8, 13
Muhereza, E.F., 161
Multi-sited ethnography, 29

N
National Afforestation Programme (NAP), 2, 5–8, 144, 148–152, 154

Index

Nature conservation, 2, 3, 7–9, 15, 23, 29, 64, 71, 77, 95, 98, 113, 118, 119, 197, 198, 206–208, 211–213, 215–218
Neumann, R., 72
Njogu, J.G., 84, 192
Nordstrom, C., 87, 198
Norgaard, R.B., 76

O
Oyono, P.R., 161

P
Park, T., 153
Pashmina, 25, 48, 52, 53, 60, 79, 81–84, 86, 96, 98–102, 104, 108, 109
Peluso, N.L., 11, 12, 72, 88, 110, 192, 198
Platteau, J., 193
Political ecology, 8, 9, 14, 15
Politics of conservation, 143
Post-colonial period, 16, 122, 133–139
Power, 2, 5–8, 10, 11, 13–16, 25–27, 33, 56, 58, 64, 67, 69, 76, 78, 79, 82, 84, 86, 87, 89, 90, 95, 96, 100, 102, 107, 110, 113, 117, 119, 122–124, 130, 137, 140, 144, 147, 151–153, 157, 160–162, 166, 169, 170, 173–175, 177–181, 183–186, 193, 194, 198–201, 205–209, 211–215, 218
Puri, B., 26, 27

Q
Qualitative research, 23

R
Ramirez, R., 84
Ramnath, M., 109
Rana, P., 198
Randeria, S., 192
Rangan, H., 13, 185
Rathore, B.M.S., 152
Rawling, C.G., 39
Rehman, S., 43, 52
Research method, 15
Resource access, 121, 144, 162
Riazuddin, A., 52
Ribot, J.C., 161
Robbins, P., 8, 9, 190, 198
Rose, N., 13

Rossi, B., 13, 84, 198, 212
Ros-Tonen, M., 185

S
Saberwal, V., 76
Saberwal, V.K., 130, 185, 189, 193, 198
Sachs, W., 10
Saito-Jensen, M., 185, 188
Sanrangi, T.K., 183
Sarin, M., 179, 192, 193
Schaller, G., 4, 75
Schaller, G.B., 42
Schonsberg, B.E., 56
Scientific forestry, 16, 122, 209
Scoones, I., 10
Sen, S., 161
Shahtoosh, 2–5, 7, 9, 14–16, 28–34, 37–61, 63–90, 95–119, 205, 207, 209–217
Sharma, Y.R., 26, 27
Shawl industry, 14, 16, 23, 25, 28, 37–39, 43, 45, 51–58, 60, 61, 68, 70, 73, 76, 79, 82, 83, 89, 98, 100, 111, 113, 119, 205, 216
Shepherd, A., 11
Showeb, M., 116
Simmel, G., 108
Singh, S., 98, 110
Sinha, H., 182
Sinha, S., 208
Sivaramakrishnan, K., 13, 122
Split role, 16, 64, 77–81, 84, 90, 96, 119, 140, 143, 174, 200, 207, 211, 212
Split role of state, 86
Springate-Baginski, O., 182, 184, 193
Strauss, L., 52
Sundar, N., 13, 151, 152, 161, 168, 180, 182, 183, 185, 189, 190

T
Tibetan antelope, 2, 3, 5, 14, 37, 65, 66, 68, 70, 71, 74, 76, 78, 85, 208
Trebeck, G., 56
Tucker, R., 131

V
Van der Ploeg, J.D., 166
Vasan, S., 185, 197
Venkatesan, S., 105, 108, 113, 115, 116
Veron, R., 193

W
Wakefield, W., 56
Walton, O., 32
Wang, L., 39
Wardell, D.A., 81, 193
Watts, M., 8, 88
Widmalm, S., 26
Wiggins, S., 198
Wilson, H.H., 40
Wollenberg, E., 84, 192

X
Xi, Z., 39

Y
Yearley, S., 10, 77
Younghusband, F.E., 56

Z
Zutshi, C., 57

Printed in the United States
By Bookmasters